Green Chemistry and Green Materials from Plant Oils and Natural Acids

Green Chemistry Series

Editor-in-chief:
James H. Clark, *Department of Chemistry, University of York, UK*

Series editors:
Catherine Birch, *AgriFood X Limited, UK*
Graham Bonwick, *AgriFood X Limited, UK*
George A. Kraus, *Iowa State University, USA*
Andrzej Stankiewicz, *Delft University of Technology, The Netherlands*

Titles in the series:
1: The Future of Glycerol: New Uses of a Versatile Raw Material
2: Alternative Solvents for Green Chemistry
3: Eco-Friendly Synthesis of Fine Chemicals
4: Sustainable Solutions for Modern Economies
5: Chemical Reactions and Processes under Flow Conditions
6: Radical Reactions in Aqueous Media
7: Aqueous Microwave Chemistry
8: The Future of Glycerol: 2nd Edition
9: Transportation Biofuels: Novel Pathways for the Production of Ethanol, Biogas and Biodiesel
10: Alternatives to Conventional Food Processing
11: Green Trends in Insect Control
12: A Handbook of Applied Biopolymer Technology: Synthesis, Degradation and Applications
13: Challenges in Green Analytical Chemistry
14: Advanced Oil Crop Biorefineries
15: Enantioselective Homogeneous Supported Catalysis
16: Natural Polymers Volume 1: Composites
17: Natural Polymers Volume 2: Nanocomposites
18: Integrated Forest Biorefineries
19: Sustainable Preparation of Metal Nanoparticles: Methods and Applications
20: Alternative Solvents for Green Chemistry: 2nd Edition
21: Natural Product Extraction: Principles and Applications
22: Element Recovery and Sustainability
23: Green Materials for Sustainable Water Remediation and Treatment
24: The Economic Utilisation of Food Co-Products
25: Biomass for Sustainable Applications: Pollution Remediation and Energy
26: From C–H to C–C Bonds: Cross-Dehydrogenative-Coupling
27: Renewable Resources for Biorefineries
28: Transition Metal Catalysis in Aerobic Alcohol Oxidation
29: Green Materials from Plant Oils

30: Polyhydroxyalkanoates (PHAs) Based Blends, Composites and Nanocomposites
31: Ball Milling Towards Green Synthesis: Applications, Projects, Challenges
32: Porous Carbon Materials from Sustainable Precursors
33: Heterogeneous Catalysis for Today's Challenges: Synthesis, Characterization and Applications
34: Chemical Biotechnology and Bioengineering
35: Microwave-Assisted Polymerization
36: Ionic Liquids in the Biorefinery Concept: Challenges and Perspectives
37: Starch-based Blends, Composites and Nanocomposites
38: Sustainable Catalysis: With Non-endangered Metals, Part 1
39: Sustainable Catalysis: With Non-endangered Metals, Part 2
40: Sustainable Catalysis: Without Metals or Other Endangered Elements, Part 1
41: Sustainable Catalysis: Without Metals or Other Endangered Elements, Part 2
42: Green Photo-active Nanomaterials
43: Commercializing Biobased Products: Opportunities, Challenges, Benefits, and Risks
44: Biomass Sugars for Non-Fuel Applications
45: White Biotechnology for Sustainable Chemistry
46: Green and Sustainable Medicinal Chemistry: Methods, Tools and Strategies for the 21st Century Pharmaceutical Industry
47: Alternative Energy Sources for Green Chemistry
48: High Pressure Technologies in Biomass Conversion
49: Sustainable Solvents: Perspectives from Research, Business and International Policy
50: Fast Pyrolysis of Biomass: Advances in Science and Technology
51: Catalyst-free Organic Synthesis
52: Hazardous Reagent Substitution: A Pharmaceutical Perspective
53: Alternatives to Conventional Food Processing: 2nd Edition
54: Sustainable Synthesis of Pharmaceuticals: Using Transition Metal Complexes as Catalysts
55: Intensification of Biobased Processes
56: Sustainable Catalysis for Biorefineries
57: Supercritical and Other High-pressure Solvent Systems: For Extraction, Reaction and Material Processing
58: Biobased Aerogels: Polysaccharide and Protein-based Materials
59: Rubber Recycling: Challenges and Developments
60: Green Chemistry for Surface Coatings, Inks and Adhesives: Sustainable Applications
61: Green Synthetic Processes and Procedures
62: Resource Recovery from Wastes: Towards a Circular Economy
63: Flow Chemistry: Integrated Approaches for Practical Applications
64: Transition Towards a Sustainable Biobased Economy

65: Transportation Biofuels: Pathways for Production: 2nd Edition
66: Challenges in Green Analytical Chemistry: 2nd Edition
67: CO_2-switchable Materials: Solvents, Surfactants, Solutes and Solids
68: Green Toxicology: Making Chemicals Benign by Design
69: Sustainable and Functional Redox Chemistry
70: Waste Valorisation Using Ionic Liquids
71: Natural Product Extraction: Principles and Applications: 2nd Edition
72: Renewable Resources for Surface Coatings, Inks and Adhesives
73: Green Corrosion Inhibition: Fundamentals, Design, Synthesis and Applications
74: Chemical Valorisation of Carbon Dioxide
75: Green Sample Preparation Techniques: Concepts, Novel Materials and Solvents, and Applications
76: Methane Conversion Routes: Status and Prospects
77: Green Gasoline: A Green Spark Transportation Fuel
78: Agri-food Waste Valorisation
79: Bioplastics and Biocomposites: A Practical Introduction
80: Green Development of Photoluminescent Carbon Dots: For Analytical, Biomedical and Environmental Applications
81: Toward a PFAS-free Future: Safer Alternatives to Forever Chemicals
82: Green Fire Retardants for Polymeric Materials
83: Green Chemistry and Green Materials from Plant Oils and Natural Acids

How to obtain future titles on publication:
A standing order plan is available for this series. A standing order will bring delivery of each new volume immediately on publication.

For further information please contact:
Book Sales Department, Royal Society of Chemistry, Thomas Graham House, Science Park, Milton Road, Cambridge, CB4 0WF, UK
Telephone: +44 (0)1223 420066, Fax: +44 (0)1223 420247
Email: booksales@rsc.org
Visit our website at books.rsc.org

Green Chemistry and Green Materials from Plant Oils and Natural Acids

Edited by

Zengshe Liu
NCAUR, ARS/USDA, USA
Email: kevin.liu@ars.usda.gov

and

George Kraus
Iowa State University, USA
Email: gakraus@iastate.edu

ROYAL SOCIETY OF **CHEMISTRY**

Green Chemistry Series No. 83

Print ISBN: 978-1-83767-080-2
PDF ISBN: 978-1-83767-159-5
EPUB ISBN: 978-1-83767-160-1
Print ISSN: 1757-7039
Electronic ISSN: 1757-7047

A catalogue record for this book is available from the British Library

The Royal Society of Chemistry is a charity, registered in England and Wales, Number 207890, and a company incorporated in England by Royal Charter (Registered No. RC000524), registered office: Burlington House, Piccadilly, London W1J 0BA, UK, Telephone: +44 (0)20 7437 8656.

For further information see our website at www.rsc.org

Printed in the United Kingdom by CPI Group (UK) Ltd, Croydon, CR0 4YY, UK

Preface

This is our second book in the Royal Society of Chemistry Green Chemistry series. As with the previous book, the central theme of this book is the use of plant materials to prepare sustainable polymers, lubricants and plasticizers. We have broadened the focus to include both oils and organic acids. The list of authors includes prominent scientists and engineers from USDA national laboratories, NREL, the Kansas Polymer Research Center and the Chinese Academy of Forestry plus researchers from major agricultural universities including Fujian Agriculture and Forestry University, the South China Agricultural University and Iowa State University.

Part 1 features the application of oils to the polymer industry. Chapters cover polymer design, polymerization in supercritical carbon dioxide and the generation of thermosets, composites, epoxy resins and polyurethanes. Part 2 focuses on the application of oils and natural acids to lubricants and additives. Chapters cover oil-based plasticizers, oils for improved lubricity and the conversion of natural acids into bio-based pyrones and many other industrial chemicals.

This book was originated and assembled by Dr Zengshe Liu, a research chemist in the bio-oils research unit of the National Center for Agricultural Utilization Research located in Peoria, Illinois, and Dr George A. Kraus, a chemistry professor at Iowa State University, who has worked on green chemistry and natural resources for so many years. Dr Liu's longstanding research interest in sustainable chemistry and his many contacts in both industry and academia were vital for the expanded focus of this book. We thank Helen Armes of the Royal Society of Chemistry for suggesting a

Green Chemistry Series No. 83
Green Chemistry and Green Materials from Plant Oils and Natural Acids
Edited by Zengshe Liu and George Kraus

Published by the Royal Society of Chemistry, www.rsc.org

second book and assisting with the book proposal. We also appreciate the efforts of Amina Headley in keeping us on track and guiding us through the publication process.

Dr George A. Kraus
Iowa State University

Dr Zengshe Liu
USDA-Peoria

Contents

Chapter 1 Strategic Approach to the Design of Polymers from Natural Oils 1
Z. S. Petrović and J. Hong

1.1 Introduction 1
1.2 Biological Oils – Composition and Specific Properties 2
1.3 Planning Polymeric Materials from Natural Oils 7
1.4 Functionalization of Vegetable Oils 8
1.5 Effect of Double Bonds on Polymer Properties 9
1.6 Influence of Oil Heterogeneity on the Properties of Polyurethanes 10
1.7 Influence of the Position of Functional Groups 11
1.8 Designer Oils by Genetic Modification 12
1.9 Bio-content in Polymeric Materials 14
1.10 Use of Oils and Fats with a High Content of Saturated FA 14
1.11 Effect of Crosslinking Density on the Properties of Thermosets 15
1.12 Technologies for Converting Oils to Polymers 16
1.13 Hydrolytic Stability of Oil Products 19
1.14 Electrical Properties of Natural Oil-based Polymers 20
References 20

Green Chemistry Series No. 83
Green Chemistry and Green Materials from Plant Oils and Natural Acids
Edited by Zengshe Liu and George Kraus

Published by the Royal Society of Chemistry, www.rsc.org

Chapter 2 Natural Epoxy Oil (Euphorbia Oil) Polymerization in Liquid Carbon Dioxide–Green Solvents 23
Shailesh N. Shah and Zengshe Liu

2.1 Introduction 23
2.1.1 Plant or Natural Oil-based Polymers 24
2.1.2 Natural Plant Oils with Epoxide Rings 24
2.2 Materials, Methods and Procedures 27
2.2.1 Materials 27
2.2.2 Procedure 30
2.3 Results and Discussion 31
2.3.1 Analysis and Composition of EuO 31
2.3.2 Effect of Initiator Concentration 31
2.3.3 Effect of Monomer Concentration 34
2.3.4 Soxhlet Extraction and Characterization of the Extracted RPEuO-soluble Substance 34
2.3.5 Thermal Stability of RPEuO 38
Acknowledgements 40
References 40

Chapter 3 Soybean Oil-based Polymers and Their Composites 42
W. Liu, T. Chen and R. Qiu

3.1 Characteristics of Soybean Oil 42
3.2 Soybean Oil-based Polymers 43
3.2.1 Polyurethane 43
3.2.2 Vinyl Resins 45
3.2.3 Epoxy Resins 48
3.3 Soybean Oil-based Biocomposites 51
3.3.1 Composites from the Soybean Oil-based Matrix 51
3.3.2 Composites Compatibilized by Soybean Oil Derivatives 52
3.4 Conclusions 53
Abbreviations 54
Acknowledgements 54
References 54

Chapter 4 Plant Oil-based Polyurethanes 59
Chaoqun Zhang

4.1 Monomers Prepared from Plant Oils 60
4.1.1 Polyols from Plant Oils 61

4.1.2 Internal Emulsifiers from Plant Oils 66
4.1.3 Epoxidation/Ring-opening Method 66
4.1.4 Thiol–Ene Photo-click Method 70
4.1.5 Chain Extenders from Plant Oils 71
4.1.6 Isocyanates from Plant Oils 73
4.2 Solvent-based Plant Oil-based Polyurethanes 74
4.2.1 Partially Bio-based Polyurethanes 74
4.2.2 Fully Bio-based Polyurethanes 75
4.3 Plant Oil-based Waterborne Polyurethane Dispersions 75
4.3.1 Cationic Polyurethane Dispersions 75
4.3.2 Anionic Polyurethane Dispersions 77
4.3.3 Non-ionic Polyurethane Dispersions 78
4.4 Non-isocyanate Polyurethanes 78
4.4.1 Carbonated Plant Oils 79
4.4.2 Solvent-based NIPUs Prepared from Plant Oils 79
4.4.3 Waterborne NIPUs Prepared from Plant Oils 81
4.4.4 NIPU Foams Prepared from Plant Oils 81
4.5 Application of Plant Oil-based Polyurethanes 84
4.5.1 Protective Coatings 84
4.5.2 Adhesives 85
4.5.3 Foams 85
4.5.4 Drug Delivery 86
4.5.5 Slow-release Fertilizers 86
4.5.6 Others 87
References 87

Chapter 5 Non-isocyanate Polyurethane from Vegetable and Microalgal Oils 92
Tao Dong, Skylar Schutter, Cheng Zhang and Jacob Kruger

5.1 Introduction 92
5.2 Materials and Methods 94
5.2.1 Materials 94
5.2.2 Short Path Distillation of *Nannochloropsis* Oil 94
5.2.3 Methylation of *Nannochloropsis* Oil 94
5.2.4 Epoxidation by H_2O_2 (Soybean, Linseed and *Schizochytrium* Oils) 94
5.2.5 Epoxidation by *m*-CPBA (*Nannochloropsis* FAME) 95

5.2.6 Methylation of Epoxidized Linseed and *Schizochytrium* Oils 95
5.2.7 Carbonation 95
5.2.8 Polymer Precursor Fractionation 95
5.2.9 Aminolysis Kinetics Study 95
5.2.10 Resin Preparation 96
5.2.11 Analytical Methods 96
5.3 Results and Discussion 96
5.3.1 Fatty Acid Profile 96
5.3.2 Short Path Distillation 97
5.3.3 Epoxidation of Soybean, Linseed and *Schizochytrium* Triglyceride Oils by H_2O_2 97
5.3.4 Epoxidation of *Nannochloropsis* Oil by *m*-CPBA 99
5.3.5 Methylation of Epoxidized Algae-DHA Triglyceride Oil 100
5.3.6 Carbonation 100
5.3.7 Polymer Precursor Fractionation 102
5.3.8 Reaction Kinetics Study 102
5.4 Conclusion 107
Acknowledgements 107
References 107

Chapter 6 Preparation and Application of Renewable Intrinsically Flame-retardant Epoxy Resins from Plant Oils 109
Wenbin Li, Xiaoan Nie and Jie Chen

6.1 Introduction 109
6.2 Preparation of Epoxy Resins from Plant Oils 112
6.3 Types of Flame-retardant Epoxy Resins from Fatty Acid Derivatives 115
6.3.1 Epoxy Resins from a Halogen-containing Fatty Acid Derivative 115
6.3.2 Epoxy Resins from a Carbon-containing Fatty Acid Derivative 118
6.3.3 Epoxy Resins from a Phosphorus-containing Fatty Acid Derivative 119
6.3.4 Epoxy Resins from a Nitrogen-containing Fatty Acid Derivative 123
6.3.5 Epoxy Resins from a Silicon-containing Fatty Acid Derivative 127

6.3.6 Epoxy Resins from Intumescent Flame Retardants (IFRs) and Nanoscale Flame Retardants 130
6.4 Conclusions 131
Acknowledgements 132
References 132

Chapter 7 Reactions of Mixed Natural Acids and Esters in Acidic Media 135
William J. Bradley, Yang Qu, Zachery Crandall, Theresa L. Windus and George A. Kraus

7.1 Introduction 135
7.2 Results and Discussion 136
7.3 Computational Evaluation 139
7.4 Materials and Methods 141
7.4.1 Procedure for the Synthesis of **4** 141
7.4.2 Procedure for the Synthesis of **6** 141
7.4.3 Methyl Coumalate from Dimethyl Malate 142
Acknowledgements 142
References 142

Chapter 8 Industrial Chemicals *Via* Decarboxylation of Natural Carboxylic Acids 144
Kenneth M. Doll and Bryan R. Moser

8.1 Plant Feedstocks for Industrial Use 144
8.1.1 Plant Oil Production and Availability 145
8.2 Removal of Oxygen from Natural Carboxylic Acids 147
8.3 Chemicals from Decarboxylation of Natural Carboxylic Acids 149
8.4 Ruthenium-based Decarboxylation 149
8.4.1 Other Reactions Observed in the Ruthenium System 149
8.5 Iridium-based Decarboxylation and Aromatization Reactions for Aviation Fuel 151
8.6 Methacrylic Acid from Selective Decarboxylation of Itaconic Acid 153
8.7 Conclusions 155
Acknowledgements 155
References 155

Chapter 9 A Novel Application of Chemically Modified Eleostearic Acid Methyl Ester (EAME) for Improving Diesel Lubricity **159**
Zengshe Liu, Jing Li, Gerhard Knothe, Brajendra K. Sharma and Jiangchung Jiang

9.1 Introduction 159
9.2 Characterization 161
9.2.1 FT-IR 161
9.2.2 NMR 161
9.3 Methods 161
9.3.1 Preparation of EAME from Tung Oil 161
9.3.2 Preparation of EAME/MA and Esterification with Methanol and Butanol 162
9.3.3 Lubricity Determination 162
9.4 Results and Discussion 163
9.4.1 Characterization of EAMA/MA/ME and EAMA/MA/BU 163
9.4.2 Lubricity 165
9.5 Conclusions 167
References 168

Chapter 10 Sustainable Lubricant Formulations from Natural Oils: A Short Review **170**
Brajendra K. Sharma, Gobinda Karmakar, Raj Shah, Pranab Ghosh, Majher I. Sarker and Sevim Z. Erhan

10.1 Introduction 170
10.2 Characteristics of Vegetable Oils 173
10.3 Chemical Modification of VOs 174
10.3.1 Transesterification 176
10.3.2 Hydrolysis 177
10.3.3 Catalytic Partial Hydrogenation 178
10.3.4 Amidation/Amination 179
10.3.5 Epoxidation and Carbonation 179
10.3.6 Elektrionization™ 181
10.3.7 Estolide Formation/Oligomerization 182
10.3.8 Polymerization 184
10.4 Lubricants (Base Oil/Additives) from Other Natural Resources 185
10.5 Conclusion 186
References 186

Chapter 11 Plant Oil-based Plasticizers **194**
Yun Hu and Puyou Jia

11.1 Introduction 194
11.2 Plant Oil-based Plasticizers 195
11.2.1 Plant Oil-based Epoxidized Plasticizer 195
11.2.2 Plant Oil-based Polyester Plasticizers and Macromolecular Plasticizers 197
11.2.3 Plant Oil-based Flame Retardant Plasticizers 201
11.2.4 Glyceryl Ester Plasticizer 206
11.2.5 Plant Oil-based Reactive Plasticizers and Internally Plasticized PVC 206
11.2.6 Other Plant Oil-based Plasticizers 207
11.2.7 Conclusion 209
References 210

Chapter 12 Genetic and Process Engineering for Select Glycolipid Biosynthesis from Plant/Algal Oils or Their Derivatives **213**
Joseph Msanne and Richard D. Ashby

12.1 Introduction 213
12.2 Synthesis and Regulation of Select Glycolipid Biosurfactants 215
12.2.1 Rhamnolipids (RLs) 215
12.2.2 Sophorolipids (SLs) 219
12.2.3 Mannosylerythritol Lipids (MELs) 224
12.2.4 Cellobiose Lipids (CLs) 227
12.2.5 Trehalose Lipids (TLs) 231
12.3 Conclusions 238
References 239

Subject Index **252**

CHAPTER 1

Strategic Approach to the Design of Polymers from Natural Oils

Z. S. PETROVIĆ*[a] AND J. HONG[a,b]

[a] Pittsburg State University, Pittsburg, KS 66762, USA; [b] Checkerspot Inc., Alameda, CA 94501, USA
*Email: zpetrovic@pittstate.edu

1.1 Introduction

Motivation for entering the field of bio-based materials is primarily related to the desire to improve ecology and sustainability, but the approach depends on the position of a person or a company in the society. Manufacturers of bio-based materials primarily want to add value to their existing products, but the issue is the direction they should choose and how to achieve the goal. Usually, they are directed towards making raw materials for potential users, but they may develop a technology for making the final product. Companies making petrochemical products may look for bio-based substitutes, to replace some components in their products with renewable materials. In that case, they would utilize components made by the first group. Scientists have fewer restrictions and may consider a range of options by selecting different raw materials with a range of chemistries, creating new chemicals, and targeting different applications that may not be restricted by economic reasons. An important aspect of their research would be to establish principles and relationships between the structure and properties to provide suggestions for designing materials and processes. In this chapter, an attempt is made to analyze the selection parameters to facilitate the

Green Chemistry Series No. 83
Green Chemistry and Green Materials from Plant Oils and Natural Acids
Edited by Zengshe Liu and George Kraus

Published by the Royal Society of Chemistry, www.rsc.org

decision of why and how to use natural oils for the preparation of polymeric materials.

1.2 Biological Oils – Composition and Specific Properties

Biological oils are generated by plants, animals, fish, algae, planktons, and bacteria. Biological oils are triglycerides (triacylglycerols), *i.e.*, esters of glycerin and different fatty acids (FA), but some non-triglyceride products such as cashew nutshell liquid are treated as oils. Tall oil is an extract from wood after digestion of wood chips with bases or acids and it is a mixture of chemicals rich in fatty acids and resin acids, which can be separated by distillation. The main value of biological oils is the possibility of their transformation into an array of chemicals, many of which are useful monomers for polymeric materials. Biological oils are usually liquid at room temperature due to the presence of double bonds. The presence of esters and double bonds is common in all oils, but some fatty acids may have a range of different functional groups. Most edible oils are triglycerides of five fatty acids: oleic, linoleic, linolenic, stearic, and palmitic acids, which are the major components but many other FA are minor components. Over 1000 fatty acids are identified but not more than 50 are technologically important.[1] Since almost every plant creates some kind of oil, the number of oils that were analyzed is enormous.[2] The largest volume of oils on the markets are palm and soybean followed by rapeseed, sunflower, palm kernel, cottonseed, olive, peanut, and coconut oil. Table 1.1 shows the composition of the ten largest volume edible vegetable oils and two important industrial oils (linseed and castor). Rapeseed oil contains unhealthy erucic acid (22:1) and it is used mainly for industrial applications. When erucic acid is removed from rapeseed oil, it is called canola oil which is a popular cooking oil. Table 1.1 shows that these oils have no fatty acids with more than three double bonds and only palm kernel and coconut oil have saturated fatty acids with less than 16 carbons. In spite of the refining processes and removal of phospholipids from oils, there are still about two to five percent of non-triglyceride impurities including wax esters, hydrocarbons, and phenolic derivatives.[3] These impurities may affect subsequent chemical transformations of oils.

Animal and fish fats have somewhat different compositions. Butter is the emulsion of milk fats (~80%) and protein in water. Lard and beef tallow have about 27% palmitic (C16:0) acid, as shown in Table 1.2.

Generally, animal fats are low in unsaturation. On the other side, fish oils may have very high unsaturation and are rich in fatty acids with five or six double bonds. They also contain a significant amount of saturates. There are high hopes for algal oil utilization for industrial applications but their presence in the market today is symbolic. The composition of algal oils depends on the type of algae and can be tailored by genetic modifications. The values in Tables 1.1 and 1.2 are indicative and not exact, due to

Table 1.1 Composition of large volume edible oils and two industrial oils.

Carbon atoms:double bonds	8:0	10:0	12:0	14:0	16:0	16:1	18:0	18:1	18:2	18:3	20:0	20:1	22:0	22:1	24:0	Iodine value	World consumption, 2013 (MMT)
Palm oil			0.1	1.0	44.4	0.2	4.1	39.3	10.0	0.4	0.3		0.1			50–55	52.4
Soybean oil				0.1	10.6	0.1	0.4	23.3	53.7	7.6	0.3		0.3			123–139	43.4
Rapeseed oil				0.1	3.8	0.3	1.2	18.5	14.5	11.0	0.7	6.6	0.5	41.1	1.0	100–115	23.8
Sunflower oil				0.1	7.0	0.1	4.5	18.7	67.5	0.8	0.4	0.1	0.7			125–140	13.4
Palm kernel oil	3.3	3.4	48.2	16.2	8.4		2.5	15.3	2.3		0.1	0.1				14–19	5.8
Cottonseed oil			0.1	0.7	21.6	0.6	2.6	18.6	54.4	0.7	0.3		0.2			98–118	5.3
Olive oil					9.0	0.6	2.7	80.3	6.3	0.7	0.4					76–88	2.9
Peanut oil				0.1	11.1	0.2	2.4	46.7	32.0		1.3	1.6	2.9		1.5	84–100	5.3
Coconut oil	7.8	6.7	47.5	18.1	8.8		2.6	6.2	1.6	0.1	0.1					7–12	3.7
Corn oil				0.1	10.7	0.2	2.0	25.4	59.6	1.2	0.4		0.1			118–128	2.0
Linseed oil					6.0		4.0	22.0	16.0	52.0	0.5					>177	0.7[a]
Castor oil					2.0		1.0	7.0	3.0							81–91	1.4

[a] 2004/5.

Table 1.2 Composition of fatty acids in animal fats and fish oil.

Carbon atoms:double bonds	14:0	16:0	16:1	18:0	18:1	18:2	20:5	Iodine value	World production, 2013 (MMT)
Butter	12	26	3	11	28			26–42	10
Lard	1–2	26	5	11	44	8–15		58	8.3
Tallow	3	27	2–4	15–23	43	2		50	8.7
Fish oil[a]	9	19	12	3	11		14	150–185	1.0
Algal oil[b]	5	37	23	0.9	11.9	1.5	15.3	103	—

[a] Menhaden oil contains 5–14% 20:6.
[b] *Nannochloropsis salina* oil.

variability of compositions within the same species and market sources when production figures are cited. Currently, vegetable oils are the most promising raw materials for new oleochemicals due to their well-defined composition and favorable price. The dilemma of food *versus* industrial applications is only pronounced in the case of biodiesel due to large volumes, whereas the volume for oleochemicals is much smaller and tolerable. The utilization of oils for non-food applications such as in soaps, coatings, and lubricants is as old as the industry and is not in question. There are many oils that are not used for food, and intensive investigations are carried out to find their best applications.[4,5] Some specialty oils offer new properties. Calendula or marigold seed oil has one of the highest iodine values (IV) of 242. It has 59% calendic acid with three conjugated double bonds. Tung oil (IV = 160–175) is also rich in conjugated triene acid (69%). Oils with conjugated trienic fatty acids dry faster than non-conjugated (linseed) ones. *Lesquerella* oil has fatty acids with hydroxyl groups. Vernonia seed oil is rich in vernolic fatty acid (72–78%), which has one epoxy group. Dominating fatty acids in standard oils have 18 and 16 carbons and *cis*-double bonds, but some oils are rich in shorter or longer fatty acids. Figure 1.1 shows the structure of some important fatty acids. Comprehensive reviews of vegetable oils and their compositions are given in many books and articles.[1,4,6–12]

Generally, natural raw materials, such as natural oils, are heterogeneous in structure and vary from molecule to molecule. This may create issues when competing with synthetic products, which are generated from high purity starting materials. The composition depends on the origin of the oil, but in the same type of oil, it may differ due to the geographical region, climatic conditions, genetic makeup of the plant, and other variable factors.

One of the important aspects of the application of oils in polymers is the distribution of fatty acids in triacylglycerols (TAG). There are around 35 types of triacylglycerols in oils with five fatty acids, such as soybean oil (SBO), even when we ignore the positional isomers. The most abundant TAGs in SBO are LLL (24.5%), OLL (21.6%), OOL (11.7%), and OOO (5.6%), where L stands for linoleic and O for oleic acid, but they may vary with the type of SBO.[2]

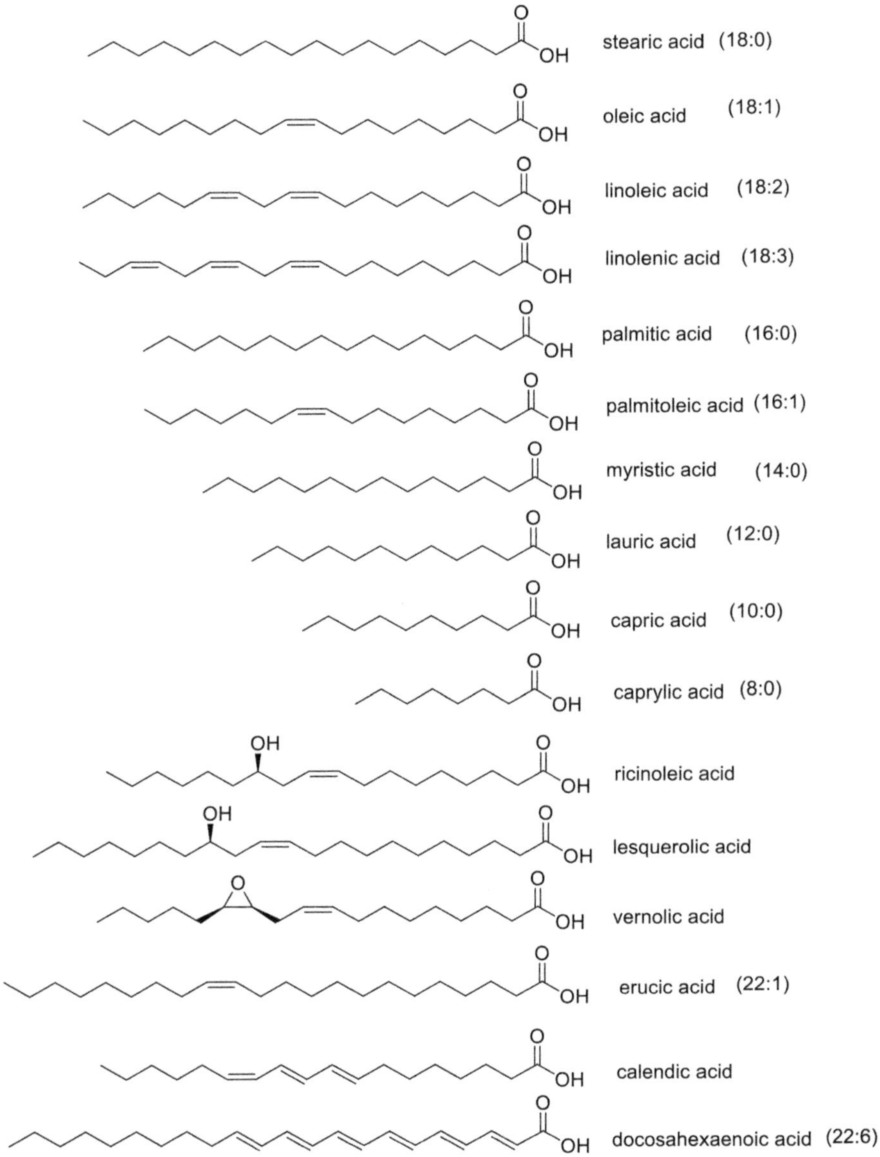

Figure 1.1 Structures of selected fatty acids.

Distribution of double bonds per TAG in SBO varies from zero to eight as illustrated in Figure 1.2, for a specific SBO.[13] It should be mentioned that the same average number of double bonds per TAG can be obtained from different TAGs. For example, three double bonds per TAG exist in LnPP, OOO, and OLP, where Ln is linolenic and P is palmitic acid, but their spatial distribution is very different. Average functionality (number of double bonds) can be expressed as the first moment of distribution

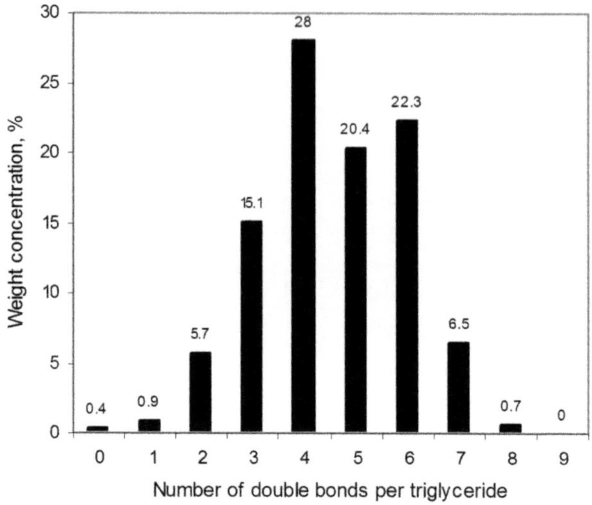

Figure 1.2 Distribution of double bonds per TAG in selected soybean oil.[4] Reproduced from ref. 4 with permission from John Wiley & Sons, Copyright © 2007 Wiley Periodicals, Inc.

(number average, f_n) or second moment (weight average, f_w), which can be calculated from known TAG compositions:

$$f_n = \sum N_i f_i \Big/ \sum N_i \text{ and } f_w = \sum N_i f_i^2 \Big/ \sum N_i f_i \tag{1.1}$$

where N_i is the number of moles of TAGs with functionality f_i.

The major components in Figure 1.2 have four, five, and six double bonds but the average content is $f_n = 4.56$ and $f_w = 5.0$. The gel point when cross-linking oils or functionalized oils is related to f_w rather than f_n.

The molar mass of oils can be calculated from their composition determined by GC/MS, assuming that oils are pure triacylglycerols (without free FA). Although the composition of oils is very heterogeneous, molar masses of standard oils vary in a narrow range, the strongest effect coming from the number of carbons in fatty acids. The molar mass of triacylglycerols consisting of FA with 18 carbon atoms lies between tristearin ($M = 891$) and trilinolenin ($M = 873$). The difference is two percent, which is within the experimental error of some experimental methods. The number average molar masses of selected oils calculated from FA composition have these approximate values: SBO glyc max (872), SBO low sat (878), sunflower (877), corn (870), canola (887), palm kernel (689), palm (864), and castor oil (925) with the experimental error below one percent. Weight average molar masses are close to M_n values, meaning that standard oils are essentially monodisperse. The GC/MS method for molar mass determination is not applicable to oils with a high free acid content, oligomerized oils, and modified oils or their mixtures. An alternative method is vapor pressure osmometry (VPO), which requires no knowledge of the composition and is particularly useful for studying mixtures. The error when applying to oils is below ten percent but could be less than two percent, and it is more precise at lower molar masses.[14] NMR was used for the calculation of molar mass, but it could be applied to TAGs of known purity.[15,16] Size exclusion chromatography (SEC) also gives approximate M_n and M_w values, although the values relative to polystyrene (typical standards) give fairly large errors. The combination of SEC with light scattering (LS) is not precise at low molar masses (below 1000). Mass spectrometry techniques are good for identifying species but are not suitable for quantitative determination of average molar masses.

Heterogeneity of oil structures arises from the fact that triglycerides are usually composed of several fatty acids which differ in length, number of double bonds, the position of double bonds in fatty acids, and geometry of DB (*cis–trans*).

Heterogeneous structures have a lower crystallization tendency and produce materials with lower glass transitions than ordered systems, which could be beneficial for extending the elastic behavior to lower temperatures.

1.3 Planning Polymeric Materials from Natural Oils

Consideration for making new polymeric materials from natural oils should include the cost, structure of available oils, and desired properties. Generally, there are two classes of polymers: thermoplastics and thermosets. Thermoplastic polymers are linear and are processed by rapid techniques such as extrusion and injection molding, but also by casting from solution. They are made from pure di-functional components generated by breaking triglyceride molecules. One of the largest oil-based commercial products by volume in this category is polyamide 11. It is obtained from castor oil in several steps involving methanolysis to obtain methyl ricinoleate, which is then thermally cracked to methyl undecylenic acid and heptaldehyde. Undecylenic acid is reacted with HBr to obtain 11-bromoundecanoic acid, which after treatment with ammonia gives the monomer-11-ω-amino undecanoic acid. Polycondensation produces polyamide 11.[5] Various diacids, hydroxy acids, or diols can be made by methods including ozonolysis,[6] metathesis,[7] thiol–ene additions to double bonds,[8] epoxidation, hydroformylation, and transesterification. Most frequently, such monomers are used for polyesters or polyester polyols for urethanes, polyamides, and epoxy resins with potential biodegradability as listed in several monographs.[9–12]

Thermosets are cross-linked systems and are natural products from modified triacylglycerols. Their applications vary from cast resins to foams, adhesives, elastomers, and coatings. Although the structure of coatings is in many ways the same as that of cast resins, elastomers, and foams, they are applied from solutions in thin layers and deserve special consideration.[13] The selection of functional groups for thermosets is wider, covering the whole of organic chemistry. Points of attack for chemical transformation are

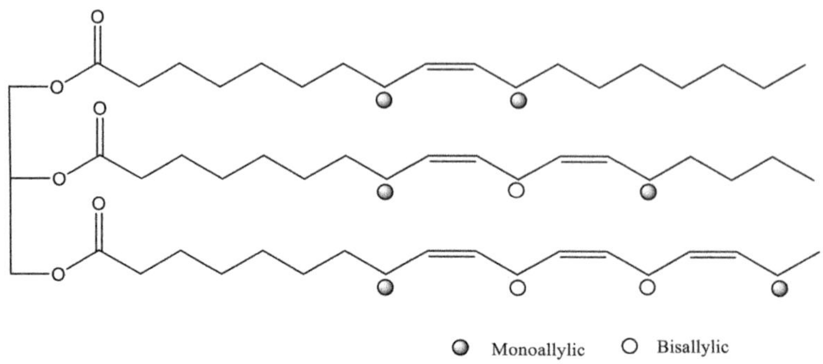

Figure 1.3 Sensitive points in an oil structure for a chemical attack.

usually double bonds (DB), ester bonds, and allylic positions, as indicated in Figure 1.3.

Converting oils directly to products is a desirable goal. Drying oils such as linseed and tung oil can be used directly as coatings. Coating films are formed by oxidative polymerization through double bonds or allylic sites. Direct polymerization of oils is carried out on an industrial scale to give higher molar mass and higher viscosity oils for the coating industry or as rubber processing aids (factice). "Blown oils" are produced by bubbling air through soybean or linseed oil at temperatures of 100–110 °C for a relatively long time (30–50 hours), while "bodied oil" is partially polymerized (linseed) oil by heating at 330 °C. Polymerization of oils can be carried out at temperatures below 100 °C in the presence of superacids such as HBF_4.[14] Although oxidative polymerization may lead to solid products, their usefulness was not confirmed. For controlled polymerization to useful products, oils must be modified by introducing reactive functional groups.

1.4 Functionalization of Vegetable Oils

The largest volume polymers from vegetable oils are alkyds and polyurethanes which are produced from polyols and isocyanates. Although there are some commercial aliphatic isocyanates based on fatty acids, the main use of oils is in polyols. Thermoplastic polyurethane elastomers are made from fatty acid-based diols (macrodiols), usually diphenylmethane diisocyanate (MDI), and a short diol (chain extender). Macrodiols are usually polyesters of molar mass from 1000 to 2000. Ricinoleic acid is a natural hydroxy containing fatty acid that would self-polymerize just by heating, but a more elegant way is to run transesterification of methyl ricinoleate. Control of molecular weight can be achieved by adding a short diol, preferably with more reactive primary hydroxyls.[15] Since the OH group in ricinoleic acid is located on the 12th carbon, the polymerized polyol will have a linear structure with a hanging six carbon chain as a side group (dangling chain). When a hydroxy fatty acid is prepared by epoxidation and ring opening of oleic acid, or by hydroformylation, dangling chains are longer: eight or nine

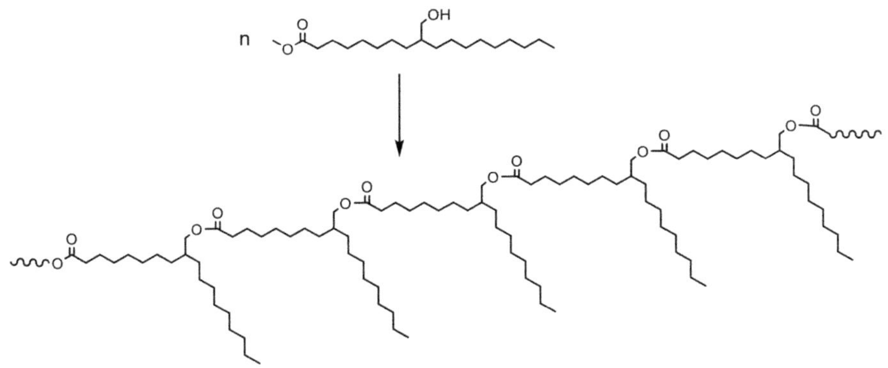

Figure 1.4 Hydroxylated methyl oleate generated by hydroformylation of methyl oleate and resulting polyester with dangling chains.

carbons as shown in the polyester in Figure 1.4. Dangling chains in polyurethanes act as plasticizers in elastomers and, when compared with petrochemical analogues, they lower and slow down the elastic recovery after deformation. If a higher elastic response is desired, then the elimination of dangling chains is preferable. However, when a memory effect in foams is desired, then polyols with dangling chains are useful.

Ozonolysis and reductive hydrogenation of triolein produce a triol with terminal hydroxyl groups, while hydroxylated triolein by epoxidation and reduction with hydrogen gives a triol with dangling chains. The effect of dangling chains on polyols and MDI cross-linked polyurethanes from polyols with and without dangling chains is illustrated in Figure 1.5.

Polyurethane with MDI without dangling chains is glassy solid, stronger, with a high modulus and a higher glass transition by about 25 °C, but with low elongation at break. Polyurethanes plasticized with dangling chains behave as elastomers at room temperature and are more suitable for low temperature applications where high elongations are required.

1.5 Effect of Double Bonds on Polymer Properties

The influence of double bonds on the properties of polyurethane networks is best illustrated with polyurethanes from castor oil and hydrogenated castor oil, as shown in Figure 1.6. Double bonds disturb the regularity of chains and lower the crystallization temperature and glass transition of polymers.

While the presence of double bonds in oils has a profound effect on the crystallization temperature, their influence on polymer properties is less pronounced. The glass transition in PU with double bonds is around 5 °C lower, and the tensile strength and modulus at 50% elongation are comparable. Elongation at break is higher in double bond containing samples, and the effect of double bonds would be higher at higher concentrations. Generally, double bonds in the chain increase the elasticity, lower the T_g, decrease the oxidative stability, and give a brownish tint to the material.

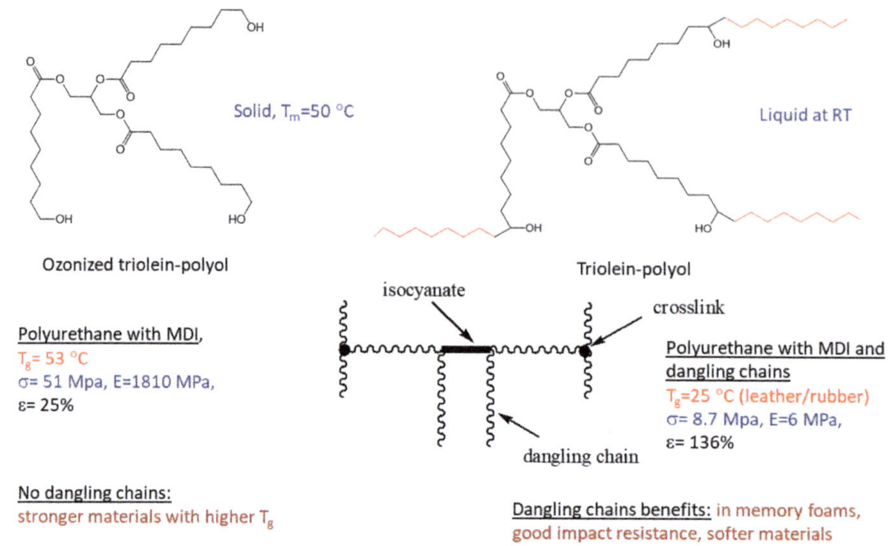

Figure 1.5 The triol obtained by ozonolysis is solid while the polyol obtained by epoxidation and ring opening with hydrogen has dangling chains (red) and is a slow crystallizing liquid.[16] Reproduced from ref. 16 with permission from American Chemical Society, Copyright 2005.

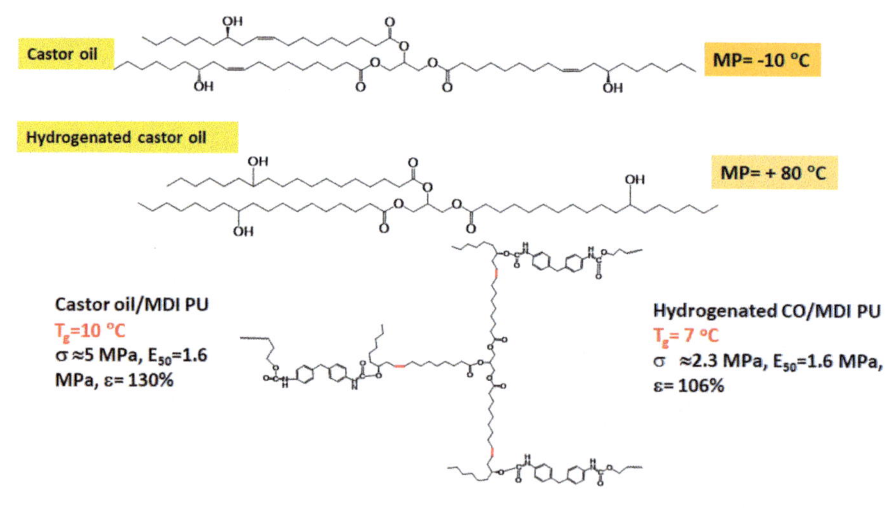

Figure 1.6 Properties of castor oil and hydrogenated castor oil and polyurethane networks with both polyols.

1.6 Influence of Oil Heterogeneity on the Properties of Polyurethanes

Table 1.3 displays the properties of the polyol from SBO with heterogeneity displayed in Figure 1.2 and is compared with those from high oleic oil,

Table 1.3 Properties of polyurethanes from the SBO polyol with a heterogeneous structure and from the triolein polyol.

	SBO polyol	Triolein polyol
Functionality, f_n	3.5	3
OH number, mg KOH per g	179	160
	Polyurethane with MDI	Polyurethane with MDI
T_g, °C	31	33
Tensile strength, MPa	20	20
Elongation at break, %	108	200
Modulus, MPa	312	190

having a structure close to triolein. The functionality of two polyols was 3.5 for the SBO polyol and around three for triolein.

Despite the lower OH number, PU from the triolein polyol gave almost identical T_g and tensile strength to those of PU from the SBO polyol (heterogeneous), but a lower modulus and higher elongation. Whereas triolein-based polyurethane contained dangling chains of eight or nine carbon atoms, SBO also had around 15% of saturated fatty acids bearing no functional groups, acting as a stronger plasticizer. SBO polyurethane made from the polyol of higher OH number also had a higher content of MDI, resulting in a higher modulus.

1.7 Influence of the Position of Functional Groups

The double bond in oleic acid is on the ninth or tenth carbon. Linoleic acid also has a double bond in that location, but also has an additional double bond between the 12th and 13th carbon. Linolenic has yet a third double bond between the 15th and 16th carbon. When functional groups are introduced at these positions, they will be inside the fatty acid chains. This has strong implications on the crosslinking density and properties. When epoxidized fatty acids are polymerized by cationic initiators, close epoxy groups (in linoleic and linolenic) may lead to cyclization and reduce the functionality of crosslinked networks or reduce the OH number in polyols made by ring opening with alcohols. Terminal OH or acid groups may be introduced by oxidizing methyl groups using metabolic engineering, although the process is tedious.[17] In standard polyurethanes from high oleic oils, the polyol has an OH group either on the ninth or on the tenth carbon, as shown in Figure 1.7. The part from the first and ninth carbon will be a part of the network chain (see Figure 1.5) that is subjected to stress in mechanical testing, whereas the pendant (dangling) chain is not, but has a plasticizing function. Moving the OH group to the left will result in shorter elastic chains and longer dangling chains. A shorter elastic chain increases glass transition, strength, and modulus but longer dangling chains act in the opposite way. The net effect should be checked experimentally.

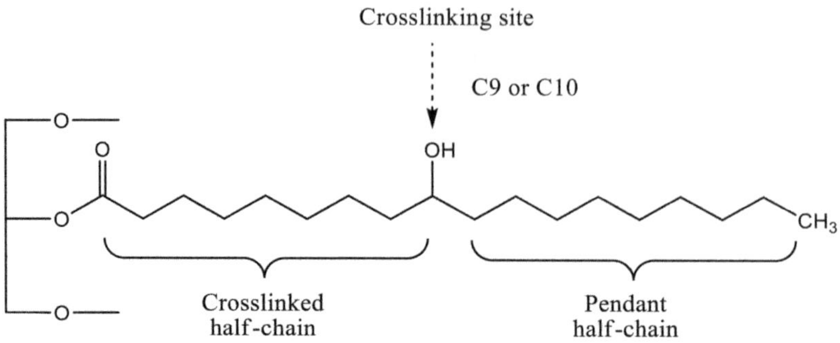

Figure 1.7 Schematic representation of the position of functional groups in a fatty acid chain.[18] Reproduced from ref. 18 with permission from John Wiley & Sons, Copyright © 2000 John Wiley & Sons, Inc.

Saturated dangling chains may crystallize when they reach a certain length. When polyitaconic acid was esterified with alcohols of different lengths, crystallization occurred when the hydrocarbon side-chain reached ten carbons.[19] Crystallization may occur even if the main chain is long enough. Saturated hydrocarbon chains from fatty acids with a terminal hydroxy group on the 18th carbon are expected to be crystalline. Polyesters and polyurethanes from nonanoic hydroxy fatty acids are highly crystalline materials and display excellent memory effects.[20]

Terminal hydroxy groups are highly desirable but there are only a few natural fatty acids of this type. Some fatty acids with functional groups at different positions in the fatty acid chain with potential for application in polyurethanes are shown in Figure 1.8.

Gunstone lists hydroxy fatty acids with different positions of the OH group in the fatty acid chain: ricinoleic with OH on the 12th carbon, lesquerolic and auricolic on the 14th, coriolic on the 13th, and helenynolic on the 9th.[21]

1.8 Designer Oils by Genetic Modification

Genetic modification of plants to produce desirable fatty acids with specific functional groups is possible.[22] Algae are especially conducive to this because of their rapid growth rate and accumulation of lipids. This could be the future in designing new molecules. Designing oils with preferable structures typically requires both the manipulation of endogenous genes as well as the introduction of heterologous constructs from new plant or algal species. The process can be complex and, in the case of soybean producing ricinoleic acid, the success was limited. The possibility of creating new functionalized oils offers exciting opportunities, however. Factors to consider are as follows: (a) terminal OH groups are desirable for utilization of the total FA length as an elastic chain for higher elasticity, higher elongation, and lower modulus for elastomers and flexible foams; (b) fatty acids with shorter chains are useful for rigid applications, while

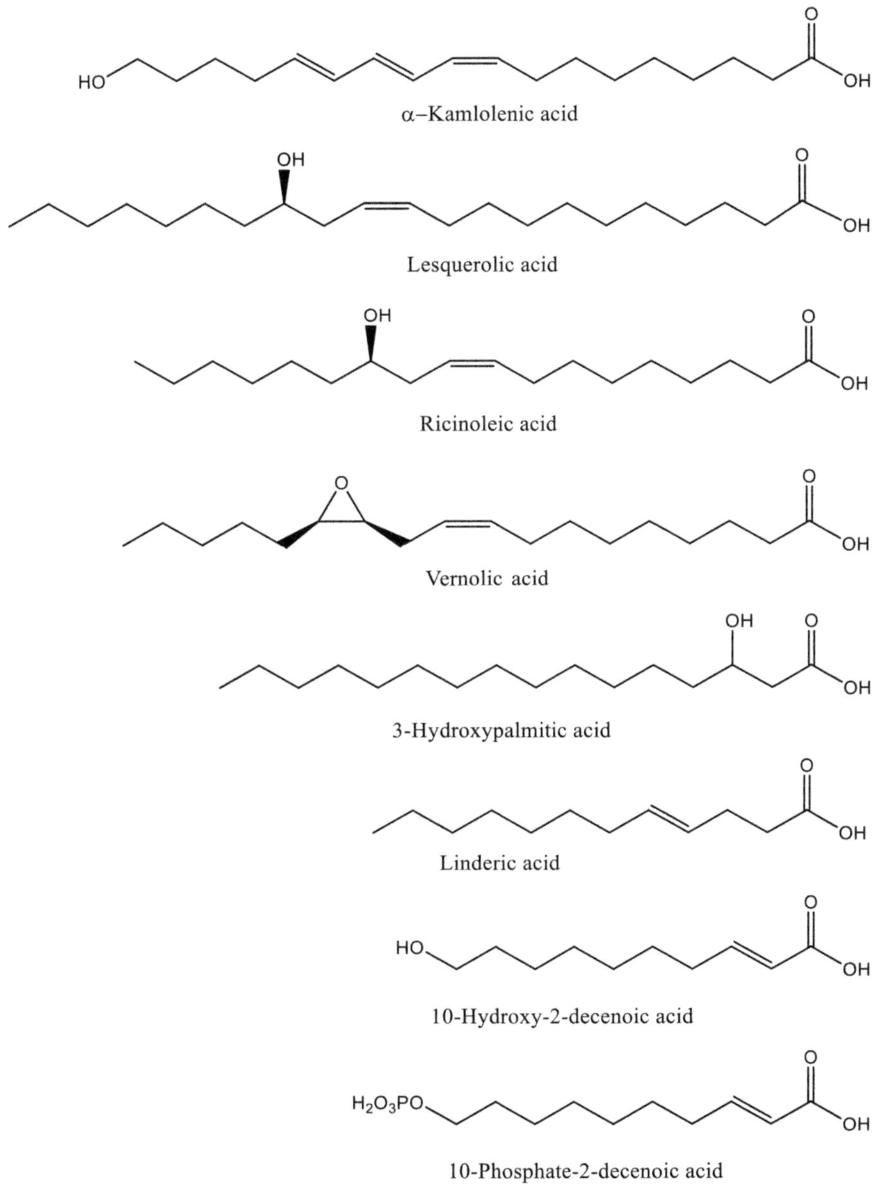

Figure 1.8 Some naturally occurring fatty acids with functional groups at different positions in the FA chain for potential use in polyurethanes.

long chains are for elastic materials; (c) the position of a functional group in the FA chain; and (d) primary *versus* secondary OH groups determine reactivity and other factors. Long chains between functional groups may allow crystallization of hydrocarbon chains, decreasing low temperature elasticity.

1.9 Bio-content in Polymeric Materials

When making thermosetting polyurethane cast resins, coatings, adhesives, and foams, oil-based polyols are reacted with diisocyanates or polyisocyanates. The ratio of polyols and isocyanates depends on the applications. For hard and rigid materials, a higher aromatic isocyanate content is required. Also, higher polyol functionality and lower molar mass are desired for higher rigidity. Typically, rigid foams have less than 50% oil-based polyols and flexible polymers have above 80%. However, bio-polyols in foams on a commercial scale are mixed with petrochemical polyols and their content generally does not exceed 30% except in applications such as carpet backing, where deterioration of properties is tolerated. Figure 1.9 shows epoxy resins and foams from epoxidized oils with 100% oil content.[23–26]

1.10 Use of Oils and Fats with a High Content of Saturated FA

Transformation of oils and fats with low unsaturation to polyols can be carried out by transesterification with polyhydric alcohols. Theoretically, transesterification of triglycerides with glycerol at a molar ratio of 1 : 2 should give monoglycerides, *i.e.*, diols. This does not occur since transesterification by chemical means is a reversible process and one obtains a mixture of glycerol, monoglycerides, diglycerides, and triglycerides (starting oil).[13,27] The mixture of non-functional oil, mono-functional diglycerides, di-functional monoglycerides, and tri-functional free glycerol would give polyols of functionality varying between two and three, but the OH value will be controlled by

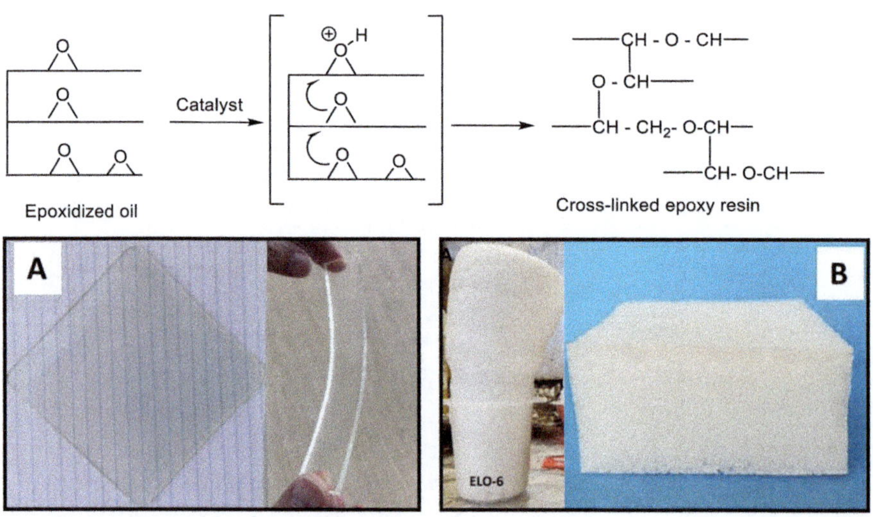

Figure 1.9 Epoxy cast resins and foams prepared entirely from epoxidized soybean (A) and linseed oils (B).

the excess of glycerol. Typically, at two moles of glycerol per mole of triglyceride, the yield is 40–60% with a base catalyst and could reach 70% with ionic liquids.[28] Using an excess of glycerol shifts the equilibrium to a higher monoglyceride content and free glycerol. Since glycerol is not compatible with oil, it creates a cloudy polyol above some critical excess. It is interesting that non-functionalized triglycerides may not be detrimental if kept within certain limits. A high monoglyceride content has been obtained by enzymatic trans-esterification.[29] Free glycerol has a higher density than oil-based polyols and may precipitate with time. Transesterification with higher alcohols such as sorbitol leads to higher functionality and higher viscosity polyols.

1.11 Effect of Crosslinking Density on the Properties of Thermosets

Thermosetting polymers are networks of chains whose properties depend on the cross-linking density and structure of chains. Generally, highly cross-linked polymer resins are rigid, strong, and often brittle materials *vs.* low crosslinked materials, which have elastomeric (rubbery) properties. The theory of networks is well-developed for low crosslink density elastomers.[30,31] The gelation theory predicts the structure of networks depending on the starting monomer functionality and concentration.[32–36] Crosslinking of oil-based polyols with isocyanates is displayed in Figure 1.10a. The functionality of networks is related to the functionality of reactants and usually they are tri- or tetra-functional defined by the number of chains emanating from the crosslink junction, Figure 1.10b. Longer chains in the networks are coiled and uncoiled during the sample elongation. However,

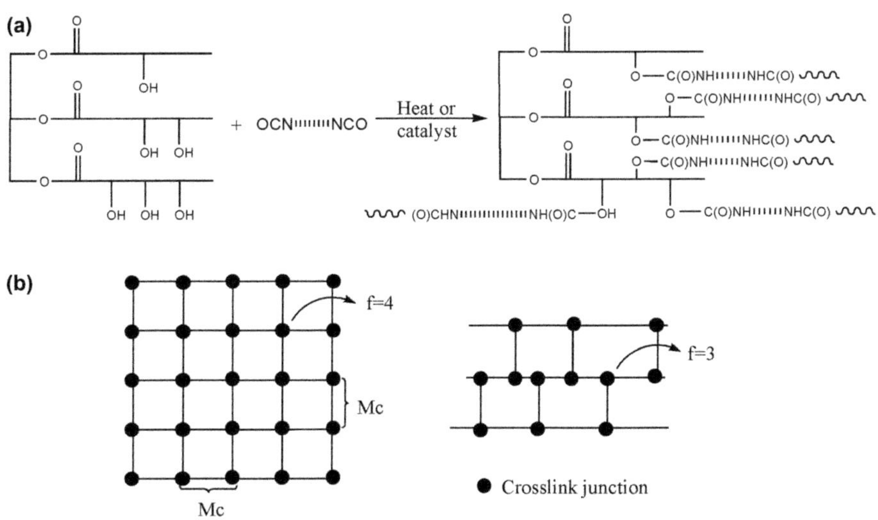

Figure 1.10 Schematic representation of the formation of polyurethane networks (a) and network types (b).

short chains are relatively stretched and do not allow high elongation. Crosslinking density is expressed either as the molar mass between crosslinks, M_c, or as the number of crosslinks per unit volume, ν. These two quantities are determined either from swelling experiments or by measuring shear modulus in the rubbery state, G, of a sample with density ρ. The relationship $\nu = \rho/M_c$ indicates that the crosslinking density decreases with increasing M_c.

$$G = \rho RT/M_c = \nu RT \tag{1.2}$$

It can be shown that the OH number in polyols is directly proportional to the crosslinking density. Glass transition, T_g, and mechanical properties increase with the crosslinking density, Figure 1.11.

Figure 1.11 shows low glass transitions for systems with OH numbers below 160 mg KOH per g that are rubbery at room temperature. Their tensile strengths were below 5 MPa. On the other hand, those with higher OH numbers were stronger with modulus values in the range of polyethylene. The crosslinking density of highly crosslinked polymer networks is better assessed by measuring glass transition. The Fox–Loshaek empirical relationship[38] predicts the effect of crosslinking density on T_g:

$$T_g = T_{g\infty} + K_1/M_c = T_{g\infty} + K_2\nu \tag{1.3}$$

Eqn (1.3) shows the linear dependence of T_g on the crosslinking density, ν. $T_{g\infty}$ represents the glass transition of the linear (uncrosslinked) polymer with infinite M_c. Constants K_1 and K_2 are determined from experiments of known crosslinking density or have been calculated.[39]

1.12 Technologies for Converting Oils to Polymers

The majority of applications of oil-based polymers in coatings, foams, and adhesives require a crosslinked structure. Economics favors processing in

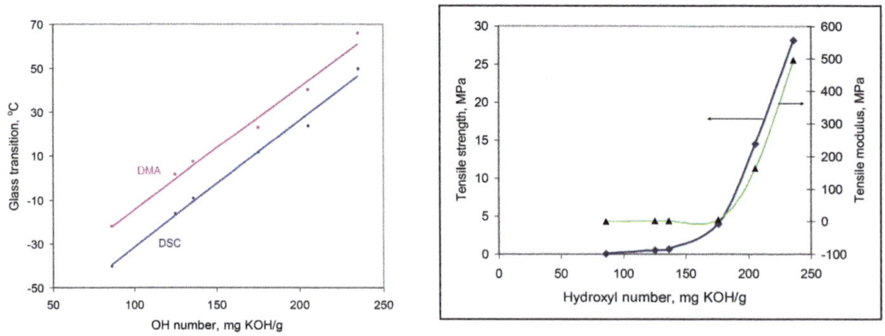

Figure 1.11 Glass transition and tensile strength and modulus *vs.* OH number (crosslinking density) in selected oil-based polyurethanes cross-linked with MDI.[37] Reproduced from ref. 37 with permission from John Wiley & Sons, Copyright © 2007 Society of Chemical Industry.

the fewest number of steps. Preparation of polyols for polyurethanes directly from oils (Schemes 1.1–1.3) is possible by thiol–ene addition of mercapto alcohols,[8] hydroformylation, and metathesis or with an extra step involving epoxidation, ring opening with alcohols, acids, hydrogen, or inorganic acids, and ozonolysis/hydrogenation.[16] Epoxidized oils can be directly polymerized to polymers using cationic catalysts.[40–44]

Radical polymerization of oils is carried out by introducing acrylic vinyl groups.[45–49] These raw materials may be co-reacted with a range of vinyl monomers for coatings or for unsaturated polyester binders for composites.[50–52] A great advantage of radical polymerization is that no by-products are generated

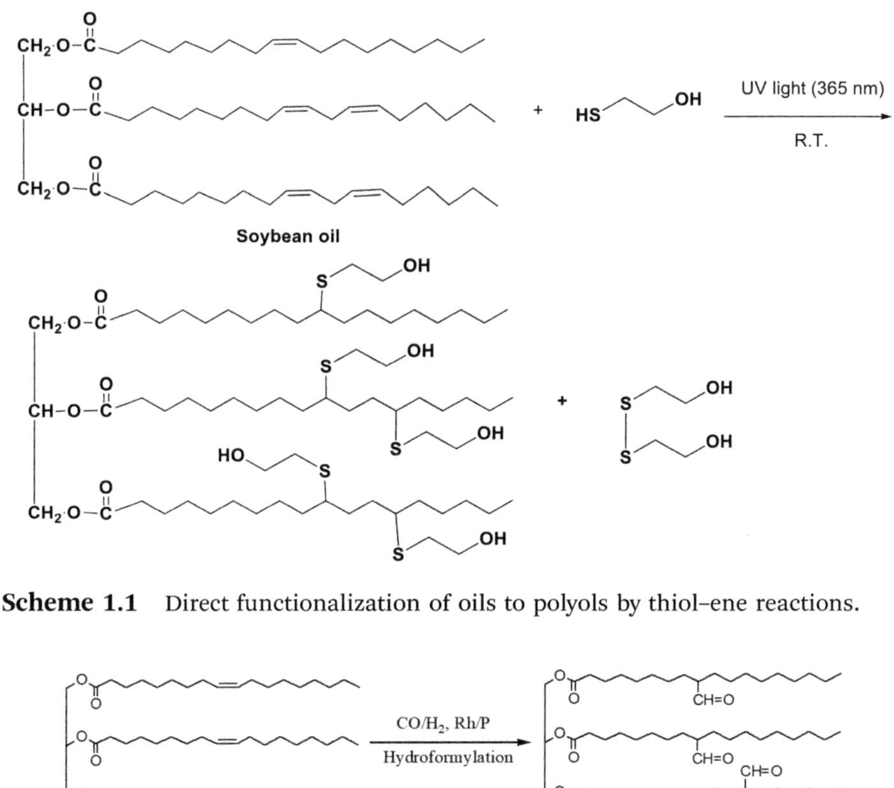

Scheme 1.1 Direct functionalization of oils to polyols by thiol–ene reactions.

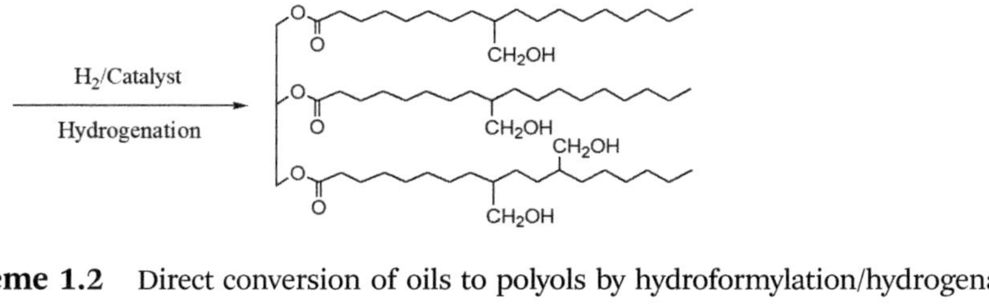

Scheme 1.2 Direct conversion of oils to polyols by hydroformylation/hydrogenation.

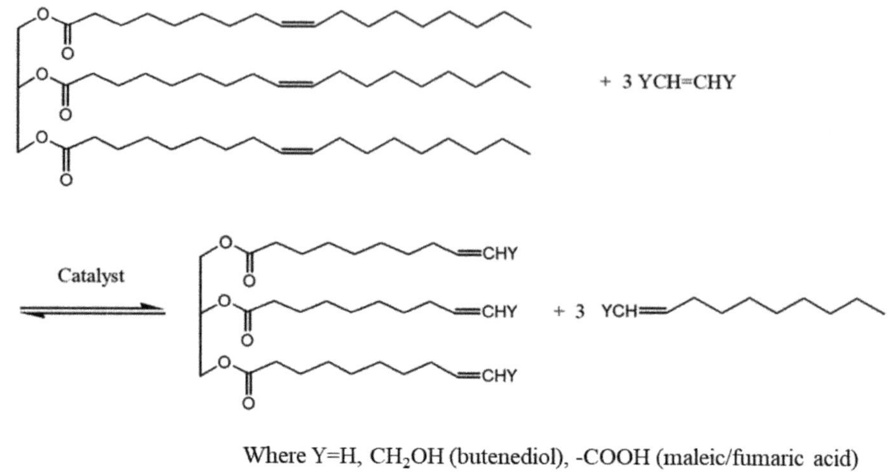

Scheme 1.3 Direct functionalization of oils by co-metathesis with ethylene, butene diol, or maleic/fumaric acids – preparation of oils with terminal double bonds, polyols, and polyacids.

as in condensation reactions, and it can be carried out at lower temperatures. However, the chain polymerization reaction may be affected by a large number of allylic hydrogens which are good radical blockers. Olefinic terminal double bonds as in oils with undecylenic fatty acid or ones obtained by co-metathesis of oils with ethylene or by ring opening of epoxy groups with allyl alcohol react with radicals to a low degree of conversion, but maleic anhydride may be a useful co-monomer for linking two double bonds. It also allows additional crosslinking through esterification with hydroxyls when they are present (dual curing mechanism) and gives excellent thermosetting resins.[53]

A new trend in using environmentally friendly technologies is the application of "click" chemistry processes such as a thermally activated azide–alkyne reaction.[54–56] Since the presence of catalysts and solvents is undesirable, useful resins were prepared just by heating azidated oil with alkynated oils at 100 °C (see Figure 1.12). The conversion of the thermally activated reaction reached 90% conversion after 24 hours.[57]

Copolymerization of functionalized oils with different reactants not only affects the polymer properties but also may reduce the bio-content. Using aromatic co-reactants introduces rigidity in the final polymer and may improve strength, as in polyurethanes cured with different isocyanates. Aliphatic diisocyanates give lower strength and more flexible materials, whereas aromatic triisocyanates were found to impart the highest strengths. Cycloaliphatic diisocyanates give similar properties to aromatic diisocyanates but have better light stability. Components with hydrogen bonds, such as polyurethanes and polyamides, usually offer greater strength and toughness. Polymers entirely made from triglycerides, such as epoxy resins from epoxidized oils, lack hydrogen bonding and have somewhat reduced strength, which is primarily controlled by the crosslinking density. This is

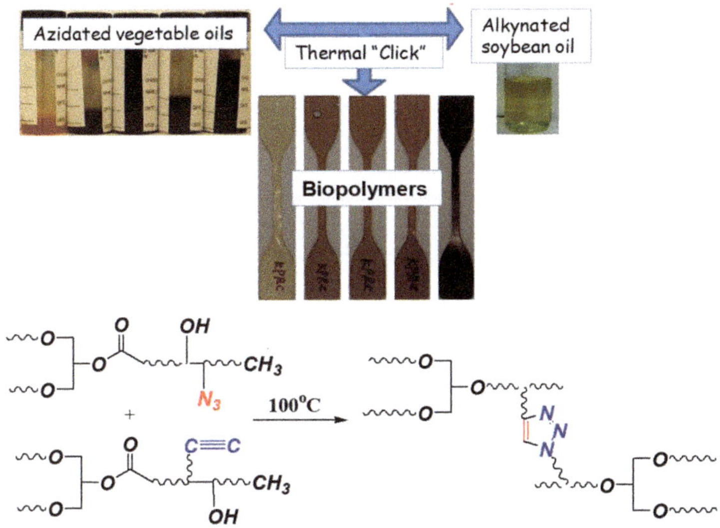

Figure 1.12 Appearance of different azidated vegetable oils and alkynated soybean oil and cast polymers with triazine rings.

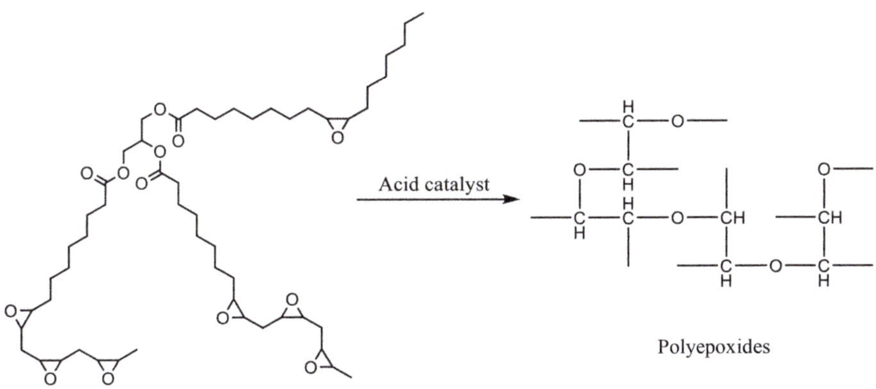

Figure 1.13 Schematic representation of epoxy resin preparation from epoxidized oils.

well-illustrated by comparing epoxy resins from linseed and soybean oils, shown in Figure 1.13 and Table 1.4.

Thus, copolymerization with different, primarily petrochemical co-reactants can be used for controlling the mechanical and other properties at the expense of bio-content.

1.13 Hydrolytic Stability of Oil Products

Ester bonds hydrolyze under conditions of high humidity and in the presence of bases. This is a serious issue in the case of thin products such as

Table 1.4 Properties of epoxidized linseed oil (ELO), epoxidized soybean oil (ESO), and their epoxy resins.

	ELO	ESO
Functionality, f_n	6–7	~4
EOC, %	9.5	7
T_g, °C	5	?
	Polymer	Polymer
T_g, °C	44	3
Tensile strength, MPa	14	3.3
Elongation at break, %	5.8	6.1
Modulus, MPa	350	34

coatings, exposed to environmental elements like heat and humidity. In the urethane industry, oil-based polyol systems may be delivered with base (amine) catalysts causing a change in properties (viscosity, acid and hydroxyl numbers) on storage with time. However, ester bonds are well-shielded by long, hydrophobic aliphatic chains and do not hydrolyze readily at room temperature in solid samples at moderate relative humidity. Rapid deterioration of cast polyurethane samples based on soybean oil polyols was found to occur in a water-saturated atmosphere above 80 °C.[58]

1.14 Electrical Properties of Natural Oil-based Polymers

Natural oils are relatively low polarity materials and have good electrical properties. Purified soybean oil is commercially used as an insulating liquid in transformers. The room temperature dielectric constant of oils (ε') is ~2.8. Epoxidized soybean oil has somewhat higher ε' of ~4.8.

Epoxy resins from epoxidized soybean oil displayed excellent insulation properties: a dielectric strength of 232 kV cm^{-1}, a volume resistivity of 1.1×10^{15} Ω cm, and a surface resistivity of 5.2×10^{15} Ω square^{-1}, comparable with those of standard polymeric insulating materials.[44] The dielectric properties of polyurethanes and other polymers are related to their composition and the presence of polar components. Generally, the permittivity (dielectric constant) of oil-based polyurethanes should be lower than those of standard polyurethanes with polyester and polyether polyols. It is not a constant and increases significantly above the glass transition temperature.

References

1. F. D. Gunstone, *The Chemistry of Oils and Fats*, CRC Press LLC, Boca Raton, FL, 2004.
2. *Physical and Chemical Characteristics of Oils, Fats, and Waxes*, ed. D. Firestone, AOCS Press, 2nd edn, 2006.
3. A. Jakab, K. Heberger and E. Forgacs, *J. Chromatogr. A*, 2002, **976**, 255–263.
4. Z. S. Petrović, L. Yang, A. Zlatanić, W. Zhang and I. Javni, *J. Appl. Polym. Sci.*, 2007, **105**, 2717–2727.

5. W. Sorenson and T. W. Campbell, *Preparative Methods of Polymer Chemistry*, Interscience Publishers, New York, London, 2nd edn, 1968.
6. I. Cvetković, J. Milić, M. Ionescu and Z. S. Petrović, *Chem. Ind.*, 2008, **62**, 319–328.
7. A. Rybak, P. A. Fokou and M. A. R. Meier, *Eur. J. Lipid Sci. Technol.*, 2008, **110**, 797–804.
8. M. Ionescu, D. Radojčić, X. Wan, Z. S. Petrović and T. A. Upshaw, *Eur. Polym. J.*, 2015, **67**, 439–448.
9. A. Gandini and T. M. Lacerda, *Polymers from Plant Oils*, Scrivener Publishing-Wiley, 2nd edn, 2018.
10. S. A. Madbouly, C. Zhang and M. R. Kessler, *Bio-Based Plant Oil Polymers and Composites*, Plastics Design Library, Elsevier/William Andrew, Amsterdam, The Netherlands, 2016.
11. Z. Liu and E. G. Kraus, *Green Materials from Plant Oils*, Royal Society of Chemistry, 1st edn, 2014.
12. D. Kyriacos, *Biobased Polyols for Industrial Polymers*, John Wiley & Sons, 2020.
13. D. G. Hayes and M.-J. Dumont, in *Industrial Oil Crops*, ed. T. A. McKeon, D. G. Hayes, D. F. Hildebrand and R. J. Weselake, AOCS Press, 2016, pp. 43–73.
14. M. Ionescu and S. Z. Petrovic, in *Soybean - Applications and Technology*, ed. T.-B. Ng, InTech, 2011.
15. O. S. Yemul and Z. S. Petrović, in *Contemporary Science of Polymeric Materials*, ed. L. Korugić-Karasz, ACS symposium series, 2010, ch. 2, vol. 1061, pp. 29–39.
16. Z. S. Petrović, W. Zhang and I. Javni, *Biomacromolecules*, 2005, **6**, 713–719.
17. S. Picataggio, T. Rohrer, K. Deanda, D. Lanning, R. Reynolds, J. Mielenz and L. D. Eirich, *Nat. Biotechnol.*, 1992, **10**, 894–898.
18. Z. S. Petrović, A. Guo and W. Zhang, *J. Polym. Sci., Part A: Polym. Chem.*, 2000, **38**, 4062–4069.
19. J. Veličković, Z. Petrović, D. Petrović-Djakov and J. Budinski, *27th Int. IUPAC Symposium sur les macromolecules Strasbourg*, 1981.
20. Z. S. Petrović, J. Milić, F. Zhang and J. Ilavsky, *Polymer*, 2017, **121**, 26–37.
21. F. D. Gunstone, *The chemistry of oils and fats, Sources, Composition, Properties and Uses*, Blackwell Publishing, Oxford, UK, 2004.
22. T. McKeon, D. Hayes, D. Hildebrand and R. Weselake, *Industrial Oil Crops*, 1st edn, 2016.
23. J. V. Crivello and K. D. Carlson, *Macromol. Rep.*, 1996, **33**, 251–262.
24. D. Radojčić, Z. S. Petrović, J. Tanasić and I. Ristić, *J. Polym. Environ.*, 2020, **28**, 1292–1301.
25. Z. S. Petrović, J. Hong, M. L. Vuković and J. Djonlagić, *Biocatal. Agric. Biotechnol.*, 2022, **39**, 102269.
26. CertainTeed, *US Pat.*, 9487654B2, 2016.
27. N. O. V. Sonntag, *J. Am. Oil Chem. Soc.*, 1982, **59**, 795A–802A.
28. H. Luo, Z. Zhai, W. Fan, W. Cui, G. Nan and Z. Li, *Ind. Eng. Chem. Res.*, 2015, **54**, 4923–4928.
29. F. O. Nitbani, P. J. P. Tjitda, B. A. Nurohmah and H. E. Wogo, *J. Oleo Sci.*, 2020, **69**, 277–295.

30. P. Flory, *Polym. J.*, 1985, **17**, 1–12.
31. B. Erman and J. E. Mark, *Structure and Properties of Rubberlike Networks*, Oxford University Press, New York, Oxford, 1997.
32. P. L. Flory, *Principles of Polymer Chemistry*, Cornell University Press, Ithaca, New York, 1953.
33. W. H. Stockmeyer, *J. Chem. Phys.*, 1943, **11**, 45–55.
34. M. Gordon and S. B. Ross-Murphy, *Pure Appl. Chem.*, 1975, **43**, 1–26.
35. K. Dušek, H. Galina and J. Mikeš, *Polym. Bull.*, 1980, **3**, 19–25.
36. K. Dušek, M. Duškova-Smrčkova, A. Zlatanić and Z. Petrović, *ACS-Polymers from Renewable Resources*, Orlando, FL, 2002.
37. Z. S. Petrović, A. Guo, I. Javni, I. Cvetković and D. P. Hong, *Polym. Int.*, 2008, **57**, 275–281.
38. T. G. Fox and S. J. Loshaek, *J. Polym. Sci.*, 1955, **15**, 371.
39. J. Bicerano, *Prediction of Polymer Properties*, Marcel Dekker, New York, 2002.
40. J. V. Crivello, R. Narayan and S. S. Sternstein, *J. Appl. Polym. Sci.*, 1997, **64**, 2073–2087.
41. D. Radojčić, J. Hong and Z. S. Petrović, *J. Polym. Environ.*, 2022, **30**, 765–775.
42. Z. Liu and A. Biswas, *Appl. Catal., A*, 2013, **453**, 370–375.
43. C. Noè, L. Iannucci, S. Malburet, A. Graillot, M. Sangermano and S. Grassini, *Macromol. Mater. Eng.*, 2021, 2100029.
44. D. Radojčić, J. Hong and Z. S. Petrović, *J. Polym. Environ.*, 2021, **30**, 765–775.
45. Inmot Corp., *USA Pat.*, 3673140, 1972.
46. Union Carbide Corp., *USA Pat.*, 4118405, 1978.
47. S. N. Khot, J. J. La Scala, E. Can, S. S. Morye, G. I. Williams, G. R. Palmese, S. H. Kusefoglu and R. P. Wool, *J. Appl. Polym. Sci.*, 2001, **82**, 703–723.
48. J. L. Scala and R. P. Wool, *J. Am. Oil Chem. Soc.*, 2002, **79**, 59–63.
49. L. Montero de Espinosa, J. C. Ronda, M. Galià and V. Cádiz, *J. Polym. Sci., Part A: Polym. Chem.*, 2009, **47**, 1159–1167.
50. Union Carbide Corp., *USA Pat.*, 3931071, 1976.
51. Union Carbide Corp., *USA Pat.*, 4220569, 1980.
52. S. Grishchuk and J. Karger-Kocsis, *eXPRESS Polym. Lett.*, 2011, **5**, 2–11.
53. Q. Luo, M. Liu, Y. F. Xu, M. Ionescu and Z. S. Petrović, *Macromolecules*, 2011, **44**, 7149–7157.
54. A. Biswas, B. K. Sharma, J. L. Willett, A. Advaryu, S. Z. Erhan and H. N. Cheng, *J. Agric. Food Chem.*, 2008, **56**, 5611–5616.
55. J. Hong, Q. Luo, X. Wan, Z. S. Petrović and B. K. Shah, *Biomacromolecules*, 2012, **13**, 261–266.
56. J. Hong, B. K. Shah and Z. S. Petrović, *Eur. J. Lipid Sci. Technol.*, 2013, **115**, 55–60.
57. J. Hong, D. Radojčić, D. Hairabedian, X. Wan and Z. S. Petrović, *Eur. J. Lipid Sci. Technol.*, 2015, **117**, 266–270.
58. I. Javni, R. Etekallapalli and Z. S. Petrović, in *Polyurethanes 2005 Technical Conference and Fair*, Houston, 2005.

CHAPTER 2

Natural Epoxy Oil (Euphorbia Oil) Polymerization in Liquid Carbon Dioxide–Green Solvents†

SHAILESH N. SHAH[a] AND ZENGSHE LIU*[b]

[a] Chemistry and Biochemistry Department, The University of Texas at Dallas, 800 W. Campbell Rd., Richardson, TX 75080, USA; [b] United States Department of Agriculture, Agricultural Research Service, National Center for Agricultural Utilization Research, Bio-Oils Research Unit, 1815 N. University St., Peoria, IL 61604, USA

*Email: kevin.liu@usda.gov

2.1 Introduction

Vegetable oils are non-toxic, biodegradable, renewable, environmentally friendly, non-polluting and non-toxic. Globally, vegetable oils contribute to approximately 80% of fat production, while animal fats contribution has declined to 20%.[1] Vegetable oils are the most essential raw materials used in the synthesis of biobased products from agricultural resources, in which plant oils, polysaccharides, sugars, wood and other materials have been used. A variety of industrial applications have been developed with soybean, sunflower and canola oils, including plastics, lubricants, adhesives, inks, fuels/biodiesel, solvents and surfactants. Coatings, printing

†DISCLAIMER: mention of trade names or commercial products in this publication is solely for the purpose of providing specific information and does not imply recommendation or endorsement by the U.S. Department of Agriculture. USDA is an equal opportunity provider and employer.

Green Chemistry Series No. 83
Green Chemistry and Green Materials from Plant Oils and Natural Acids
Edited by Zengshe Liu and George Kraus

Published by the Royal Society of Chemistry, www.rsc.org

inks and varnishes have been synthesized using linseed oil and tung oil for generations.[2–10]

There has been a continuous worldwide growth in vegetable oil production in the past few decades, though not all oil production is increasing equally. In the 2021–2022 crop year, vegetable oil production surpassed 200 million metric tons worldwide. At 75.5 million metric tons, palm oil ranked first among the major categories of vegetable oils. In 2021–2022, palm oil represented more than 35% of vegetable oil production, followed by soybean oil (28%), rapeseed oil (13%) and sunflower oil (10%). Palm oil takes the top spot globally. The second place is occupied by soybean oil. Corn is the largest crop in the United States, followed by soybeans (https://www.statista.com/statistics/263933/production-of-vegetable-oils-worldwide-since-2000/).

2.1.1 Plant or Natural Oil-based Polymers

In recent years, several research studies have been reported for the synthesis of polymers from vegetable oils,[11–17] in which the conversion of soybean oil (SBO) and other vegetable oils into conjugated oils and polymers and their physicochemical properties have been discussed. Epoxidation of plant oils can lead to the synthesis of polymeric materials through a variety of chemical reactions. A plant oil polymer can be soft, rubbery or a stiff thermoset, depending on its stoichiometry and the type of oil used.

According to Cowan *et al.*[18] and Erhan *et al.*,[19] vegetable oils can be polymerized at high temperatures (300 °C) and extended reaction times to form polymers with relatively low molecular weights (MWs, 8900 g mol^{-1}).

AESO is an acrylated epoxidized soybean oil resin[20–23] that can be combined with natural fibre mats made from flax, cellulose, pulp and hemp to produce composites. Recent studies have shown that polyurethanes can be generated from vegetable oil-based polyols. A review has been recently published by Kaikade *et al.*[24] SBO was also extensively investigated for its ability to produce polyols and polyurethanes.[25]

2.1.2 Natural Plant Oils with Epoxide Rings

Epoxidation of oils can lead to the synthesis of polymeric materials through a variety of chemical reactions.

Epoxidized soybean oil (ESO) is manufactured *via* the Prileschajew reaction using peracids.[26] During the Prileschajew reaction, alkenes are chemically reacted with alkyl acids to form epoxides, and this reaction is also referred to as the Prileschajew epoxidation by Li and Corey,[27] as shown in Figure 2.1.

Thus, to build an epoxide (oxirane) ring across the conjugated bond of a fatty acid of a natural plant oil, chemical modification of the conjugated bond by peracids was carried out. According to Wai *et al.*,[28] degradation of the oxirane ring is one of the unwanted side reactions of epoxidation of plant

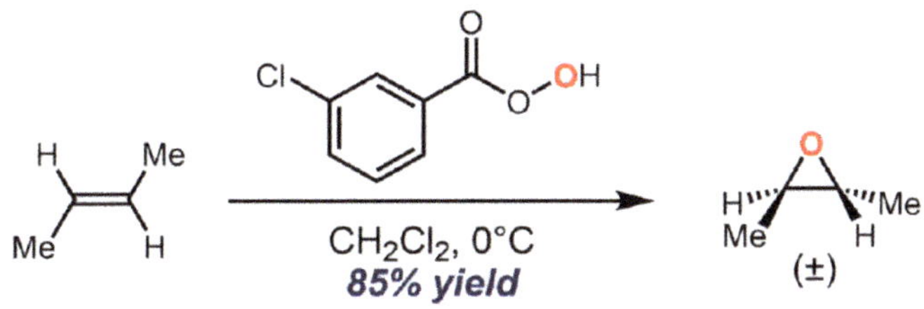

Figure 2.1 Epoxidized double bonds in soybean oil synthesis *via* the Prileschajew reaction using peracids. Reproduced from https://en.wikipedia.org/wiki/Prilezhaev_reaction, under the terms of the CC BY-SA 4.0 license https://creativecommons.org/licenses/by-sa/4.0/.

oils. To overcome the above-mentioned degradation of the oxirane ring, researchers looked for naturally occurring oils that have an inbuilt epoxy ring. By using direct polymerization of this naturally epoxidized oil, new materials can be synthesized.

Recently, ESO has been polymerized directly in supercritical carbon dioxide ($scCO_2$) with the catalyst boron trifluoride diethyl etherate ($BF_3{\cdot}OEt_2$) resulting in the formation of a polymer as reported by Liu *et al.*[29]

Furthermore, Liu *et al.*[30] reported that $BF_3{\cdot}OEt_2$ catalysed ring-opening polymerization of epoxidized soybean oil (RPESO) in liquid CO_2. These studies reveal that polymer MWs can be controlled by the initiator concentration, CO_2 pressure and change in the temperature. In the presence of 65.5 bar of subcritical CO_2, RPESO occurred under mild conditions at room temperature. As a result of RPESO, highly cross-linked polymers are formed. These polymers have glass transition temperatures (T_g) between −11.9 °C and −24.1 °C. They are thermally stable at temperatures below 200 °C.

Similar experiments were conducted by Liu and Shah[31] with jojoba oil, resulting in oligomerization. Its low MW and solid nature make it suitable for cosmetics, pharmaceuticals, dermatological products, lubricants, nano-lubricants, non-caloric fat-free food additives and plasticizers.

The purpose of this chapter is to discuss the polymerization of the naturally occurring epoxide ring in EuO as shown in Figure 2.2. EuO, extracted from *Euphorbia resinifera* seeds and rich in natural epoxy rings, has been exploited for the synthesis of polymeric materials through a variety of chemical reactions. By using direct polymerization, natural epoxy rings can be converted into new polymeric materials for different applications in the lubricant and coating sectors.

A native of Spain, *Euphorbia lagascae* (Spreng.) is a drought-tolerant plant. Euphorbia seed contains up to 45–50% oil, most of which (60–65%) is vernolic acid.[32] Vernolic acid is a C:12 epoxy fatty acid as shown in Figure 2.3 and it has potential use as a drying solvent in alkyd resin paints. In conventional paints, vaporization of alkyd resins in the paint results in the release of volatile organic compounds (VOCs). Paints formulated with vernolic acid would contain significantly reduced VOCs which are found in air pollution.

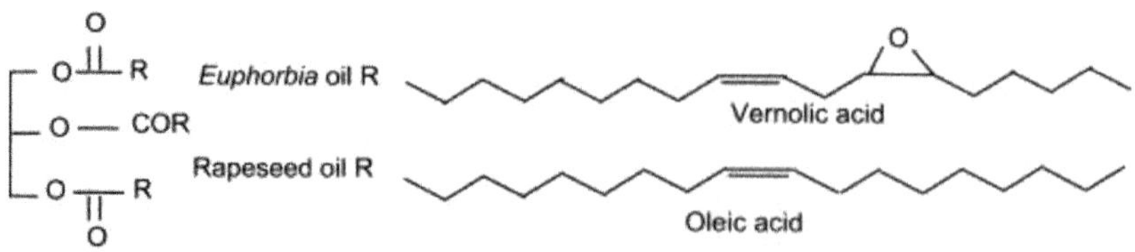

Figure 2.2 Chemical structure of EuO. Reproduced from ref. 50 with permission from Elsevier, Copyright 2013.

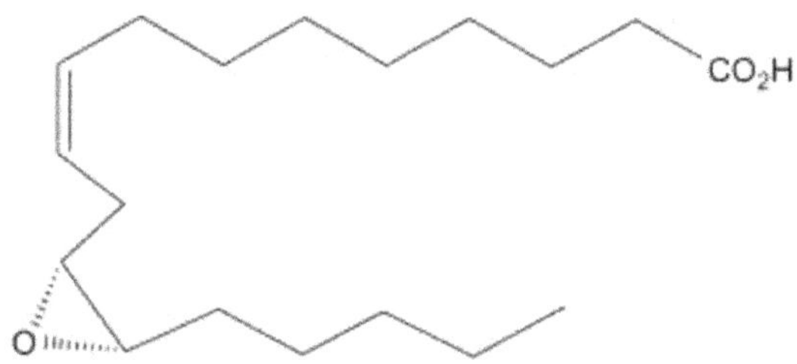

Figure 2.3 Chemical structure of vernolic acid. Reproduced from ref. 50 with permission from Elsevier, Copyright 2013.

Another most important natural plant oil with epoxy groups is vernonia oil, which contains a high proportion of vernolic acid. Ncube *et al.*[33] reported that vernonia seeds contain about 72–80% of the constituent acids of the oil, comprising an epoxy acid commonly known as vernolic acid (*cis*-12,13-epoxy-*cis*-9-octadecenoic acid, as shown in Figure 2.3). Vernonia oil has many properties like ESO and linseed oil. As a natural epoxide, it is transparent and homogeneous and exhibits low viscosity at room temperature. It can also be poured at temperatures below 0 °C.[34]

In comparison with vernonia oil and commercial epoxidized oils, EuO has a lower oxirane content especially when compared to commercial epoxidized oils and natural epoxy oils. Krewson *et al.*[35] and Riser *et al.*[36] used derivatives of these oils as plasticizers and stabilizers for polyvinyl chloride. The epoxy groups in EuO make it a potential monomer for making biopolymers, as shown in Figures 2.2 and 2.3.

Environmentally friendly alternatives to VOCs have gained attention over the past several years.[37–41] $scCO_2$ is one promising candidate.[42] There are numerous advantages of using $scCO_2$, such as its low cost, availability and non-flammability. For the synthesis of polymers and processing of biomaterials,[43–46] $scCO_2$ is an appealing medium due to its low toxicity and lack of toxic solvent residues.

In the current chapter, we discuss in detail how the natural oxirane ring in EuO is opened and polymerized using $scCO_2$ as a green solvent. CO_2 has a critical pressure of 73.8 bar and a critical temperature of 31.1 °C. It may be challenging to obtain a high P_c even though the T_c is low. However, liquid CO_2 provides many of the advantages of $scCO_2$ without the additional pressure and temperature.

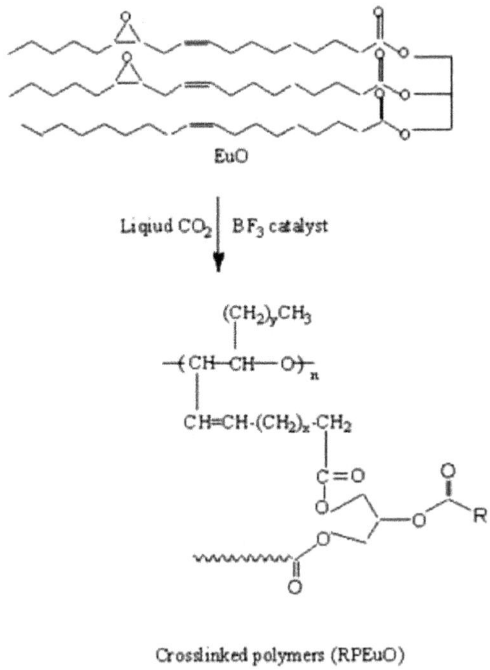

Figure 2.4 Synthesis route for RPEuO. Reproduced from ref. 50 with permission from Elsevier, Copyright 2013.

This chapter describes the boron trifluoride (Lewis's acid)-catalysed RPEuO in a liquid CO_2 medium, as shown in Figure 2.4.

Polymers formed by this process are referred to as RPEuO polymers. A study was conducted to investigate the effects of initiator and monomer concentrations on the thermal properties of RPEuO polymers. A number of spectroscopy techniques were used to characterize and confirm the structure of the RPEuO polymers, including Fourier transform infrared (FTIR), ^{1}H-NMR, ^{13}C-NMR and solid-state ^{13}C-NMR spectroscopy. Thermogravimetric analysis (TGA) and differential scanning calorimetry (DSC) were used to study the thermal properties of RPEuO polymers. The MW of the extracted soluble substances from RPEuO polymers was determined by gel permeation chromatography (GPC).

2.2 Materials, Methods and Procedures

2.2.1 Materials

2.2.1.1 Euphorbia Seeds

Dr Richard Roseburg, Oregon State University, Klamath Basin Research Extension Center, Klamath Falls, OR, USA, offered the euphorbia seeds (51% oil content).

2.2.1.2 EuO Extraction

A hydraulic press (Model 3851, Carver, Inc., Wabash, IN, USA) equipped with a cage apparatus was utilized to extract the oil from the seeds (Carver Catalog No. 2094). Euphorbia seeds in 3.81 cm-thick layers were placed inside a slotted cage cylinder (8.89 cm inner diameter × 19.05 cm high), which was divided by filter pads and separator discs. The seeds were compressed by gradually increasing the plunger's pressure (up to 151.7 bar) until the oil stopped dripping from the cage's side. Whatman filter paper no. 4 was used to filter EuO which has a light amber colour.

2.2.1.3 Chemicals

Aldrich Chemical Inc. (Milwaukee, WI, USA) supplied BF_3-OEt_2 that had been purified and redistilled. Carbon dioxide (>99.8%) was acquired from Linde Gas LLC (Independence, OH, USA). Bicarbonate, hexane (98.5%, A.C.S. quality) and tetrahydrofuran (THF, A.C.S. quality) were purchased from Aldrich Chemical Inc.

2.2.1.4 Fatty Acid Composition Determination by Gas Chromatography (GC)

The fatty acid composition of EuO was examined by using GC. Using a Polymer Laboratories (Long Beach, CA, USA) 8400 GC with a flame ionization detector and a Supelco (Bellefonte, PA, USA) Model 2380 column (30 m, 0.25 mm i.d., 0.20 m film thickness), fatty acids were separated (triplicates, means stated). Hexane at 1 mL min^{-1} served as the carrier gas. The oven's temperature was originally maintained at 150 °C for 15 min, then increased to 210 °C at 2 °C min^{-1} and 220 °C at 50 °C min^{-1} and then held at 220 °C for 10 min. Temperatures for the injector and detector were 240 °C and 270 °C, respectively. By comparing the peaks' retention periods to those of reference standards, the peaks were identified.

2.2.1.5 Spectroscopic Characterization

Spectroscopic characterization, FTIR, GC, ^{1}H-NMR and ^{13}C-NMR were carried out using the following spectroscopic instrumentation.

2.2.1.6 Fourier-transform Infrared Spectroscopy (FTIR)

A Thermo Nicolet Nexus 470 FTIR system (Madison, WI, USA) coupled with Smart ARK accessory for liquid samples in the scanning range of 650–4000 cm^{-1} for 32 scans at a spectral resolution of 4 cm^{-1} was employed to record the FTIR spectra. Using the FTIR system with the Smart Orbit attachment, solid sample spectra were captured.

2.2.1.7 Nuclear Magnetic Resonance (NMR) Spectroscopy

A Bruker (Rheinstetten, Germany) ARX-500 NMR spectrometer operating at frequencies of 500.13 and 125.77 MHz, respectively, was used to record the ^{1}H-NMR and ^{13}C-NMR spectra for extracted soluble compounds from RPEuO samples. The probe employed was a 5 mm inverse Z-gradient probe in $CDCl_3$ (Cambridge Isotope Laboratories, Andover, MA, USA). Using a Bruker ARX-300 NMR spectrometer, solid-state ^{13}C-NMR spectra for extracted insoluble compounds were recorded.

2.2.1.8 High-performance Liquid Chromatography (HPLC)

An HPLC system from Waters Corporation (Milford, MA, USA) which included a 1515 isocratic HPLC pump, a 717 plus automated injector, a column heater and controlled with Breeze software was used to obtain a GPC profile. Polymer Laboratories (Long Beach, CA, USA) provided the MIXED-E 3 m and Guard 5 m columns (part numbers PL1110-6300 and PL1110-1520, respectively) used for separation (Varian, Inc., Amherst, MA, USA). Astra V macromolecular characterization software (Wyatt Technology Corporation, Santa Barbara, CA, USA) was used to process signals generated by mini-DAWN TREOS triple-angle light scattering and Optilab rEX refractive index detectors. A flow rate of 1 mL min^{-1} of THF was used in the mobile phase, and the columns were maintained at 40 °C. Fisher Scientific (Suwanee, GA, USA) supplied THF stabilized with butylated hydroxytoluene at a known concentration near 4.00×10^{-3} g mL^{-1} for solution preparation of the liquid phase samples. From a 1 mL sample vial, 100 μL was injected using the Waters Autosampler. MWs of all polymers of RPEuO were calibrated using linear polystyrene standards (Polymer Laboratories, Long Beach, CA, USA) with $M_n = 580–100k$ and $M_w/M_n = 1$. Calculation of MW was performed with Astra V software.

2.2.1.9 Thermal Properties of Polymers

TGA and DSC techniques were employed to examine the thermal properties of the polymers.

2.2.1.10 Thermogravimetric Analysis (TGA)

The weight loss of the RPEuO samples was measured using a TA Instruments (New Castle DE, USA) Q500 thermogravimeter with an autosampler under a flowing nitrogen atmosphere. To measure the thermal properties of RPEuO, 20 mg of sample was used in TGA. Weight loss was measured as a function of temperature for samples heated at a rate of 10 °C min^{-1} from 30 °C to 600 °C.

2.2.1.11 *Differential Scanning Calorimetry (DSC)*

A TA Instruments Q2000 model thermogravimeter with an autosampler was utilized to record the DSC thermograms of the test materials. The RPEuO sample was carefully weighed at 10 mg and then placed in an aluminium pan sealed with pin-perforated lids. To remove thermal history and any moisture, the DSC oven was ramped at 10 °C min^{-1} to 110 °C. The sample was equilibrated at −60 °C from 110 °C using a refrigerated chilling system at a rate of 5 °C min^{-1}. Data were gathered when the oven temperature was increased by 5 °C min^{-1}, from −60 °C to 150 °C. DSC was run under an inert atmosphere by purging the oven with nitrogen at a rate of 50 mL min^{-1}. Thermal Advantage and Universal Analysis software (TA Instruments) was used to analyse the data.

2.2.1.12 *MW Determination of Polymers*

Gel permeation chromatography (GPC) was used to determine the MW of soluble compounds that were extracted from polymers.

2.2.2 Procedure

2.2.2.1 *EuO Natural Epoxy Ring Opening Polymerization in CO_2*

In a Parr Instrument Company (Moline, IL, USA) 4560 mini benchtop unit equipped with a Parr 4843 controller and thermocouple, ring-opening polymerization reactions were carried out in a 300 mL high Parr pressure reactor. Quartz viewing windows were installed in the reactor body. Figure 2.5 illustrates the experimental setup used for ring-opening polymerization of EuO. To charge the reactor with CO_2, a Teledyne Isco, Inc. (Lincoln, NE, USA) Model 260D high

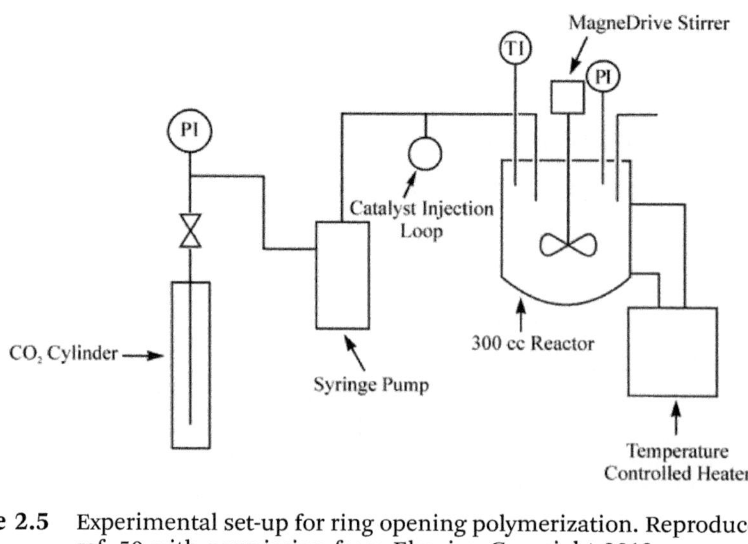

Figure 2.5 Experimental set-up for ring opening polymerization. Reproduced from ref. 50 with permission from Elsevier, Copyright 2013.

Figure 2.6 Starting material for EuO (liquid) and RPEuO polymers (solid). Reproduced from ref. 50 with permission from Elsevier, Copyright 2013.

pressure syringe pump was used. The reactor is typically filled with 30 g of EuO and then sealed. The reactor was purged with nitrogen for 5 min. Then CO_2 was injected into the reactor to a pressure of 62.1 bar. A Parr 4843 controller controlled the temperature. BF_3-OEt_2 was added to the reactor using a Rheodyne injector (Rheodyne LLC, Rohnert Park, CA, USA) after the reactor was heated to 25 °C. At a pump pressure of 65.5 bar, CO_2 was then injected into the injection loop. To deactivate the catalyst after the reaction had been going for 3 h, 2 mL of ethanol/water (1 : 1) was added to the reactor. The white polymers (RPEuO) were repeatedly washed with water, 5% aqueous sodium bicarbonate and water. At 60 °C, the RPEuO polymer was dried in a vacuum. RPEuO sample weighing between 28 g and 30 g was collected. The liquid precursor EuO and RPEuO polymer (white solid sample) are shown in Figure 2.6. The soluble material was removed from the RPEuO using the Soxhlet extraction method, with hexane as the reflux solvent for MW and spectroscopic (FTIR, ^{1}H-NMR and ^{13}C-NMR) characterization. After Soxhlet extraction, the residual insoluble materials were subjected to TGA, DSC and solid-state ^{13}C-NMR spectroscopic studies.

2.3 Results and Discussion

2.3.1 Analysis and Composition of EuO

The fatty acid composition and physical characteristics of EuO determined by GC are summarized in Tables 2.1 and 2.2 respectively. It is noticeable that 60% of the EuO composition is vernolic acid.

2.3.2 Effect of Initiator Concentration

Our previous studies showed that as the ring-opening polymerization of ESO progressed from 20–350 °C, the T_g of the formed polymers slightly decreased.[29]

Table 2.1 Fatty acid composition (wt%) of euphorbia oil (EuO). Reproduced from ref. 50 with permission from Elsevier, Copyright 2013.

Fatty acid[a]	EuO
C10:0	0.19
C12:0	0.12
C14:0	1.08
C16:0	4.33
C18:0	1.54
C18:1 9c	20.99
C18:1 11c	0.70
C18:2 9c, 12c	9.35
C18:3 9c, 12c, 15c	0.30
C20:0	
C20:1 11c	1.07
C22:3 13c, 16c, 19c	
C24:1 15c	
Vernolic acid	**60.34**

[a] For example, C18:1 9c signifies an 18-carbon fatty acid chain with one *cis* (c) double bond located at carbon 9.

Table 2.2 Physical characteristics of euphorbia oil (EuO). Reproduced from ref. 50 with permission from Elsevier, Copyright 2013.

Physical properties	EuO
Cloud point, °C	−22 (1)
Pour point, °C	−32 (1)
Oxidation stability index, 110 °C, h	9.4 (0.2)
ν, 25 °C, $mm^2\,s^{-1}$	113.65
ν, 40 °C, $mm^2\,s^{-1}$	59.72
ν, 100 °C, $mm^2\,s^{-1}$	10.87
Viscosity index	177
Specific gravity, 25 °C	0.953
Specific gravity, 40 °C	0.943
Lubricity, 60 °C, μm	125
Acid value, mg KOH per g	17.5
Iodine value	90.26

There are several factors that determine whether cyclic monomers can be converted into polymers, but thermodynamics is the most significant. However, the reaction is controlled by the relative stability of the cyclic monomer and polymer structure.[47,48] Chanda[49] reports that ring-opening polymerization is more favourable due to its thermodynamic nature (ΔG, free-energy change is negative).

The standard Gibbs free energy change (ΔG) in the current ring-opening polymerization indicates that it is the thermodynamic favourability of a physical or chemical process. When $\Delta G < 0$, the process is thermodynamically favoured. For a current polymerization process, the value of ΔG can be calculated directly from the values of ΔH and ΔS using the following equation (following the thermodynamic relation):

$$\Delta G = \Delta H - T\Delta S,$$

where ΔH is the enthalpy change, ΔS is the entropy change, T is the temperature (°K), the temperatures used in this system, and ΔH is the major factor for determining ΔG for a 3-membered ring, while ΔS would be more important for the 5- and 6-membered ring-opening reactions. The 3-membered ring opening reactions are determined by ΔG, while the 5- and 6-membered rings are determined by ΔS. Sawada[48] reported that 3-membered rings have a strongly negative value for ΔH, which is -113.0 kJ mol^{-1}, while for the 5- and 6-membered ring-opening reactions, ΔS is an especially important factor.

Since EuO molecules contain multiple epoxy groups, ring-opening polymerization facilitated the formation of cross-linked polymers. In this ring-opening polymerization reaction, temperature (T) has no significant effect. Consequently, 250 °C is used as the T for RPEuO since it is convenient.

To determine the effects of initiator loading on the thermal properties of the generated RPEuO polymers, initiator loadings between 1.87 mmol and 4.66 mmol were evaluated. For example, Figure 2.7 shows the determination of T_g by DSC of RPEuO-I (insoluble fraction after extraction). Table 2.3 summarizes all the T_g data for polymers. As the initiator loading increases, T_g increases as well. It is expected that these results will be achieved because

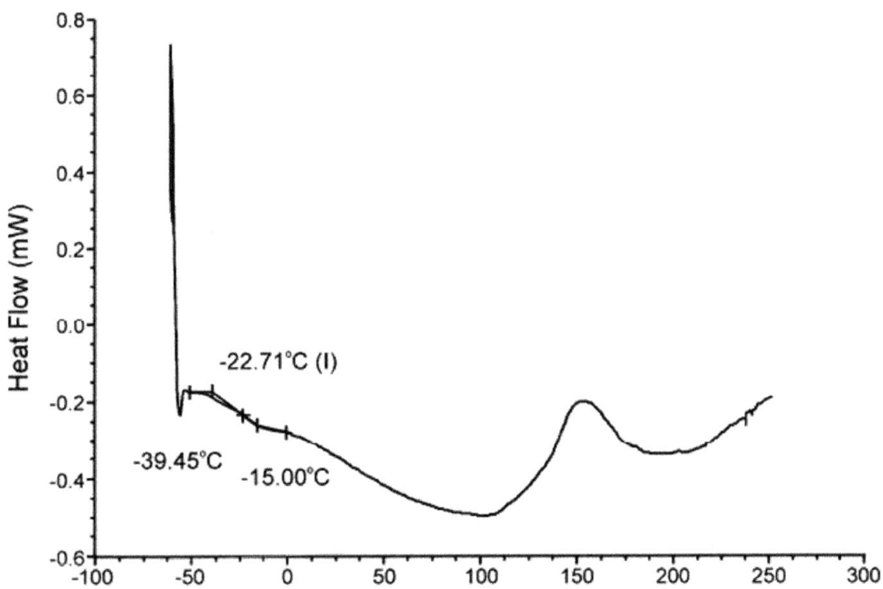

Figure 2.7 Glass transition temperature (T_g) thermogram of RPEuO-I by DSC. Reproduced from ref. 50 with permission from Elsevier, Copyright 2013.

Table 2.3 The glass transition temperatures (T_g) of RPEuO prepared at various initiator concentrations. Reproduced from ref. 50 with permission from Elsevier, Copyright 2013.

Sample entry	Polym. temp. (°C)	Initiator (mmol)	T_g (°C)
RPEuO-I	25	1.87	−22.7
RPEuO-II	25	2.79	−15.9
RPEuO-III	25	4.66	−15.0

the MW of a linear polymer decreases with increasing initiator concentration, and thus their T_g decreases similarly as well. Based on a comparison of RPEuO-I and RPEuO-III, it is shown that increasing the initiator loading from 1.87 mmol to 4.66 mmol results in a change of T_g from −22.7 °C to −15.0 °C in the resultant polymers. This point establishes that initiator concentration influences the MW of the resulting polymers which affects their thermal properties. As shown in Table 2.5, the current results are in agreement with the results of the crosslink density study. Upon hexane extraction, RPEuO-I showed a soluble content of 39%, RPEuO-II showed a soluble content of 29% and RPEuO-III showed a soluble content of 14%, which will be discussed in the following sections. Our previous research[29] on polymers derived from ESO revealed a different trend in T_g for RPEuO influenced by the initiator loading. Since ESO contains a much higher content of epoxy compounds than EuO, its polymer chains are more cross-linked than those of RPEuO, resulting in a different T_g value. Moreover, vernolic acid contains some double bonds. Movement of the side group is easier in RPEuO than in RPESO. As a result of the smaller content of epoxy groups present in EuO, the resultant RPEuO polymers are more linear.

2.3.3 Effect of Monomer Concentration

The thermal properties of RPEuO polymers were investigated with monomer loading between 30 g and 50 g, and Table 2.4 summarizes the results. It was observed that increasing the monomer concentration resulted in a slight decrease in T_g. As anticipated, the MW of RPEuO polymers decreased with an increase in monomer concentration. It is obvious that a polymer with lower MW exhibits lower T_g. Once again, Table 2.5 indicates that the RPEuO polymer exhibits a lower crosslinking density when derived from higher monomer concentrations. This is demonstrated by a higher percentage of the weight which was extracted by hexane.

2.3.4 Soxhlet Extraction and Characterization of the Extracted RPEuO-soluble Substance

A hexane refluxed Soxhlet extractor was used for 16 h to extract all RPEuO samples. As shown in Table 2.5, the soluble substances extracted from the

Table 2.4 The T_g of RPEuO polymers prepared at various monomer concentrations. Reproduced from ref. 50 with permission from Elsevier, Copyright 2013.

Sample entry	Polym. temp. (°C)	Initiator (mmol)	Monomer (mol)	T_g (°C)
RPEuO-III	25	4.66	0.03	−15.0
RPEuO-IV	25	4.66	0.04	−15.1
RPEuO-V	25	4.66	0.05	−15.3

Table 2.5 Data related to the extracted soluble substances from RPEuO polymers. Reproduced from ref. 50 with permission from Elsevier, Copyright 2013.

Sample entry	Polymer temp. (°C)	Initiator (mmol)	Monomer (g)	Soxhlet extraction (wt%)	
				Soluble	Insoluble
RPEuO-I	25	1.87	30	30	70
RPEuO-II	25	2.79	30	29	71
RPEuO-III	25	4.66	30	14	86
RPEuO-IV	25	4.66	40	22	78
RPEuO-V	25	4.66	50	35	65

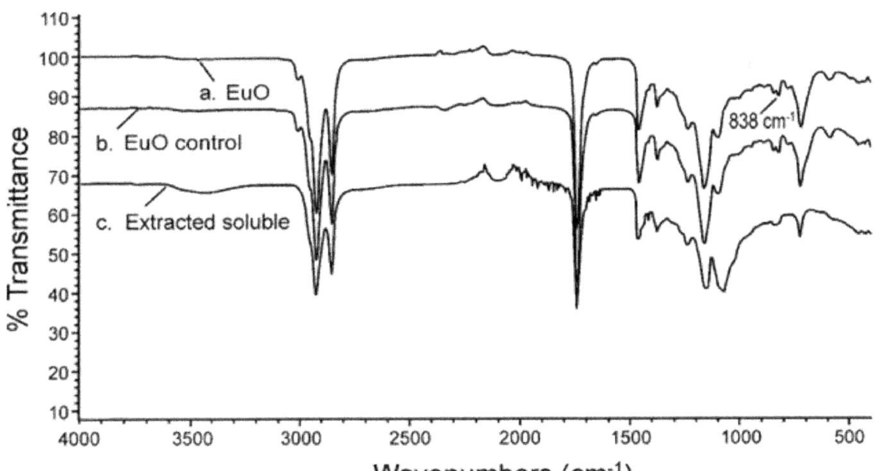

Figure 2.8 FTIR spectra of EuO, EuO (control) and RPEuO (soluble extract). Reproduced from ref. 50 with permission from Elsevier, Copyright 2013.

RPEuO samples ranged between 14 wt% and 35 wt%. The spectroscopic analytical methods mentioned above were used to characterize the soluble and insoluble components of the extract. The FTIR spectra are shown in Figure 2.8 for EuO, EuO control (experiment without catalyst) and hexane-extracted soluble substances. EuO and EuO control IR spectra clearly show absorption at 838 cm^{-1} for the oxirane group. A similar peak can also be observed in the extracted soluble substance from RPEuO, but with reduced strength, indicating that EuO underwent ring-opening polymerization, but there are no EuO monomers or small molecules such as dimers and trimers. Further analysis was conducted by using NMR and GPC.

Figure 2.9 shows the FTIR spectra of EuO and the extracted RPEuO-insoluble substance. The 838 cm^{-1} absorption peak is absent in the RPEuO polymer, which indicates that the hexane-extracted RPEuO-insoluble substances are the ring-opened products of EuO. These FTIR spectra demonstrated that the hexane-extracted RPEuO-insoluble substances are the ring-opening products of EuO because of the absence of the 838 cm^{-1} absorption peak in the RPEuO polymer.

Figure 2.10 shows the ^{1}H-NMR spectra of EuO and the RPEuO-soluble substances extracted with hexane. In both the EuO and RPEuO extracted soluble sample spectra, there are peaks related to epoxy protons at 2.7–2.9 ppm. The glycerol backbone methine (–CH_2–CH–CH_2–) protons and methylene (–CH_2–CH–CH_2–) protons of EuO are present at δ 5.1–5.3 ppm and δ 4.1–4.3 ppm, respectively, which indicates that the triglyceride structure of EuO does not change during the reaction.

The ^{13}C-NMR spectra of EuO and RPEuO-soluble substances extracted with hexane are shown in Figure 2.11. As seen in the ^{13}C-NMR spectra, the peaks at δ 54–57 ppm correspond to epoxy carbons. CH and CH_2 carbons of the –CH_2–CH–CH_2–glycerol backbone are also observed at δ 69 ppm and δ 63 ppm, respectively. As demonstrated in the ^{13}C-NMR spectra, epoxy carbons have peaks at δ 54–57 ppm.

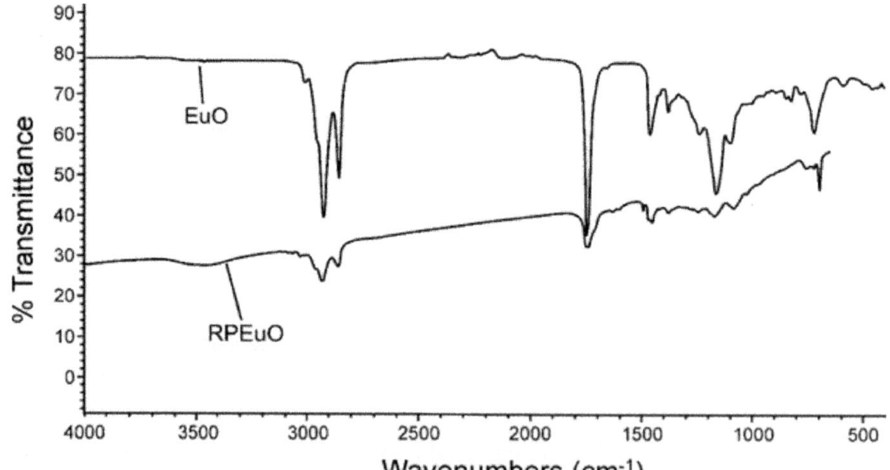

Figure 2.9 FTIR spectra of EuO and the RPEuO (insoluble extract) polymer. Reproduced from ref. 50 with permission from Elsevier, Copyright 2013.

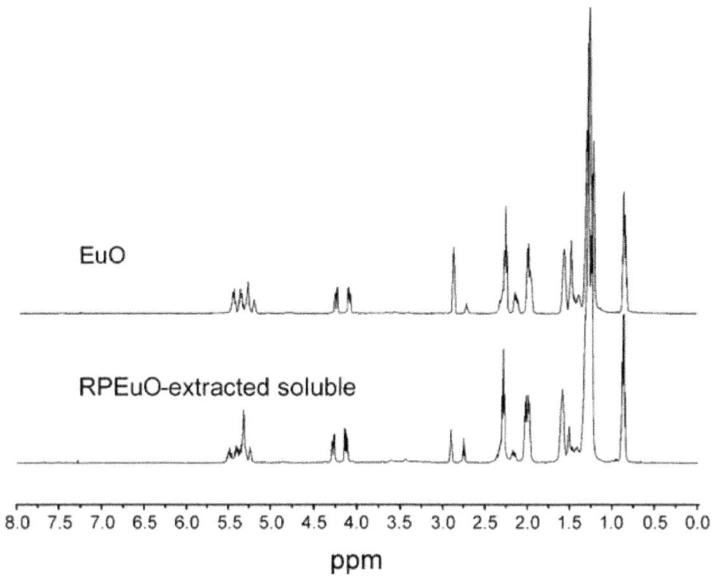

Figure 2.10 Proton NMR spectra of EuO and RPEuO (soluble extract). Reproduced from ref. 50 with permission from Elsevier, Copyright 2013.

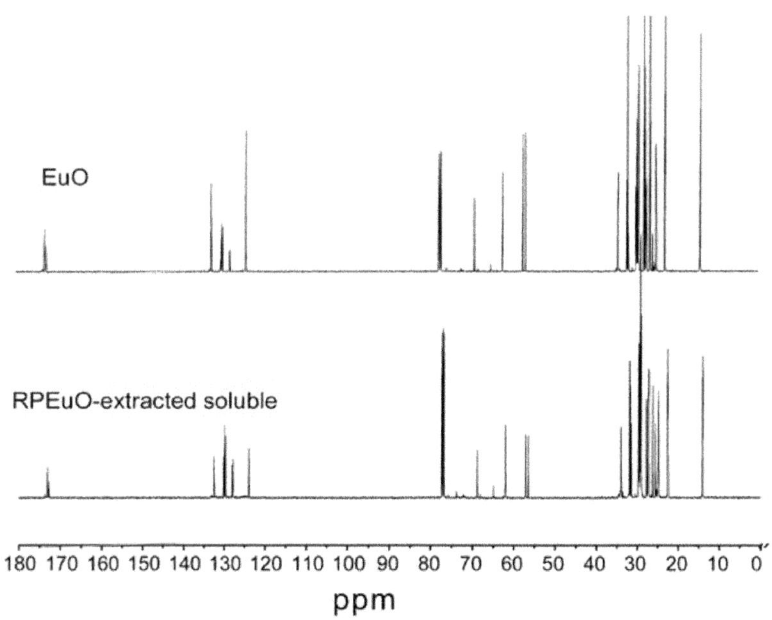

Figure 2.11 ^{13}C-NMR spectra of EuO and RPEuO (soluble extract). Reproduced from ref. 50 with permission from Elsevier, Copyright 2013.

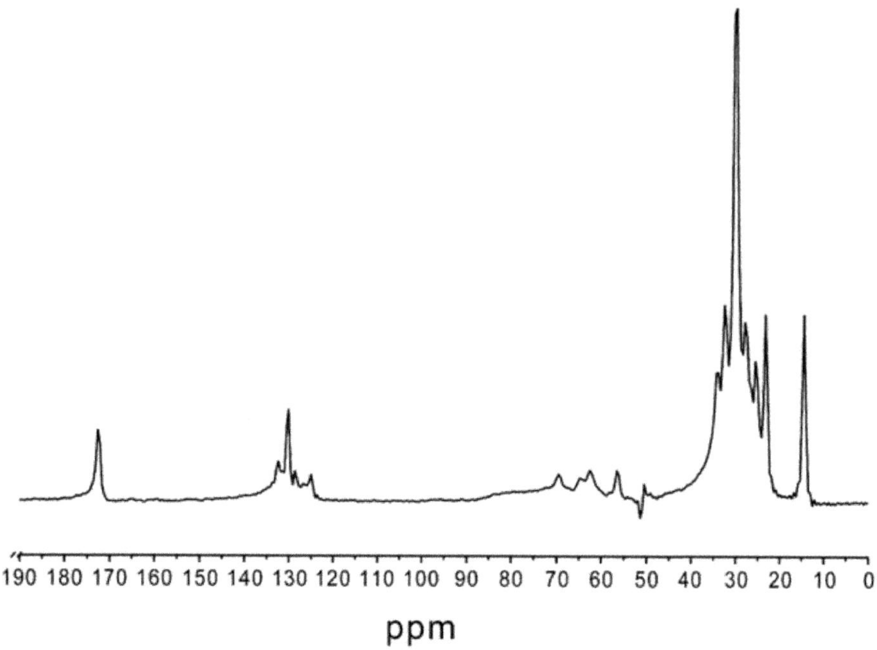

Figure 2.12 Solid-state ^{13}C-NMR spectrum of EuO (low MW) and RPEuO. Reproduced from ref. 50 with permission from Elsevier, Copyright 2013.

Based on the ^{1}H-NMR and ^{13}C-NMR spectra, we can conclude that the extracted RPEuO-soluble substances are lower MW EuO molecules. As revealed previously in the FTIR analysis section, the soluble fraction of EuO is composed of monomers as well as dimers and trimers. GPC analysis validates this result. After Soxhlet extraction of RPEuO-insoluble substances in hexane, the solvent yields cross-linked polymers. Solvents such as CH_2Cl_2, $CHCl_3$ and THF do not dissolve these cross-linked insoluble polymeric materials. The solid-state ^{13}C-NMR spectrum of the insoluble polymeric material is shown in Figure 2.12. Using solid-state ^{13}C-NMR spectra, valuable information was collected on these insoluble polymeric materials. The spectrum strongly displays the ester carbonyl carbon of the oil triglycerides at δ 173 ppm. The absence of the signal at δ 52 ppm strongly indicates that the carbon–carbon epoxy bond has been transferred to ether and establishes that EuO has undergone a ring-opening polymerization process and formed the cross-linked polymer.

2.3.5 Thermal Stability of RPEuO

The decomposition behaviour of RPEuO polymers was investigated using TGA under nitrogen conditions to learn more about their thermal properties.

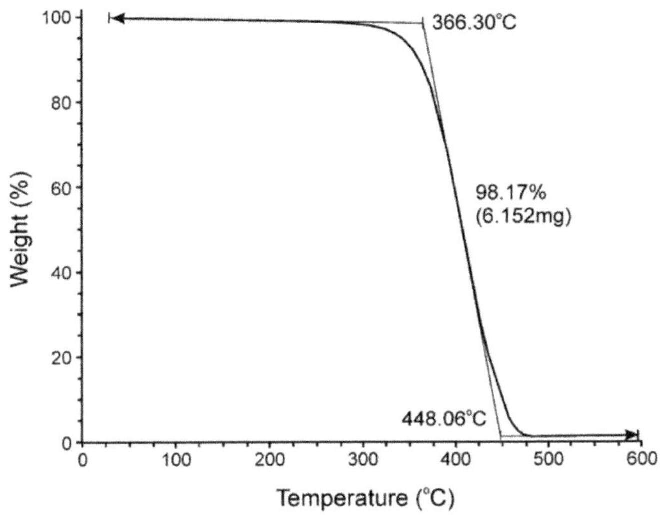

Figure 2.13 TGA spectra of RPEuO-I. Reproduced from ref. 50 with permission from Elsevier, Copyright 2013.

Table 2.6 Thermal stability data of the RPEuO polymers. Reproduced from ref. 50 with permission from Elsevier, Copyright 2013.

Samples	200–220 (°C)	240–320 (°C)	340–450 (°C)
RPEuO-I	Stable	2 wt% loss	87 wt% loss
RPEuO-II	Stable	3 wt% loss	85 wt% loss
RPEuO-III	Stable	4 wt% loss	85 wt% loss
RPEuO-IV	Stable	3 wt% loss	84 wt% loss
RPEuO-V	Stable	3 wt% loss	88 wt% loss

The TGA spectra of RPEuO-I are shown in Figure 2.13. RPEuO-I has remarkably high thermal stability below 200 °C, which is evident from the TGA spectrum in Figure 2.13. RPEuO-I samples experience weight loss at two different temperature ranges, which are 240–320 °C and 340–450 °C. At temperatures between 240 and 320 °C, the material slowly loses about 2% of its weight, followed by an almost immediate 80% loss at 340–450 °C. At temperatures between 240 °C and 450 °C, TGA measurements revealed a 98% weight loss. The TGA curves of other RPEuO samples obtained at various initiator and monomer concentrations are like those of RPEuO-I. It was observed that less than 5 wt% loss was observed around 240 °C for some samples. The TGA results for RPEuO polymers are summarized in Table 2.6. RPEuO polymers lose about 5 wt% of their weight in the first temperature region and 80 wt% of their weight in the second temperature region.

Acknowledgements

The authors gratefully acknowledge Mr Daniel Knetzer for help in GPC, DSC and TGA experiments, Erin Walter for help in the oil physical properties study and Dr Karl Vermillion for collecting the NMR spectra. This work is supported in part by the U.S. Department of Agriculture.

References

1. J. O. Metzger and U. Bornscheuer, *Appl. Microbiol. Biotechnol.*, 2006, **71**, 13.
2. A. Cunningham and A. Yapp, *US Pat.*, 3827993, 1974.
3. G. W. Bussell, *US Pat.*, 3855163, 1974.
4. L. E. Hodakowski, C. L. Osborn and E. B. Harris, *US Pat.*, 4119640, 1975.
5. D. J. Trecker, G. W. Borden and O. W. Smith, *US Pat.*, 3979270, 1976.
6. D. J. Trecker, G. W. Borden and O. W. Smith, *US Pat.*, 3931075, 1976.
7. D. K. Salunkhe, R. N. Adsule, J. K. Chavan and S. S. Kadam, *World Oilseeds: Chemistry, Technology and Utilization*, Van Nostrand Reinhold, New York, 1992.
8. C. G. Force and F. S. Starr, *US Pat.*, 4740367, 1988.
9. S. N. Shah, S. K. Mendon and S. F. Thames in *Green Materials from Plant Oils*, ed. Z. S. Liu and G. Kraus, The Royal Society of Chemistry, UK, 2015, p. 293.
10. N. Patel and S. N. Shah, in *Food, Energy, and Water, The Chemistry Connection*, ed. S. Ahuja, Elsevier, USA, 2015, p. 278.
11. R. P. Wool, S. H. Kusefoglu, G. R. Palmese, S. N. Khot and R. Zhao, *US Pat.*, 6121398A, 2000.
12. R. C. Larock, X. Dong, S. Chung, C. K. Reddy and L. E. Ehlers, *J. Am. Oil Chem. Soc.*, 2001, **78**, 447.
13. Z. S. Liu and S. Z. Erhan, *J. Polym. Environ.*, 2010, **18**, 243.
14. Y. Xia and R. C. Larock, *Green Chem.*, 2010, **12**, 1893.
15. G. Lligadas, J. C. Ronda, M. Galià and V. Cádiz, *Mater. Today*, 2013, **16**(9), 337.
16. F. W. Shaarani and J. J. Bou, *Sci. Total Environ.*, 2017, **598**, 931.
17. J.-G. Rosenboom, R. Langer and G. Traverso, *Nat. Rev. Mater.*, 2022, **7**, 117.
18. J. C. Cowan, W. C. Ault and H. M. Teeter, *Ind. Eng. Chem.*, 1946, **38**, 1138.
19. S. Z. Erhan and M. O. Bagby, *J. Am. Oil Chem. Soc.*, 1994, **71**, 1223.
20. W. Thielemans, E. Can, S. S. Morye and R. P. Wool, *J. Appl. Polym. Sci.*, 2002, **83**, 323.
21. W. Thielemans and R. P. Wool, *Polym. Compos.*, 2005, **26**, 695.
22. A. Bakthavachalam, K. B. Ayalew and S. Pilla, *ACS Omega*, 2019, **4**(26), 21799.
23. S. Abdollahi, B. M. Ebrahimi and M. Khorasani, *Polym. Test.*, 2022, **115**, 107727.
24. D. S. Kaikade and A. S. Sabnis, *Polym. Bull.*, 2023, **80**, 2239–2261.
25. A. Guo, Y. Cho and Z. S. Petrovic, *J. Polym. Sci., Part A: Polym. Chem.*, 2000, **38**, 3900.

26. N. Prileschajew, Oxydation ungesättigter Verbindungen mittels organischer Superoxyde [Oxidation of unsaturated compounds by means of organic superoxides], *Ber. Dtsch. Chem. Ges.*, 1909, **42**(4), 4811, https://en.wikipedia.org/wiki/Prilezhaev_reaction#cite_note-JJL-1.
27. J. J. Li and E. J. Corey, in *Name Reactions of Functional Group Transformations*, ed. J. J. Li and E. J. Corey, John Wiley & Sons, Hoboken, NJ, 2007, p. 274.
28. P. T. Wai, P. Jiang, Y. Shen, P. Zhang, Q. Gu and Y. Leng, *RSC Adv.*, 2019, **9**, 38119.
29. Z. S. Liu, B. K. Sharma and S. Z. Erhan, *Biomacromolecules*, 2007, **8**, 233.
30. Z. S. Liu, K. M. Doll and R. A. Holser, *Green Chem.*, 2009, **11**, 1774.
31. Z. S. Liu and S. N. Shah, *US Pat.*, 8742148B1, 2014.
32. R. L. Evangelista, T. A. Isbell, R. J. Roseberg and S. C. Cermak, Presented in part at the Association for the Advancement of Industrial Crops Annual Conference, Sonoma, CA, 2012.
33. I. Ncube, J. S. Read, P. Adlercreutz and B. Mattiasson, *Phytochemistry*, 1998, **47**(5), 723.
34. T. Abraham and R. Höfer in *Polymer Science: A Comprehensive Reference*, ed. K. Matyjaszewski, M. Moller, J. E. McGrath, M. A. Hickner and R. Hofer, Elsevier, Amsterdam, 2012, p. 15.
35. C. F. Krewson, G. R. Riser and W. E. Scott, *J. Am. Oil Chem. Soc.*, 1966, **43**, 377.
36. G. R. Riser, R. W. Riemenschneider and L. P. Witnauer, *J. Am. Oil Chem. Soc.*, 1996, **43**, 456.
37. J. A. Darr and M. Poliakoff, *Chem. Rev.*, 1999, **99**, 495.
38. J. Dupont, C. S. Consorti and J. Spencer, *J. Braz. Chem. Soc.*, 2000, **11**, 337.
39. E. J. Beckman, *J. Supercrit. Fluids*, 2004, **28**, 121.
40. S. Bektesevic, A. M. Kleman, A. E. Marteel-Parrish and M. A. Abraham, *J. Supercrit. Fluids*, 2006, **38**, 232.
41. Ž Knez, M. Pantić, D. Cör, Z. Novak and M. Knez Hrnčič, *Chem. Eng. Process.*, 2019, **141**, 107532.
42. T. Wu and B. Han in *Encyclopedia of Sustainability Science and Technology*, ed. R. Meyers, Springer, New York, 2019, p. 1.
43. R. Butler, C. M. Davies and A. I. Cooper, *Adv. Mater.*, 2001, **13**, 1459.
44. A. I. Cooper, *Adv. Mater.*, 2003, **15**, 1049.
45. C. D. Wood, B. Tan, H. Zhang and A. I. Cooper in *Thermodynamics, Solubility and Environmental Issues*, ed. T. Letcher, Elsevier, 2007, p. 383.
46. P. L. Pavlova, A. V. Minakov, D. V. Platonov, V. A. Zhigarev and D. V. Guzei, *Sustainability*, 2022, **14**, 698.
47. H. R. Allcock, *J. Macromol. Sci., Revs. Macromol. Chem.*, 1970, **C4**, 149.
48. H. Sawada, *Thermodynamics of Polymerization*, Marcel Dekker Inc., New York, 1976, p. 403.
49. M. Chanda, *Advanced Polymer Chemistry: A Problem-Solving Guide*, Marcel Dekker Inc., New York, 2000.
50. Z. Liu, S. N. Shah, R. L. Evangelista and T. A. Isbell, *Ind. Crops Prod.*, 2013, **41**, 10–16.

CHAPTER 3

Soybean Oil-based Polymers and Their Composites

W. LIU, T. CHEN AND R. QIU*

College of Transportation and Civil Engineering, Fujian Agriculture and Forestry University, Fuzhou, 350108, P. R. China
*Email: renhuiqiu@fafu.edu.cn

3.1 Characteristics of Soybean Oil

As an important oil and grain resource, soybean is extracted from soybean seeds by a pressing process. The production of soybean oil (SO) has greatly increased with the development of transgenic technology. It reached 61.3 million tons globally in 2021, which accounts for more than 30% of all vegetable oils.[1] China is a major SO producer in the world. The output of SO in China has increased rapidly in the past decades, reaching 17.74 million tons in 2020.[1] Concerning the nonrenewability of petroleum resources, SO has become an important chemical raw material for application in biofuels, coatings, plasticizers, polymers, and composites.

SO is a kind of semi-dry oil. Crude SO contains 95–97% triglycerides by mass fraction. Triglycerides are molecules of glycerol connected with three long fatty acids containing 12 to 22 carbon chains (Figure 3.1). The main fatty acid composition of SO includes 53.3% linoleic acid (two C=C bonds), 23.4% oleic acid (one C=C bond), 11% palmitic acid, 7.8% linolenic acid (three C=C bonds), and 4% stearic acid.[2] Therefore, SO triglycerides have many reactive sites such as ester groups, C=C bonds, allyl carbon, and ester α-carbon. Through various chemical modification strategies, a variety of SO derivatives can be obtained, and they have been widely used in various applications because of the advantages of SO including its abundance, low

Green Chemistry Series No. 83
Green Chemistry and Green Materials from Plant Oils and Natural Acids
Edited by Zengshe Liu and George Kraus

Published by the Royal Society of Chemistry, www.rsc.org

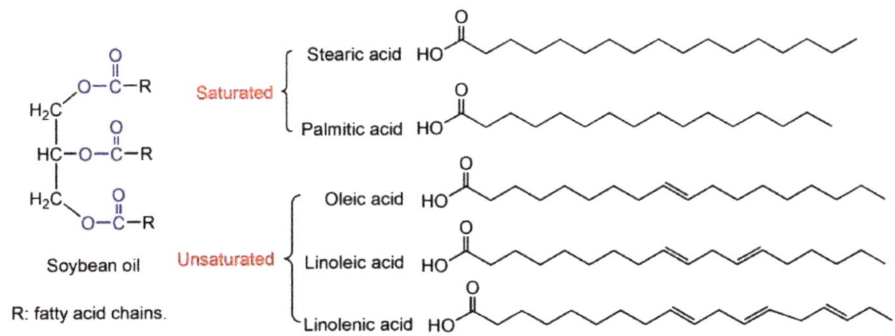

Figure 3.1 Chemical structure of SO.

toxicity, and biodegradability.[3] SO-based thermosets such as polyurethane, acrylates, and epoxy resins have been widely developed to replace their petroleum-based counterparts.

3.2 Soybean Oil-based Polymers

3.2.1 Polyurethane

Polyurethane is usually prepared by the addition reaction of polyisocyanates or diisocyanates and polyols (polyester or polyether). The type of isocyanate is relatively limited, but polyols are available in a variety of ways. Therefore, the choice of polyols determines the properties and application of polyurethane to a certain extent. Vegetable oils (VOs) contain flexible fatty acid chains, and thus VO-based polyols are commonly used to form amorphous regions of polyurethanes, providing the necessary flexibility for the resulting polyurethanes. Excluding a few VOs such as tung oil (which naturally contains hydroxyl groups), most of the VOs including soybean oil require the modification of triglycerides to synthesize polyols. As shown in Figure 3.2, SO-based polyols can be synthesized from SO through epoxide-ring opening, transesterification or amidation, ozonation hydrogenation, hydroformylation, and a thiol–ene click reaction.

Epoxide-ring opening is the most commonly used method for the synthesis of SO-based polyols, which consists of two processes: one is the epoxidation of unsaturated C=C bonds in fatty acid chains, and the other includes the use of nucleophiles such as amines, carboxylic acids, and alcohols to open the epoxy ring. Yang *et al.* synthesized SO-based polyols using diethanolamine as a ring-opening agent for epoxidized soybean oil (ESO), and polyurethanes were prepared from the polyols with commercial PL-5601 polyester polyols and toluene diisocyanate.[4] In the work of Miao *et al.*, isopropanolamine was used to react with ESO.[5] Both ester and epoxy groups of ESO can react with the amino group of isopropanolamine to yield a polyol with a high hydroxyl value of 317 mg KOH per g. The prepared polyurethane obtained a tensile strength of 2.7–27.8 MPa. Wang *et al.* synthesized a series of

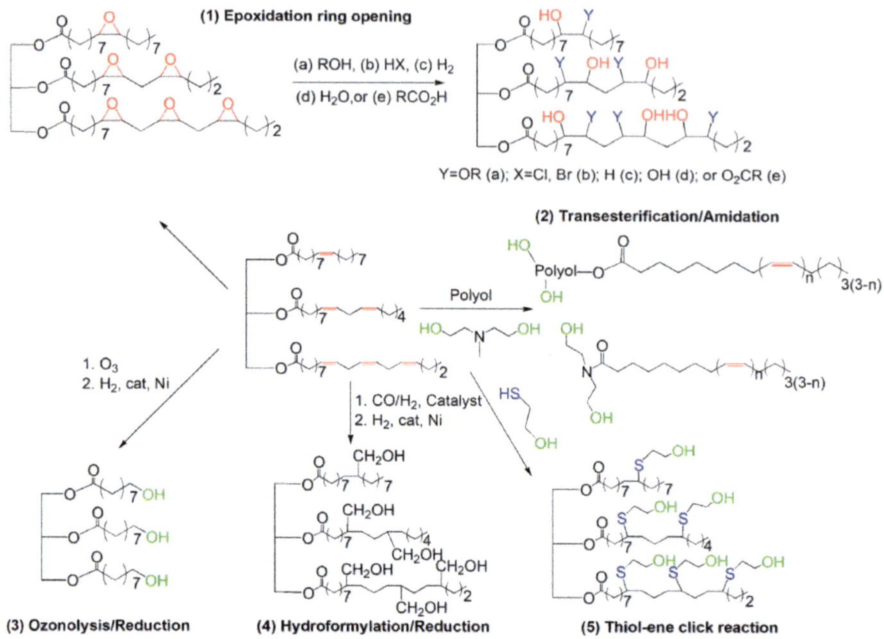

Figure 3.2 Synthesis routes of SO-based polyols. Reproduced from ref. 10 with permission from Elsevier, Copyright 2017.

polyols with different numbers of hydroxyl groups (ranging from 2.6 to 4.9 per molecule) using methanol, glycerol, and propylene glycol as the ring-opening agents for ESO, respectively.[6] The mechanical properties and heat resistance of the obtained polyurethane were significantly related to the hydroxyl value of the polyols. Zhang *et al.* synthesized polyols from ESO by reacting it with the carboxylic group of hydroxyl group-containing fatty acids from tung oil, soybean oil, or linseed oil, respectively.[7,8] The cross-linking density and glass transition temperature (T_g) of the polyurethane increased significantly with the increase of the hydroxyl value of the polyols. Dodangeh *et al.* used citric acid as a ring-opening agent for ESO to synthesize a polyol that was further reacted with tetraethyl orthosilicate and polydiphenylmethane diisocyanate to prepare polyurethane wood adhesives.[9] The bonding strength between the adhesives and wood was up to 6.5 MPa.

The reactivity of the secondary hydroxyl group obtained by the epoxide-ring opening method is lower than that of the primary hydroxyl group. SO-based polyols containing a primary hydroxyl group can be obtained by ozonolysis, which results in the fatty acid chain break from the unsaturated C=C bonds to form a hydroxyl group in the end, and hence the formed polyols have a small molecular weight and high reactivity.[11] Petrovic *et al.* synthesized polyols with functional degrees of 2.8 and 2.5 from rapeseed oil and SO by ozonolysis, respectively, and the corresponding polyurethane had T_gs of 36 °C and 22 °C.[12] SO-based polyols containing primary hydroxyl groups can also be synthesized by hydroformylation. The C=C bonds in the fatty acids are first converted to aldehyde groups with a rhodium or cobalt catalyst, and then the hydroxyl group is obtained by hydrogenation. Guo *et al.* compared the catalytic efficiencies of the rhodium or cobalt catalyst for synthesizing SO-based polyols.[13] It was found that the rhodium catalyst had a higher catalytic efficiency than the cobalt catalyst and the conversion rate for the rhodium catalyst was up to 95%. On the other hand, the cobalt catalyst is relatively cheap, but its catalytic efficiency (65%) is low and it is easy to cause the isomerization of C=C bonds.

A thiol–ene click reaction could directly convert the unsaturated C=C bonds from fatty acid chains into hydroxyl groups and has the advantages of mild reaction conditions, no by-product generation, and high yields.[14] The reaction involves three processes: the generation of free-radicals under thermal- or photo-initiation, the interaction of free-radicals with thiols to generate sulfur radicals, and the connection of sulfur radicals onto the C=C bonds *via* an addition reaction.[15] In the work of Gaillol *et al.*, polyols were synthesized from SO and thiols *via* a thiol–ene click reaction and then used for reacting with methylene diphenyl diisocyanate to prepare polyurethane.[16] Using a micro-fluidic technique, Feng *et al.* prepared SO-based polyols using a solvent-free thiol–ene click reaction and studied the effects of photo-initiators, reaction time, and ratios of the raw compositions on the reaction efficiency.[17] The T_g, tensile strength, and elongation at break of the prepared polyurethanes were 41.3 °C, 15.7 MPa, and 471%, respectively. To improve the reactivity of C=C bonds in the middle of fatty acid chains, Lonescu *et al.* used thiol SO to react with allyl monomers containing C=C bonds to synthesize polyols.[18] This method can also be used to prepare other functional monomers such as amines, isocyanates, and epoxy.

In addition, SO-based polyols can be synthesized by the transesterification of triglycerides with alcohols or the amidation of amines using a base or an enzyme as the catalyst. Glycerol is the most commonly used polyol for transesterification. The reaction between SO and glycerol is usually above 200 °C, and the synthesized polyols can be used in the preparation of polyurethane elastomers and rigid foams.[19]

3.2.2 Vinyl Resins

Vinyl resins are often used to prepare high-performance composites because of their high T_g, mechanical strength, thermal stability, and corrosion resistance.[20] Although there are a certain number of C=C bonds in the fatty acid chains of triglycerides, most of the C=C bonds are non-conjugated and do not exhibit enough reactivity for free-radical or cationic polymerization.[21] Therefore, chemical modification is a common method to prepare VO-based vinyl resins, which is generally the grafting of acrylic or methacrylic groups on the fatty acid chains.

Acrylation or methacrylation of VOs usually refers to two reactions: one is the epoxidation of C=C bonds and another is the use of acrylic acid or

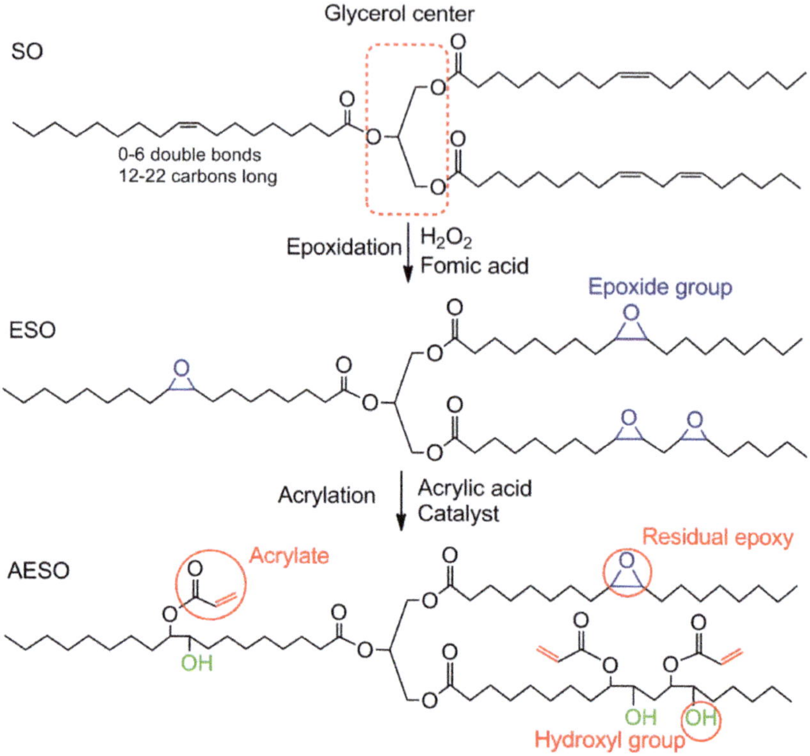

Figure 3.3 Synthesis of AESO.

methacrylic acid for epoxide-ring opening. For example, the synthesis of acrylated epoxidized soybean oil (AESO) is given in Figure 3.3.[22] Wool *et al.* studied the acrylation kinetics of epoxidized VOs, showing that the acrylation reaction was a first-order kinetic reaction, which was mainly related to the concentration of epoxy groups.[23] As the average number of epoxy groups per fatty acid chain increases, the reaction rate decreases due to the steric and electronic hindrance caused by the C=C bonds being located in the middle of the fatty acid chains. Zhang *et al.* prepared SO-based acrylates with a 76% C=C conversion rate by one-step acrylation of SO using boron trifluoride diethyl etherate (BF_3Et_2O) as a catalyst.[24] Wool *et al.* reported the modification of AESO with maleic anhydride, and the modified resins had higher thermo-mechanical properties due to the introduction of carboxylic groups.[25] In the work of Adekunle *et al.*, maleic anhydride was used to modify methacrylated ESO, and the obtained resins were further used for composite preparation *via* a resin transfer molding technique.[26]

AESO is a commercially available VO derivative (Figure 3.3). AESO has three flexible long molecular chains, a high molecular weight, and high viscosity, and thus the cured AESO resin exhibits poor physical and mechanical properties due to the lack of a rigid structure and low cross-linking density. Therefore, small molecular monomers are usually added as reactive diluents to reduce the resin viscosity to satisfy the processing requirement and to increase the cross-linking density and rigidity of the cured resins. Styrene is a traditionally perfect reactive diluent. But styrene is limited by its high volatility, inhalability, and potential carcinogenicity. Many petroleum-based reactive diluents such as divinylbenzene, *N*-vinyl pyrrolidone, 1,4-butanediol dimethacrylate, and trimethylolpropane trimethacrylate can be used to replace styrene for co-polymerizing with AESO.[27] These monomers have many reactive sites, and the resulting resins exhibit a fast curing rate and a high cross-linking density. However, they are still toxic and environmental pollutants.

Concerning the nonrenewability of petrochemical resources and the increasingly serious environmental pollution caused by petrochemicals, current research on AESO resins mostly focuses on the development of biobased substitutes for styrene to prepare styrene-free AESO resins and their composites. The recently developed biobased comonomers are shown in Figure 3.4. VO-based derivatives with a low molecular weight such as methacrylated fatty acid esters (MFA) can be directly used as comonomers

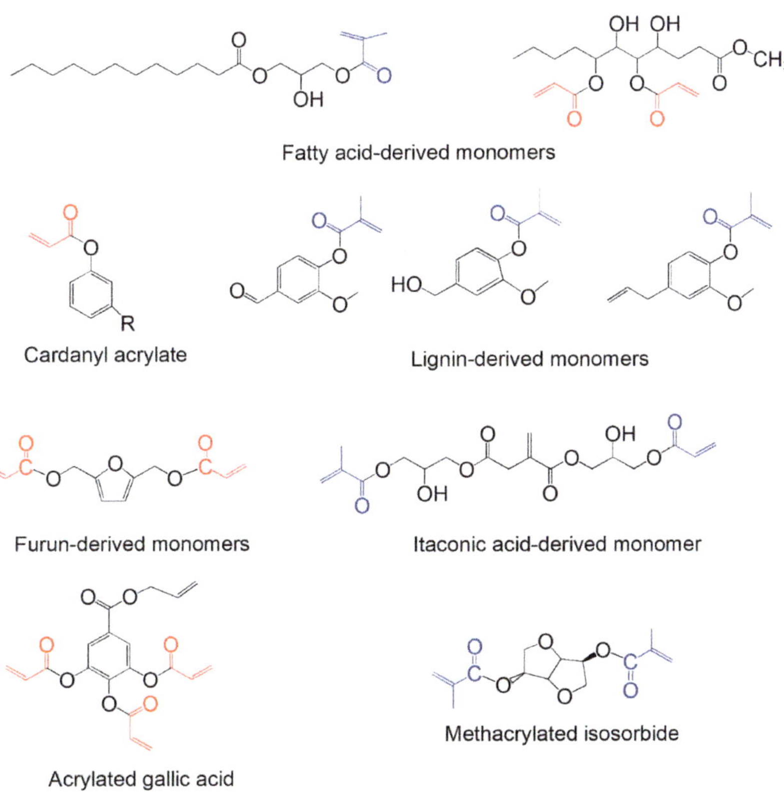

Figure 3.4 Chemical structures of biobased comonomers.

for AESO. However, MFA still have the problems of high viscosity and poor thermal stability and need to be shared with styrene.[28–30] Cardanol is a kind of natural polyphenol with saturated or unsaturated long chains. It can not only participate in the cross-linking of resins, but also provide necessary mechanical and thermal properties endowed by its rigid benzene ring structure.[31,32] Lignin is the second most abundant natural compound in nature. Monomers derived from lignin, such as vanillin, eugenol, guaiacol, catechol, and cinnamic acid, can be grafted with acrylic or methacrylic groups for being used as reactive diluents.[33–37] These monomers have rigid benzene rings and a low molecular weight. AESO resins using these monomers achieve the viscosity requirements of coating and composites during processing, meanwhile having excellent thermal and mechanical properties. Carbohydrates, including sucrose and cellulose, contain five-carbon or six-carbon sugar structures and have the advantages of low cost, abundance, renewability, and biodegradability. The use of furan derivatives including tetrahydrofurfuryl acrylate and furan diacrylate as cross-linkers significantly improved the tensile strength of the AESO resins used as photo-curable coatings.[38,39] Direct co-polymerization of AESO with dibutyl itaconate reduced the T_g of the resins, while glycidyl methacrylate modified itaconate exhibited higher photo-curing reactivity, and thus the resulting resins obtained higher mechanical properties, hardness, adhesion strength, and solvent resistance.[40,41] Isosorbide is one of the top twelve most promising renewable building blocks. It is derived from glucose, which is widely available from the de-polymerization of biomass-based materials including cellulose and starch. Isosorbide dimethacrylate (IM) was obtained from the reaction of isosorbide with methacrylate anhydride. IM has a unique bicyclic ring structure, and the polymerized IM-AESO network exhibited superior mechanical and thermal properties.[42,43] Tannic acid is a water-soluble polyphenolic compound. Hyper-branched tannic acid methacrylate cross-linked AESO resins have a high biobased content and excellent mechanical properties, but the resin showed high viscosity and is difficult to process.[44] Gallic acid can be obtained from the hydrolysis of tannic acid. Acrylated gallic acid has four C=C bonds, a rigid benzene ring, and a low molecular weight, and thus is an excellent reactive diluent for AESO resins.[45]

3.2.3 Epoxy Resins

Epoxy resin is a prominent thermosetting resin due to its excellent mechanical properties, chemical corrosion resistance, electrical insulation, and adhesive properties. Epoxy resins are usually obtained from the polymerization of monomers or oligomers containing epoxy groups with curing agents including amines, anhydrides/acids, and isocyanates. At present, bisphenol A-based epoxy occupies 80% of the market, but the health problem associated with its use is of concern. Epoxidized VOs are traditionally used as plasticizers for thermoplastics including polyvinyl chloride and poly(lactic acid) (PLA). VOs naturally do not contain epoxy groups, and thus

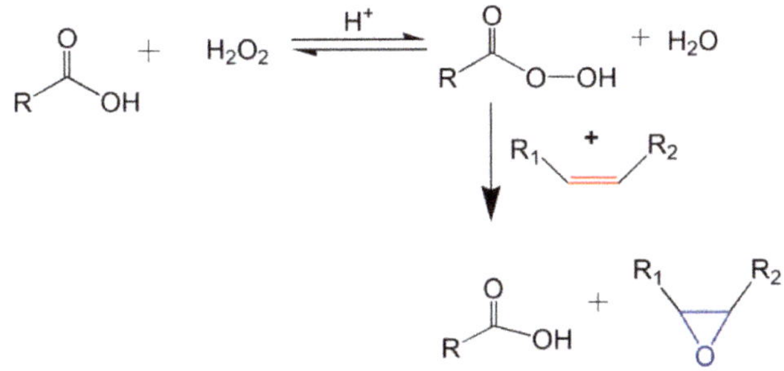

Figure 3.5 Epoxidation of vegetable oils.

epoxidized VOs are obtained by the chemical or enzymatic hydrolysis of C=C bonds *via* epoxidation. The most efficient method in industry is peroxylic acid epoxidation (Figure 3.5).[46] Wool *et al.* studied the epoxidation kinetics of different fatty acids, showing that oleic acid and linoleic acid had similar reactivities, but they showed much lower reactivity than linolenic acid.[47] Epoxidation of VOs can be catalyzed by transition metal complexes such as those of molybdenum and rhenium. Maritana *et al.* used *tert*-butyl hydrogen peroxide as an oxidant and molybdenum acetylacetone as a catalyst to synthesize epoxidized soybean oil (ESO), and the optimal epoxidation conditions were 110 °C and 2 h.[48,49] The activity of the C=C bonds in the middle of fatty acid chains was lower than that of the C=C bonds at the end. A lipase-catalyzed chemical epoxidation method has the advantages of mild reaction conditions, high yield, and few side reactions.[50] Klaas *et al.* used hydrogen peroxide as an oxidant and commercial Novozym®435 enzyme as a catalyst to prepare epoxidized soybean oil, sunflower oil, rapeseed oil, and linseed oil, with yields of more than 90%.[51]

ESO can directly undergo cationic or anionic self-polymerization in the presence of a thermal initiator. Poly(ESO) was prepared using BF_3Et_2O as a catalyst, and the T_g of the resin was between −16 °C and −48 °C.[52] The T_g and thermal deformation temperature of the poly(ESO) catalyzed by magnesium stearate were −27 °C and 90 °C, respectively.[53] However, the epoxy group of ESO is located in the middle of the molecular chain, which has low reactivity. Moreover, due to the lack of a necessary rigid structure, poly(ESO) exhibits poor mechanical performance and is normally used with traditional petroleum-based resins to adjust their toughness and biobased contents. Amines are the commonly used curing agents for epoxy resins because of their high reactivity and low curing temperature. Epoxy resins were prepared from ESO and bisphenol A using isophorone diamine as a curing agent, and the mechanical strength of the resins decreased significantly with the addition of flexible ESO, while the elongation at break and impact toughness increased.[54] Robertson *et al.* studied the hydrothermal degradability of the mixed ESO and bisphenol A resin with 4,4′-diaminodiphenyl methane as a curing agent and found that the T_g of the resins decreased with the increase of ESO usage.[55] When the ESO content was above 60%, the resin was completely degraded in sodium hydroxide solution. Acids/anhydrides are also common curing agents for epoxy resins. Misra *et al.* used sebacic acid as a biobased curing agent for bisphenol A and ESO resins.[56] Compared to amines, sebacic acid significantly increased the release enthalpy of the resin curing. The addition of ESO increased the enthalpy, maximum exothermic temperature, and activation energy of the reaction. Chow *et al.* studied the curing kinetics of ESO with methyl hexahydrobenzene anhydride using 2-ethyl-4-methylimidazole (EMI) as a catalyst.[57] The results showed that the resins exhibited auto-catalytic behaviour under isothermal conditions. With the increase in EMI usage, the activation energy of the reaction decreased and the conversion rate increased. Balart *et al.* prepared biobased epoxy resins from ESO and maleic anhydride (MA).[58] When the mass ratio of ESO to MA was 1 : 1, the resins had the highest cross-linking density and mechanical properties. Samper *et al.* prepared epoxy resins from ESO, epoxidized linseed oil, and maleic anhydride.[59] It was found that with the increase of the ESO amount, the T_g and flexural strength of the resins decreased, while the linear expansion coefficient increased, which was due to the lower epoxy value of ESO. Stefani *et al.* prepared ESO and bisphenol A resins using methyl hexahydrobenzene anhydride as a curing agent.[60] They found that the storage modulus and T_g of the polymers decreased with the increase of the ESO content.

Vitrimer is a dynamic covalent three-dimensional network that is self-healable, recyclable, and reprocessable, as defined by Leibler and his colleagues in 2011.[61] A β-hydroxyl ester based on carboxylic acid and epoxy is the most typical dynamic covalent bond for vitrimer. Recently, vitrimers prepared using ESO as raw materials have attracted much attention from researchers. The commonly used curing agents for ESO vitrimers are shown in Figure 3.6. Water-soluble citric acid was used to cure ESO and the obtained thermoset showed excellent self-healing ability without any catalyst due to the presence of the β-hydroxyl ester.[62] Wu *et al.* prepared an ESO vitrimer based on β-hydroxyl esters using biobased glycyrrhizic acid as a curing agent.[63] The unique structure of glycyrrhizic acid endowed the resin with excellent mechanical properties and thermal stability. Zhang *et al.* used biobased D-camphor acid as a curing agent to prepare an ESO vitrimer for carbon fibre composites.[64] Fumaric acid derived from rosin was used with ESO to obtain a vitrimer that can be recycled *via* ethanol degradation.[65] Mija *et al.* studied epoxidized VO-based vitrimers using carboxylic acid monomers containing disulfide bonds as the curing agents.[66–68] The results indicated that the incorporation of dynamic disulfide bonds greatly promoted the repairability and recyclability of the resins. The type and amount of a transesterification catalyst significantly affect the exchange rate of β-hydroxyl ester. Liu *et al.* introduced dynamic disulfide bonds into amine-cured ESO, and the resin had excellent mechanical properties, weldability, and recyclability.[69] Furthermore, a phenolic curing agent containing a dynamic Schiff base was synthesized from vanillin and *p*-aminophenol.[70] The cured ESO resin can be used to prepare carbon fibre reinforced composites,

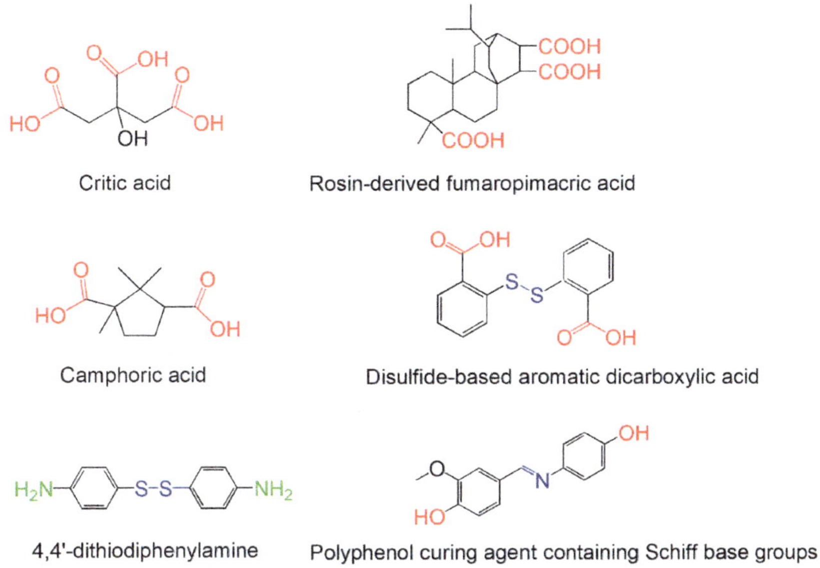

Figure 3.6 Curing agents for ESO-based vitrimers.

and the degradability of the resins enabled recovery of the carbon fibres without damage.

3.3 Soybean Oil-based Biocomposites

3.3.1 Composites from the Soybean Oil-based Matrix

ESO and AESO are the two commercially available SO derivatives for composite applications. ESO is mostly used as a toughening agent for conventional epoxy resins due to its low reactivity of epoxy groups. Biocomposites were prepared from ESO and regenerated cellulose fibres using terpenyl anhydride as the curing agent for ESO.[71] The obtained composites had a tensile strength of 65 MPa and a tensile modulus of 2.3 GPa. Microcrystalline cellulose (MCC) and dialdehyde cellulose (DAC) were used as reinforcements for the ESO matrix to prepare biocomposites, respectively.[72] The results showed that the mechanical properties of DAC/ESO composites were much higher than those of MCC/ESO composites. Liu *et al.* prepared flax fibre reinforced biobased epoxy resin composites using the blended ESO and 1,1,1-tri(4-hydroxyphenyl) ethyl triglycidyl ether resins as the matrix.[73,74] The flax fibre characteristics significantly affected the mechanical properties of the composites. The higher ESO content in the matrix resulted in the poorer flexural properties of the composites. The water resistance and mechanical properties of the composites were significantly improved after the milk protein gel coating and acylation of the fibres. Sahoo *et al.* studied the mechanical properties of ESO toughened sisal fibre/epoxy resin composites.[75] It was found that the tensile strength and modulus of the composites reinforced with 15 wt% sisal fibres were significantly higher than those of the biobased epoxy resin blends. Alkali treatment was applied for jute fibres to improve their interfacial bonding with an ESO-based epoxy matrix.[76]

More attention has been paid to the AESO composites because the C=C bonds of AESO are highly reactive. In the work of Ramamoorthy *et al.*, ramie, regenerated cellulose fibres (Lyocell and viscose fibres), and glass fibres were used as the reinforcements for AESO resins to prepare biocomposites.[77] Compared to the composites with only one fibre, the hybrid fibre-reinforced composites obtained significantly improved mechanical properties. Wool *et al.* prepared flax fibre reinforced AESO composites *via* a resin transfer molding technique.[78,79] The results showed that the mechanical properties of the composites reached the highest with a 34 wt% fibre content. The incorporation of glass fibres as hybrid reinforcements significantly increased the mechanical properties of the composites. Adekunle *et al.* mixed flax fibres with Lyocell or polyester fibres to strengthen AESO resins, and the fibre mixing process significantly affected the mechanical properties of the composites.[80,81] The effects of alkali or silane treatments on the properties of sisal fibre reinforced AESO foam composites were investigated by Wu *et al.*, indicating that the interfacial bonding of the composites was significantly improved after fibre modification.[82] The surface modification of ramie fibres with acrylic acid, silane, and peroxide treatments was performed to improve the interfacial adhesion of ramie fibre/AESO composites.[83] The mechanical properties of the composites were significantly improved after fibre modification, and the composites with silane treatment showed the best performance. Advances in reactive diluents of AESO resin and plant fibre reinforced composites have been achieved recently.[84–87] Concerning the problems of high viscosity and low cross-linking density of AESO resins, *N*-vinyl pyrrolidone (NVP) was used as a reactive diluent to prepare hemp fibre reinforced styrene-free AESO composites.[84] The addition of NVP significantly reduced the viscosity of AESO resins and improved the reactivity of the resins. The mechanical properties of the composites were much higher than those of styrene. AESO contains the hydroxyl and epoxy groups and hence isocyanate monomers were further used as the modifiers. In the composite system, isocyanates can act as both a cross-linker for the resin matrix and an interface compatibilizer for composites, and therefore, the mechanical properties and water resistance of the composites are greatly improved.[85,86] Biobased isosorbide methacrylate was also synthesized as a reactive diluent for AESO resins and their composites.[87] The mechanical properties and water resistance of bamboo and hemp fibre reinforced composites were significantly related to the fibre type and resin composition.

3.3.2 Composites Compatibilized by Soybean Oil Derivatives

SO derivatives can significantly increase the surface hydrophobicity of plant fibres and thus improve the interfacial bonding between the fibres and the hydrophobic matrix. Both ESO and AESO contain reactive functional groups such as hydroxyl or epoxy groups, which provides the possibility to react with the hydroxyl groups of cellulosic fibres. Mahmud *et al.* used ESO as a reactive plasticizer for waste cellulose fibre/PLA composites, and the addition of ESO significantly increased the elongation at break, flexural strain, and impact strength of the composites.[88] However, the reactivity of ESO is limited, and the T_g and mechanical strength of the composites were reduced after the incorporation of ESO. Huang *et al.* grafted a polymerized ESO onto a cellulosic fibre surface through epoxide-ring opening polymerization of ESO using stannic chloride as a catalyst, where the hydroxyl group on the fibres was used as the initiator, and the polymerized ESO was connected to the fibres *via* ether bonds.[89] Due to the introduction of long hydrophobic fatty acid chains, the water contact angle of the fibres reached 145.1° and the surface free energy was -13.09 mJ m^{-2}. Dai *et al.* modified maleated MCC with ESO, and the modified MCC was used to prepare PLA composites.[90] The elongation at break and impact strength of the composites increased significantly due to the formation of the flexible interface layer between cellulose and the PLA matrix. Liu *et al.* used rigid tannic acid as a cross-linking agent for ESO.[91,92] The phenolic hydroxyl groups of tannic acid could react with the epoxy group of ESO to induce the polymerization of ESO; meanwhile, tannic acid could react with the carboxylic groups of PLA to establish a chemical bond between the ESO phase and the PLA matrix. The toughness of PLA blends was significantly improved, while the strength and modulus of PLA still remained at a high level. Considering the principle that tannic acid could bond with the substrate surface through polyphenol chemistry, tannic acid-ESO oligomers were designed to distribute in bamboo fibre/PLA composites. The oligomers played the effects of interfacial modification and matrix toughening in the composites. Similarly, Song *et al.* prepared ESO oligomers using kailic acid as a ring opening agent for ESO, and the oligomers were further used to toughen bamboo fibre/PLA composites[93] The impact toughness of the composites increased from 5.8 kJ m^{-2} to 14.55 kJ m^{-2}. In the work of Liu *et al.*, AESO was grafted on bamboo fibres *via* a diisocyanate monomer that can react with the hydroxyl groups of the fibres and AESO.[94] The grafted fibres containing C=C bonds from AESO would participate in the free-radical polymerization of unsaturated polyester resins, hence forming an AESO-based interfacial layer between the fibres and the matrix. The interfacial bonding between the fibres and the matrix was greatly increased after modification, leading to significantly improved mechanical properties and water resistance of the composites.

3.4 Conclusions

Much progress in the development of SO-based polymers and composites has been made in the last decade, which will undoubtedly continue for the coming decades. Numerous efforts have been done in developing SO-based polyurethanes, vinyl and epoxy resins, and their composites with natural fibres or petroleum-based synthetic fibres. SO can be involved directly in the synthesis of polymers using their own functionalities such as ester groups and C=C bonds. Modified SO could be used in cationic polymerization, free-radical polymerization, ring-opening polymerization, and condensation polymerization to prepare various SO-based thermosets. The obtained SO polymers can be used in the applications of coatings, adhesives, plasticizers, and composites, proving to be excellent candidates for the replacement of petroleum-based materials. Fibre-reinforced composites using SO-based resins are one of the main applications. However, there are challenges to achieving SO-based polymers and composites with satisfactory performance because of the flexible molecular structure of SO. Although many novel chemical modification strategies for SO have been developed, it is still difficult to control the structures and properties of the products precisely. The developed strategies also have drawbacks including using petroleum-based resources, harsh reaction conditions, and environmental issues caused by toxic and volatile chemicals. Therefore, developing highly efficient and green methods for converting SO to monomers and polymers is still desired.

Abbreviations

AESO	Acrylated epoxidized soybean oil
BF_3Et_2O	Boron trifluoride diethyl etherate
DAC	Dialdehyde cellulose
EMI	2-Ethyl-4-methylimidazole
ESO	Epoxidized soybean oil
NVP	*N*-Vinyl pyrrolidone
SO	Soybean
T_g	Glass transition temperature
VOs	Vegetable oils

Acknowledgements

The authors gratefully acknowledge the funding from the National Natural Science Foundation of China (Grant numbers: 32171697, 32071699, 31800486, and 31670568), the Fujian Provincial Administration of Forestry, China (Grant number: 2021FKJ02), and the Fujian Agriculture and Forestry University, China (Grant numbers: KXJQ20013, 1122YB067, and 1122YB019).

References

1. https://www.indexmundi.com/agriculture.
2. F. Zaaboul, Q. Zhao, Y. Xu and Y. Liu, *Food Hydrocolloids*, 2022, **124**, 107296.
3. Y. Xia and R. C. Larock, *Green Chem.*, 2010, **12**, 1893–1909.
4. L.-T. Yang, C.-S. Zhao, C.-L. Dai, L.-Y. Fu and S.-Q. Lin, *J. Polym. Environ.*, 2012, **20**, 230–236.

5. S. Miao, S. Zhang, Z. Su and P. Wang, *J. Appl. Polym. Sci.*, 2013, **127**, 1929–1936.
6. C. S. Wang, L. T. Yang, B. L. Ni and G. Shi, *J. Appl. Polym. Sci.*, 2009, **114**, 125–131.
7. C. Zhang, Y. Li, R. Chen and M. R. Kessler, *ACS Sustainable Chem. Eng.*, 2014, **2**, 2465–2476.
8. R. Q. Chen, C. Q. Zhang and M. R. Kessler, *J. Appl. Polym. Sci.*, 2015, **132**, 41213.
9. F. Dodangeh, M. S. Seyed Dorraji, M. H. Rasoulifard and H. R. Ashjari, *Composites, Part B*, 2020, **187**, 107857.
10. C. Zhang, T. F. Garrison, S. A. Madbouly and M. R. Kessler, *Prog. Polym. Sci.*, 2017, **71**, 91–143.
11. T. S. Omonov, E. Kharraz and J. M. Curtis, *J. Am. Oil Chem. Soc.*, 2011, **88**, 689–705.
12. Z. S. Petrović, W. Zhang and I. Javni, *Biomacromolecules*, 2005, **6**, 713–719.
13. A. Guo, D. Demydov, W. Zhang and Z. S. Petrovic, *J. Polym. Environ.*, 2002, **10**, 49–52.
14. Y. H. Xu, X. Q. Xiong, L. Cai, Z. K. Tang and Z. J. Ye, *Prog. Chem.*, 2012, **24**, 385–394.
15. A. B. Lowe, *Polym. Chem.*, 2010, **1**, 17–36.
16. S. Caillol, M. Desroches, S. Carlotti, R. Auvergne and B. Boutevin, *Green Mater.*, 2013, **1**, 16–26.
17. Y. Feng, H. Liang, Z. Yang, T. Yuan, Y. Luo, P. Li, Z. Yang and C. Zhang, *ACS Sustainable Chem. Eng.*, 2017, **5**, 7365–7373.
18. M. Ionescu, D. Radojčić, X. Wan, Z. S. Petrović and T. A. Upshaw, *Eur. Polym. J.*, 2015, **67**, 439–448.
19. A. A. Lubguban, Y.-C. Tu, Z. R. Lozada, F.-H. Hsieh and G. J. Suppes, *J. Appl. Polym. Sci.*, 2009, **112**, 19–27.
20. Y. Lu and R. C. Larock, *ChemSusChem*, 2009, **2**, 136–147.
21. J.-M. Raquez, M. Deléglise, M.-F. Lacrampe and P. Krawczak, *Prog. Polym. Sci.*, 2010, **35**, 487–509.
22. F. Habib and M. Bajpai, *Chem. Chem. Technol.*, 2011, **5**, 317–326.
23. J. L. Scala and R. P. Wool, *J. Am. Oil Chem. Soc.*, 2002, **79**, 59–63.
24. P. Zhang and J. Zhang, *Green Chem.*, 2013, **15**, 641–645.
25. J. Lu, S. Khot and R. P. Wool, *Polymer*, 2005, **46**, 71–80.
26. K. Adekunle, D. Åkesson and M. Skrifvars, *J. Appl. Polym. Sci.*, 2010, **115**, 3137–3145.
27. Y. Wu, M. Fei, R. Qiu, W. Liu and J. Qiu, *Polymers*, 2019, **11**, 1815.
28. A. Campanella, J. J. La Scala and R. P. Wool, *J. Appl. Polym. Sci.*, 2011, **119**, 1000–1010.
29. J. J. La Scala, J. M. Sands, J. A. Orlicki, E. J. Robinette and G. R. Palmese, *Polymer*, 2004, **45**, 7729–7737.
30. A. Campanella, J. J. L. Scala and R. P. Wool, *Polym. Eng. Sci.*, 2009, **49**, 2384–2392.
31. Y. Hu, Q. Shang, J. Tang, C. Wang, F. Zhang, P. Jia, G. Feng, Q. Wu, C. Liu and L. Hu, *Ind. Crops Prod.*, 2018, **117**, 295–302.

32. Y. Hu, Q. Q. Shang, C. N. Wang, G. D. Feng, C. G. Liu, F. Xu and Y. H. Zhou, *Polym. Adv. Technol.*, 2018, **29**, 1852–1860.
33. J. F. Stanzione III, P. A. Giangiulio, J. M. Sadler, J. J. La Scala and R. P. Wool, *ACS Sustainable Chem. Eng.*, 2013, **1**, 419–426.
34. W. Thielemans and R. P. Wool, *Biomacromolecules*, 2005, **6**, 1895–1905.
35. N. Auclair, A. Kaboorani, B. Riedl and V. Landry, *Ind. Crops Prod.*, 2015, **76**, 530–537.
36. Y. Zhang, Y. Li, L. Wang, Z. Gao and M. R. Kessler, *ACS Sustainable Chem. Eng.*, 2017, **5**, 8876–8883.
37. Y. Zhang, V. K. Thakur, Y. Li, T. F. Garrison, Z. Gao, J. Gu and M. R. Kessler, *Macromol. Mater. Eng.*, 2018, **303**, 1700278.
38. J. Wu, Y. Qian, C. A. Sutton, J. J. La Scala, D. C. Webster and M. P. Sibi, *ACS Sustainable Chem. Eng.*, 2021, **9**, 15537–15544.
39. N. R. Jang, H. R. Kim, C. T. Hou and B. S. Kim, *Polym. Adv. Technol.*, 2013, **24**, 814–818.
40. O. Goerz and H. Ritter, *Polym. Int.*, 2013, **62**, 709–712.
41. J. Dai, S. Ma, Y. Wu, L. Han, L. Zhang, J. Zhu and X. Liu, *Green Chem.*, 2015, **17**, 2383–2392.
42. J. M. Sadler, A.-P. T. Nguyen, F. R. Toulan, J. P. Szabo, G. R. Palmese, C. Scheck, S. Lutgen and J. J. L. Scala, *J. Mater. Chem. A*, 2013, **1**, 12579–12586.
43. W. Liu, T. Xie and R. Qiu, *ACS Sustainable Chem. Eng.*, 2017, **5**, 774–783.
44. R. Liu, J. Zhu, J. Luo and X. Liu, *Prog. Org. Coat.*, 2014, **77**, 30–37.
45. S. Ma, Y. Jiang, X. Liu, L. Fan and J. Zhu, *RSC Adv.*, 2014, **4**, 23036–23042.
46. B. Rangarajan, A. Havey, E. A. Grulke and P. D. Culnan, *J. Am. Oil Chem. Soc.*, 1995, **72**, 1161–1169.
47. J. L. Scala and R. P. Wool, *J. Am. Oil Chem. Soc.*, 2002, **79**, 373–378.
48. M. Farias, M. Martinelli and D. P. Bottega, *Appl. Catal., A*, 2010, **384**, 213–219.
49. M. Farias, M. Martinelli and G. K. Rolim, *Appl. Catal., A*, 2011, **403**, 119–127.
50. T. Vlček and Z. S. Petrović, *J. Am. Oil Chem. Soc.*, 2006, **83**, 247–252.
51. S. Warwel, *Ind. Crops Prod.*, 1999, **9**, 125–132.
52. Z. Liu and S. Z. Erhan, *J. Am. Oil Chem. Soc.*, 2010, **87**, 437–444.
53. B.-J. K. Ahn, S. Kraft and X. S. Sun, *J. Agric. Food Chem.*, 2012, **60**, 2179–2189.
54. P. Czub, *Macromol. Symp.*, 2006, **242**, 60–64.
55. G. Yang, B. J. Rohde and M. L. Robertson, *Green Mater.*, 2013, **1**, 125–134.
56. G. M. Roudsari, A. K. Mohanty and M. Misra, *ACS Sustainable Chem. Eng.*, 2014, **2**, 2111–2116.
57. S. Tan, Z. Ahmad and W. Chow, *Ind. Crops Prod.*, 2013, **43**, 378–385.
58. J. M. España, L. Sánchez-Nacher, T. Boronat, V. Fombuena and R. Balart, *J. Am. Oil Chem. Soc.*, 2012, **89**, 2067–2075.
59. M. Samper, V. Fombuena, T. Boronat, D. Garcia-Sanoguera and R. Balart, *J. Am. Oil Chem. Soc.*, 2012, **89**, 1521–1528.
60. F. I. Altuna, L. H. Espósito, R. A. Ruseckaite and P. M. Stefani, *J. Appl. Polym. Sci.*, 2011, **120**, 789–798.

61. D. Montarnal, M. Capelot, F. Tournilhac and L. Leibler, *Science*, 2011, **334**, 965–968.
62. F. I. Altuna, V. Pettarin and R. J. Williams, *Green Chem.*, 2013, **15**, 3360–3366.
63. J. Wu, X. Yu, H. Zhang, J. Guo, J. Hu and M.-H. Li, *ACS Sustainable Chem. Eng.*, 2020, **8**, 6479–6487.
64. W. Zhang, J. Wu, L. Gao, B. Zhang, J. Jiang and J. Hu, *Green Chem.*, 2021, **23**, 2763–2772.
65. X. Yang, L. Guo, X. Xu, S. Shang and H. Liu, *Mater. Des.*, 2020, **186**, 108248.
66. C. Di Mauro, S. Malburet, A. Genua, A. Graillot and A. Mija, *Biomacromolecules*, 2020, **21**, 3923–3935.
67. T.-N. Tran, C. D. Mauro, A. Graillot and A. Mija, *Macromolecules*, 2020, **53**, 2526–2538.
68. C. Di Mauro, T.-N. Tran, A. Graillot and A. Mija, *ACS Sustainable Chem. Eng.*, 2020, **8**, 7690–7700.
69. Y.-Y. Liu, J. He, Y.-D. Li, X.-L. Zhao and J.-B. Zeng, *Ind. Crops Prod.*, 2020, **153**, 112576.
70. Y.-Y. Liu, J. He, Y.-D. Li, X.-L. Zhao and J.-B. Zeng, *Compos. Commun.*, 2020, **22**, 100445.
71. T. Takahashi, K.-i. Hirayama, N. Teramoto and M. Shibata, *J. Appl. Polym. Sci.*, 2008, **108**, 1596–1602.
72. M. He, J. Zhou, H. Zhang, Z. Luo and J. Yao, *J. Appl. Polym. Sci.*, 2015, **132**, 42488.
73. Z. Liu, S. Z. Erhan, D. E. Akin and F. E. Barton, *J. Agric. Food Chem.*, 2006, **54**, 2134–2137.
74. Z. Liu, S. Z. Erhan, D. E. Akin, F. E. Barton, C. Onwulata and T. A. McKeon, *Compos. Interfaces*, 2008, **15**, 207–220.
75. S. K. Sahoo, S. Mohanty and S. K. Nayak, *J. Nat. Fibers*, 2017, **14**, 357–367.
76. P. Niedermann, G. Szebényi and A. Toldy, *Polym. Compos.*, 2017, **38**, 884–892.
77. S. K. Ramamoorthy, Q. Di, K. Adekunle and M. Skrifvars, *J. Reinf. Plast. Compos.*, 2012, **31**, 1191–1200.
78. A. O'donnell, M. A. Dweib and R. P. Wool, *Compos. Sci. Technol.*, 2004, **64**, 1135–1145.
79. S. S. Morye and R. P. Wool, *Polym. Compos.*, 2005, **26**, 407–416.
80. K. Adekunle, S.-W. Cho, C. Patzelt, T. Blomfeldt and M. Skrifvars, *J. Reinf. Plast. Compos.*, 2011, **30**, 685–697.
81. D. Åkesson, M. Skrifvars and P. Walkenström, *J. Appl. Polym. Sci.*, 2009, **114**, 2502–2508.
82. S. P. Wu, J. F. Qiu, M. Z. Rong, M. Q. Zhang and L. Y. Zhang, *Polym. Int.*, 2009, **58**, 403–411.
83. T. S. Lee, H. Y. Choi, H. N. Choi, K.-Y. Lee, S.-H. Kim, S. G. Lee and D. K. Yong, *J. Adhes. Sci. Technol.*, 2013, **27**, 1335–1347.
84. W. Liu, T. Chen, T. Xie and R. Qiu, *Composites, Part A*, 2016, **82**, 1–7.
85. W. Liu, T. Xie and R. Qiu, *Composites, Part A*, 2016, **90**, 278–285.

86. W. Liu, M.-E. Fei, Y. Ban, A. Jia, R. Qiu and J. Qiu, *Compos. Sci. Technol.*, 2018, **160**, 60–68.
87. W. Liu, T. Chen, M.-E. Fei, R. Qiu, D. Yu, T. Fu and J. Qiu, *Composites, Part B*, 2019, **171**, 87–95.
88. S. Mahmud, Y. Long, J. Wang, J. Dai, R. Zhang and J. Zhu, *Fibers Polym.*, 2020, **21**, 2949–2961.
89. X. Huang, A. Wang, X. Xu, H. Liu and S. Shang, *ACS Sustainable Chem. Eng.*, 2017, **5**, 1619–1627.
90. X. Dai, Z. Xiong, H. Na and J. Zhu, *Compos. Sci. Technol.*, 2014, **90**, 9–15.
91. W. Liu, J. Qiu, L. Zhu, M.-E. Fei, R. Qiu, E. Sakai, K. Ito, G. Song and G. Tang, *Polymer*, 2018, **148**, 109–118.
92. W. Liu, J. Qiu, T. Chen, M. Fei, R. Qiu and E. Sakai, *Compos. Sci. Technol.*, 2019, **181**, 107709.
93. X. Song, C. Zhang, Y. Yang and Y. Weng, *J. Appl. Polym. Sci.*, 2022, **139**, 51583.
94. W. Liu, T. Xie and R. Qiu, *Cellulose*, 2016, **23**, 2501–2523.

CHAPTER 4

Plant Oil-based Polyurethanes

CHAOQUN ZHANG[a,b]

[a] Key Laboratory for Biobased Materials and Energy of Ministry of Education, College of Materials and Energy, South China Agricultural University, 483 Wushan Road, Guangzhou 510642, China; [b] Guangdong Laboratory for Lingnan Modern Agriculture, Guangzhou 510642, China
Emails: zhangcq@scau.edu.cn; nwpuzcq@gmail.com

Polyurethanes are one of the most versatile polymers containing repeated carbamate segments (–NHCOO–) in their main chain, which is formed by the addition of polyisocyanates and polyols. As shown in Figure 4.1, polyurethanes have become the fifth largest polymeric materials and have been widely used in various fields, such as adhesives, coatings, foams, elastomers, sealants, slow-release fertilizers, *etc.* It was reported that the global polyurethane market size is expected to grow from $143.8 billion in 2027 at a rate of 13.0% per year.[1]

The rapid development of polyurethanes is attributed to the simple preparation procedure and excellent properties of these polymeric materials, such as wear resistance, high mechanical strength, high elasticity, aging resistance, corrosion resistance, *etc.* These excellent properties are closely related to the molecular structure of polyurethane materials: the main chain of polyurethane molecules is formed by flexible long-chain polyols and rigid isocyanate blocks, which can effectively disperse stress effects, and the coexistence of polar and non-polar chain segments also improves the chemical stability of polyurethanes, while the extensive hydrogen bonding within the polymer further improves the mechanical properties of the materials.

Until now, most of the monomers for the development of polyurethanes have been derived from fossil feedstock, including isocyanates, polyols,

Green Chemistry Series No. 83
Green Chemistry and Green Materials from Plant Oils and Natural Acids
Edited by Zengshe Liu and George Kraus

Published by the Royal Society of Chemistry, www.rsc.org

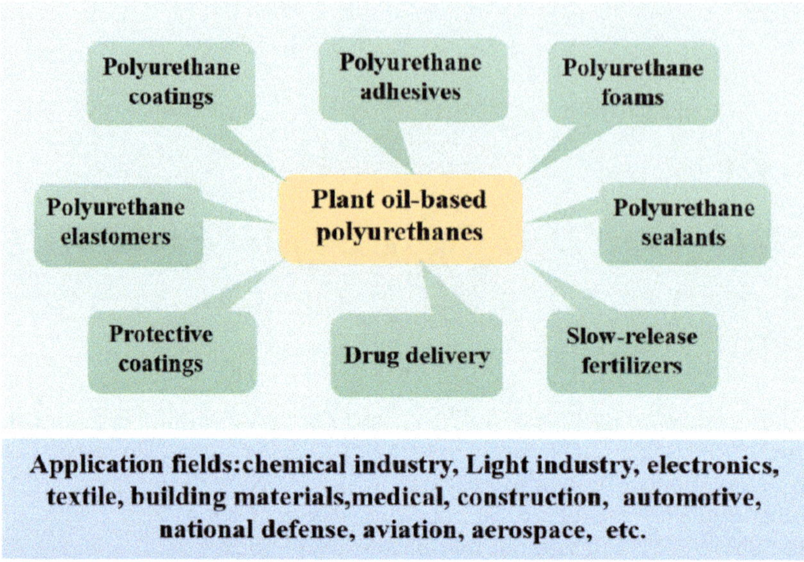

Figure 4.1 Plant oil-based polyurethanes.

chain extenders, *etc.* With the increasing global concern about the depletion of fossil fuels and the environmental impact of fossil-based compounds, exploring bio-based chemicals and monomers from a renewable resource for bio-based polyurethanes has attracted much attention from academia and industry. The current global carbon peak carbon neutrality policy also pushed towards more sustainable and greener approaches in the development of environmentally friendly polyurethanes. Among renewable resources, plant oils are a promising option because of their abundant production (about 140 million tons per year worldwide), biodegradability, and renewable origin. On the other hand, their unique triglyceride structures and long hydrophobic fatty acid chains can endow the resulting polyurethanes with enhanced thermal stability, water resistance, and solvent resistance. In this chapter, monomers and their resulting polyurethanes prepared from typical plant oils will be summarized. In addition, the application of these plant oil-based polyurethanes will also be reviewed.

4.1 Monomers Prepared from Plant Oils

The main components of plant oils are esters generated from glycerol and fatty acids containing 8–24 carbons and 0–7 carbon–carbon double bonds, such as erucic acid, tung oil acid, and ricinoleic acid. The molecular structure formulas of typical plant oils are shown in Figure 4.2. The triglyceride structure and unsaturated bonds (carbon–carbon double bonds) of plant oils can be converted into functional groups to prepare plant oil-based monomers, such as

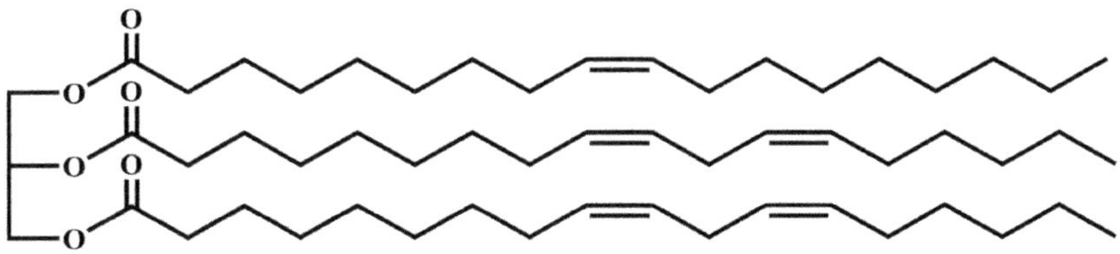

Figure 4.2 The molecular structure of typical plant oils.

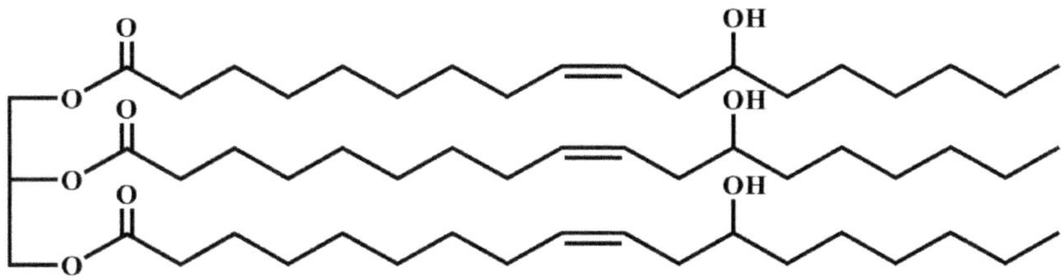

Figure 4.3 The molecular structure formula of castor oil.

polyols, internal emulsifiers, chain extenders, and isocyanates, which are necessary for the preparation of bio-based polyurethanes.

4.1.1 Polyols from Plant Oils

As far as we know, castor oil is the only plant oil that contains naturally occurring hydroxyl groups (Figure 4.3). Castor oil is a pale-yellow viscous oil and possesses an average of 2.7 hydroxyl groups (a hydroxyl value of about 164 mg KOH per g) and approximately 3.04 unsaturation/triglycerides per triglyceride molecule. The unique chemical structure and renewable characteristics of castor oil make it a promising candidate as a cross-linking monomer for synthesizing polyurethane products which could be a replacement for petro-derived counterparts, such as coatings, adhesives, foams, sealants, and so on. Therefore, castor oil has attracted significant interest for direct utilization prior to the synthesis of bio-based polyols from other plant oils by chemical or biosynthetic approaches.

Generally, most plant oils do not contain any hydroxyl group, which is the functional group of polyols. However, after specific modification on active sites, such as carbon–carbon double bonds and ester groups, plant oils could be transformed into bio-based polyols as a replacement for petroleum-based polyols. Various methods have been developed and will be described below in detail, including epoxidation/oxirane ring opening, hydroformylation/reduction, thiol–ene coupling reaction, ozonolysis, and transesterification.

4.1.1.1 Epoxidation/Oxirane Ring Opening

The epoxidation/oxirane ring opening method refers to the epoxidation of unsaturated double bonds of plant oils under the action of strong oxidizing agents, followed by an oxirane ring opening reaction initiated by some small molecule compounds (organic alcohols, organic acids, inorganic acids, amine compounds, *etc.*). The preparation diagram is shown in Figure 4.4. For example, Polaczek *et al.* synthesized palm oil-based polyols using diethylene glycol as the ring-opening agent and fluoroboric acid as the catalyst.[2] The hydroxyl value of the resulting polyols ranged from 150 to 156 mg KOH per g. Using a similar method, Zhou *et al.* designed and synthesized fully plant oil-based polyols with a low hydroxyl value and low viscosity using epoxy soybean oil and oleic acid as raw materials and tetrabutylammoniumchloride as a catalyst.[3] The hydroxyl value of the soybean oil-based polyols was 95 mg KOH per g and the viscosity was 2236 mPa s. Kessler *et al.* compared the catalytic effect of 1,8-diazabicyclo(5.4.0)-undec-7-ene and pyridine in the ring-opening reaction between epoxidized soybean oil and castor oil-based fatty acids in a solvent-free process.[4] Furthermore, a series of fully plant oil-based polyols with pre-crosslinked structures were successfully synthesized.

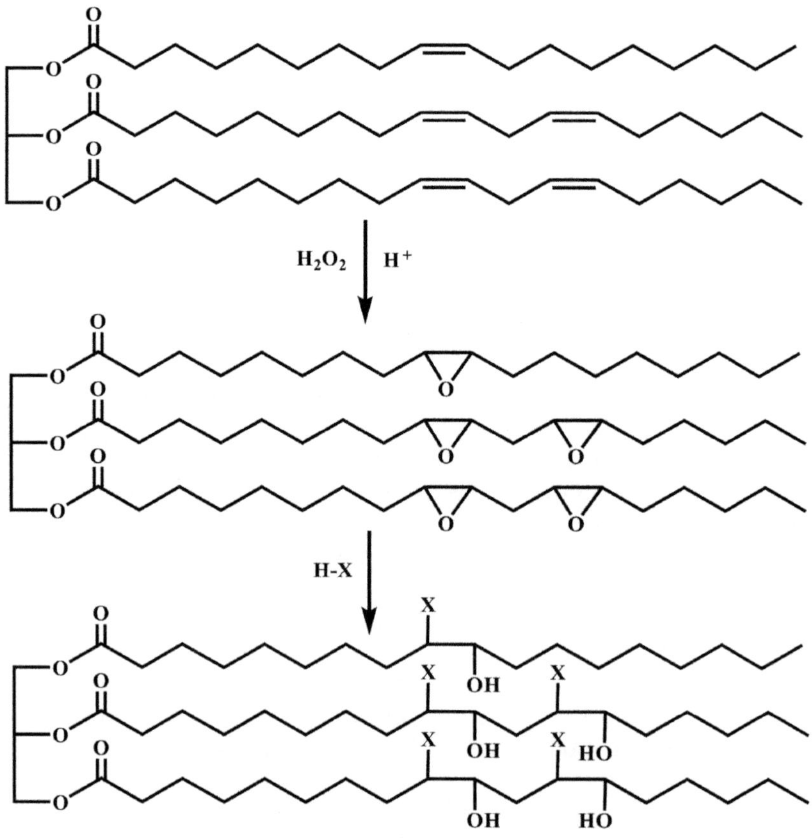

Figure 4.4 Schematic of the epoxidation/oxirane ring opening method for plant oil-based polyols.

4.1.1.2 Hydroformylation/Reduction

The hydroformylation/reduction method involves hydroformylation of the unsaturated double bonds of plant oil molecules to aldehydes using syngas (a mixture of H_2 and CO) in the presence of a Rh or Co catalyst, followed by reduction to primary hydroxyl groups in the presence of H_2 (Figure 4.5). For example, Hong *et al.* prepared rubber seed oil-based polyols by hydroformylation/hydrogenation using $Rh(CO)_2$ as a catalyst. The hydroxyl value of the resulting product was 240 mg KOH per g and the acid value was 21 mg KOH per g.[5] Tilloy *et al.* reported the synthesis of polyhydroxylated derivatives by hydrohydroxymethylation of plant oils through Rh-catalyzed hydroformylation followed by the hydrogenation of formyl groups. Under optimized reaction conditions, the yields of the polyols prepared from castor oil and ethyl castor oleate reached 74% and 80%, respectively.[6] It should be

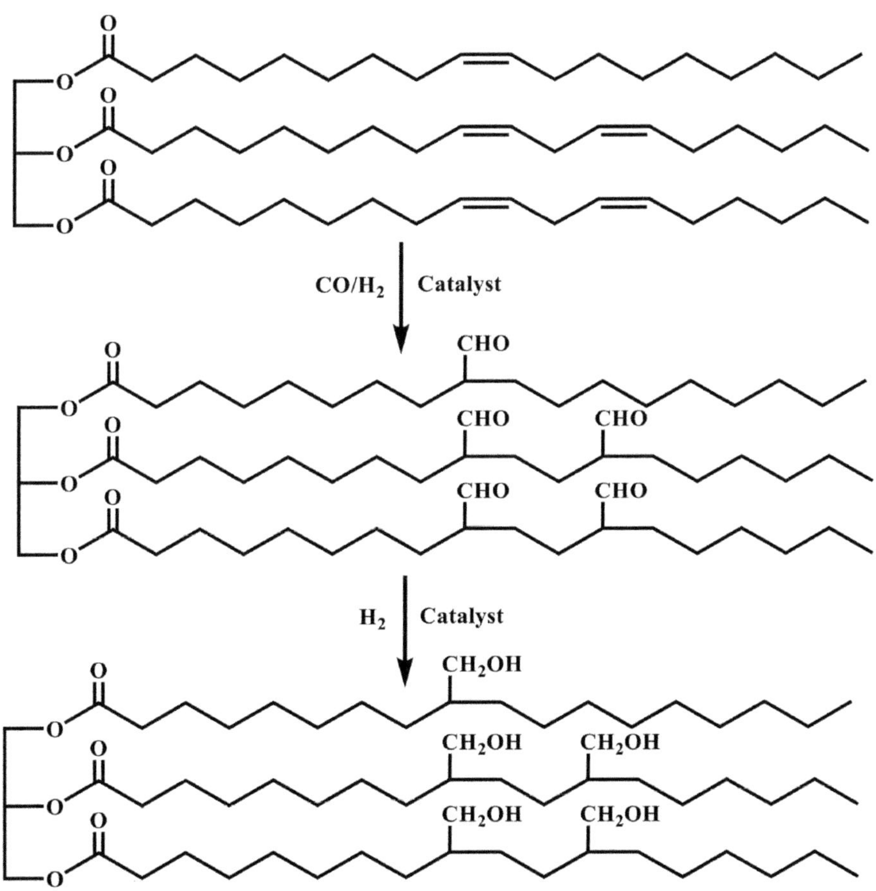

Figure 4.5 Schematic of the hydroformylation/reduction method for plant oil-based polyols.

mentioned that the primary hydroxyl groups of the polyols prepared using the hydroformylation/reduction method showed much higher reactivity than the secondary hydroxyl groups of castor oils and those prepared by other methods.

4.1.1.3 Thiol–Ene Coupling Reaction

The thiol–ene click coupling reaction has been applied to the preparation of plant oil-based polyols due to several advantages, such as its simplicity, efficiency, and controllable characteristics. The method involves the introduction of hydroxyl groups into the molecular structure of plant oils by means of a click reaction that can occur between the mercaptan group and the carbon–carbon double bonds of plant oils under the irradiation of UV light or heat and other energy resources. Among them, the initiation by UV light is most frequently used and studied.

As shown in Figure 4.6, Hong *et al.* investigated the efficient preparation of plant oil-based polyols containing primary hydroxyl groups using castor oil and soybean oil as raw materials using a thiol–ene photo-click method. The results showed that plant oil-based polyols with hydroxyl values over 200 mg KOH per g could be prepared by adjusting the reaction conditions such as the concentration of thiol and temperature.[7] Zhang *et al.* applied the thiol–ene photo-click reaction to develop a series of bio-based polyols from typical plant oils with different carbon–carbon double bonds (olive oil, grape seed oil, castor oil, rapeseed oil, rice bran oil, *etc.*) and revealed the relationship between carbon–carbon double bonds in the backbone of plant oils and the functionalities of the resulting polyols.[8]

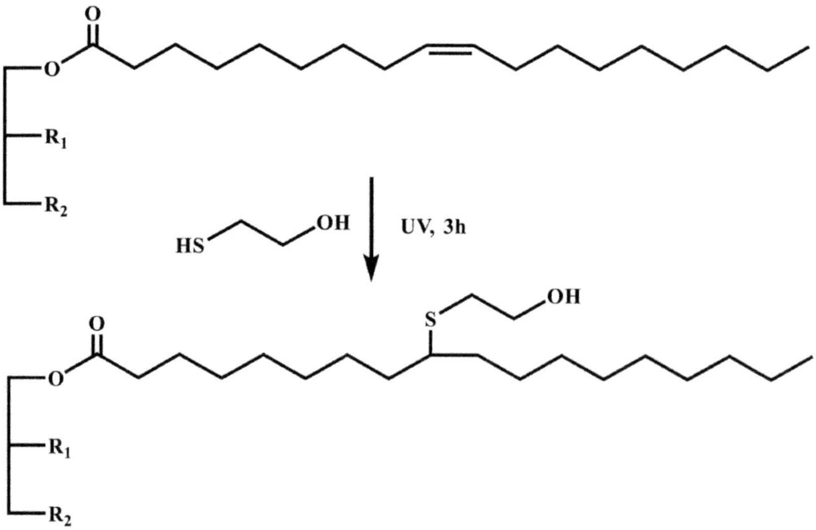

Figure 4.6 Synthesis route of plant oil-based polyols using the thiol–ene photo-click reaction (R_1 and R_2 are fatty acid chains).

4.1.1.4 Ozonolysis/Reduction

The ozonolysis/reduction method involves the oxidation of the carbon–carbon double bonds on the molecular structure of plant oils into aldehyde by ozone, followed by reduction into hydroxyl groups for plant oil-based polyols. This method usually has a high conversion rate (up to 97%) of double bonds and also generates primary hydroxyl groups. For example, Dumont *et al.* reported for the first time that 10 kinds of North American plant oils were converted to plant oil-based polyols by the ozonolysis–hydrogenation process.[9] In detail, plant oils were first converted into ozonolysis products (aldehyde oil) by the ozonolysis method, and then the aldehyde oil was hydrogenated with nickel as a catalyst for plant oil-based polyols. The hydroxyl value of the obtained polyols ranged from 163 to 210 mg KOH per g.

4.1.1.5 Transesterification/Amidation

Transesterification is a method in which the triglyceride structure of plant oils is esterified with small molecule alcohols or amines in the presence of a catalyst (strong acid or base) to produce monoglycerides or diglycerides. Common alcohols used in this method mainly include methanol, glycerol, pentaerythritol, trimethylolpropane, *etc.* However, this method involves a reversible reaction and the conversion rate of the reaction is dependent on many factors, such as temperature, catalyst type, alcohol–oil molar ratio, and free fatty acid content. For example, Fridrihsone *et al.* synthesized plant oil-based polyols with a hydroxyl value of 374 mg KOH per g by transamidation of rapeseed oil and triethanolamine. Jayavani *et al.* synthesized two types of bio-based polyols through glycerolysis of two non-edible plant oils (chaulmoogra oil and Alexandrian laurel seed oil)[10] in the presence of an environmentally friendly heterogeneous base catalyst, CaO. The hydroxyl values of the chaulmoogra oil-based polyol and the Alexandrian laurel seed oil-based polyol were 189 mg KOH per g and 200 mg KOH per g, respectively (Figure 4.7).[11] Yunus *et al.* established a kinetic/mathematical model for the synthesis of polyol esters by chemical transesterification of palm oil

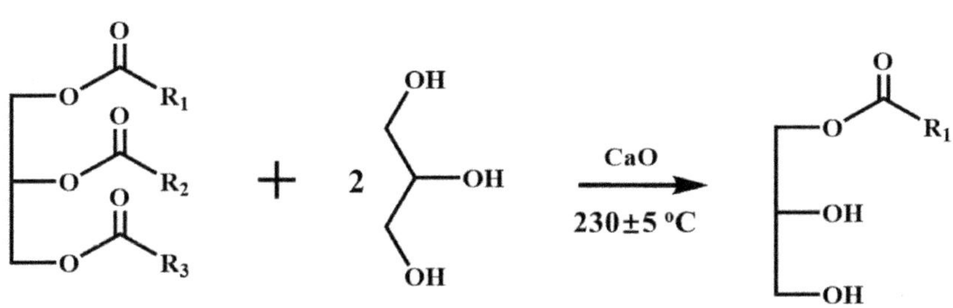

Figure 4.7 Schematic of the transesterification method for plant oil-based polyols (R_1, R_2, and R_3 are fatty acid chains).[11] Adapted from ref. 11 with permission from Elsevier, Copyright 2016.

methyl ester with trimethylopropane using liquid sodium methoxide as the catalyst.[12]

4.1.2 Internal Emulsifiers from Plant Oils

An internal emulsifier is another important raw material for the development of plant oil-based waterborne polyurethanes. In order to act as an emulsifying agent, the internal emulsifier should contain hydroxyl groups which react with isocyanates for the chain extension and carboxylic acid groups which act as a hydrophilic group to promote the dispersion of polyurethanes in the water phase. The most widely used petroleum-based internal emulsifiers are 2,2-dimethylolpropionic acid or 2,2-dimethylolbutanoic acid. Therefore, simultaneous introduction of hydroxyl groups and carboxylic acid groups into the fatty acid chains of plant oils is the key to preparing plant oil-based hydrophilic internal emulsifiers. Currently, there are two main methods for such purposes: the epoxy/ring-opening method and the thiol–ene photo-click method.

4.1.3 Epoxidation/Ring-opening Method

Different strategies have been applied to introduce hydroxyl groups and carboxylic acid groups into the fatty acid chains of plant oils through the epoxidation/ring-opening method. A typical method is as follows: (1) introducing two or more hydroxyl groups on the carbon–carbon double bonds of plant oils through the epoxy/ring-opening reaction and then introducing a certain number of carboxylic acid groups through a saponification reaction or an esterification reaction with acid anhydride and (2) directly using a small molecule acid (phosphoric acid, polycarboxylic acid, *etc.*) to carry out the ring-opening reaction on epoxidized plant oils to form new hydroxyl groups and maintain a certain number of residual carboxylic acid groups.

Much effort has been dedicated to developing new internal emulsifiers prepared from plant oils. For example, Palanisamy *et al.* prepared a bio-based internal emulsifier by esterification of hydroxylated cottonseed oil with maleic anhydride.[13] In detail, the epoxidized cottonseed oil was subjected to *in situ* ring-opening hydroxylation under the action of formic acid, and finally it underwent an esterification reaction with maleic anhydride in the presence of tin chloride catalysis to obtain an internal emulsifier with both hydroxyl (OH value: 95 mg KOH per g) and carboxyl groups (acid value: 68 mg KOH per g) (Figure 4.8a). The waterborne polyurethane prepared using this internal emulsifier exhibits excellent thermal stability. Nabid *et al.* proposed a new scheme for the preparation of plant oil-based internal emulsifiers by saponification after *in situ* ring-opening epoxy sunflower oil under the action of methanol.[14] Different from other bio-based internal emulsifiers in which hydroxyl and carboxyl groups were directly introduced into the carbon–carbon double bonds of plant oils, the resulting internal emulsifier is a single fatty acid chain structure, and the hydroxyl groups of

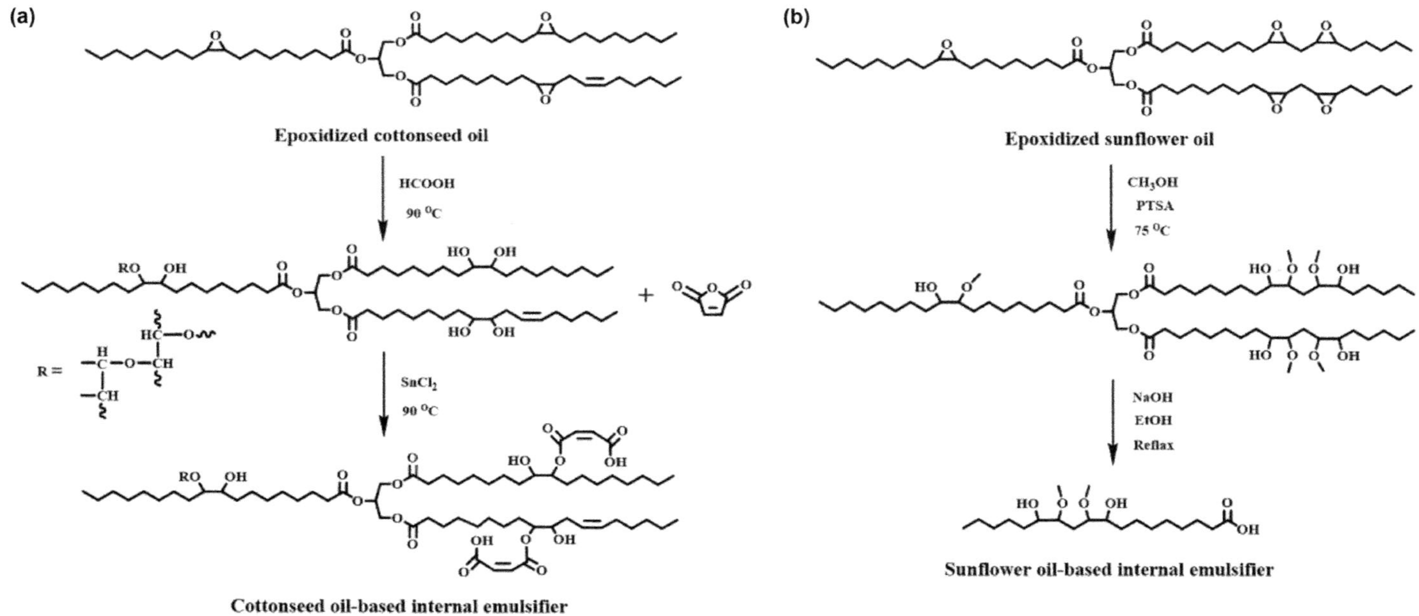

Figure 4.8 Synthesis route of the bio-based internal emulsifier from cottonseed (a)[13] and sunflower (b).[14] (a) Adapted from ref. 13 with permission from Springer Nature, Copyright 2016. (b) Adapted from ref. 14 with permission from Elsevier, Copyright 2016.

this internal emulsifier came from the ring-opening of epoxy sunflower oil by methanol, and the hydrophilic carboxylic acid group was derived from the hydrolysis of a triglyceride of the sunflower oil (Figure 4.8b). The particle size of the waterborne polyurethane dispersions prepared using the emulsifier was between 80 and 130 nm, and the obtained films exhibited excellent mechanical properties.

Compared with the step-by-step introduction of hydroxyl and hydrophilic carboxyl groups into plant oils to synthesize bio-based internal emulsifiers, it is simpler and more efficient to obtain plant oil-based internal emulsifiers directly through ring-opening of epoxy plant oils with small molecular acids. For example, Palanisamy *et al.* synthesized a novel phospholipid internal emulsifier through the ring-opening hydrolysis of epoxidized cottonseed oil in the presence of *ortho*-phosphoric acid[15] (Figure 4.9a). The obtained phospholipid internal emulsifier is a liquid at 30 °C, and its hydroxyl value and acid value are 117 mg KOH per g and 68 mg KOH per g, respectively. The particle size of the waterborne polyurethane dispersion prepared using the internal emulsifier was between 32 and 68 nm, and the waterborne polyurethane could be used for anti-corrosion coatings. Zhang *et al.* reported a solvent-free and self-catalyzed epoxy/ring-opening method for the preparation of novel fully bio-based emulsifiers from epoxidized soybean oil and glutaric acid[16] (Figure 4.9b). The effects of carboxyl/epoxy molar ratios and reaction time on the chemical structures of the resulting internal emulsifiers have been systematically studied. The results show that a molar ratio of the epoxidized soybean oil epoxy group to glutaric acid of 4 : 1 and a reaction time of 90 min were the optimal reaction conditions, resulting in the highest OH number (OH value: 99.58 mg KOH per g) and acid number (acid value: 119.15 mg KOH per g) of the emulsifier, as well as the highest atomic and energy efficiency. This novel plant oil-based internal emulsifier was a liquid at 30 °C and exhibited a chemical structure similar to those of 2,2-dimethylolpropionic acid and 2,2-dimethylolbutanoic acid, with the carboxylic acid group serving as an ionic segment and the hydroxyl groups acting as crosslinking agents. This liquid plant oil-based emulsifier has a lower melting point than 2,2-dimethylolpropionic acid and 2,2-dimethylolbutanoic acid, resulting in good compatibility with bio-based polyols endowing the resulting dispersion with excellent storage stability, a high solid content (up to 45%), and a high bio-based content (up to 74%). Comparing the waterborne polyurethane films prepared from 2,2-dimethylolpropionic acid or 2,2-dimethylolbutanoic acid and the traditional solvent-based polyurethane films, the waterborne polyurethane films prepared from the obtained internal emulsifier exhibited higher toughness and thermal stability. In addition, natural diacids (adipic acid and pimelic acid) were used as a replacement for glutaric acid to prepare plant oil-based internal emulsifiers with different hydroxyl numbers and acid numbers for the synthesis of pressure-sensitive adhesives.[17]

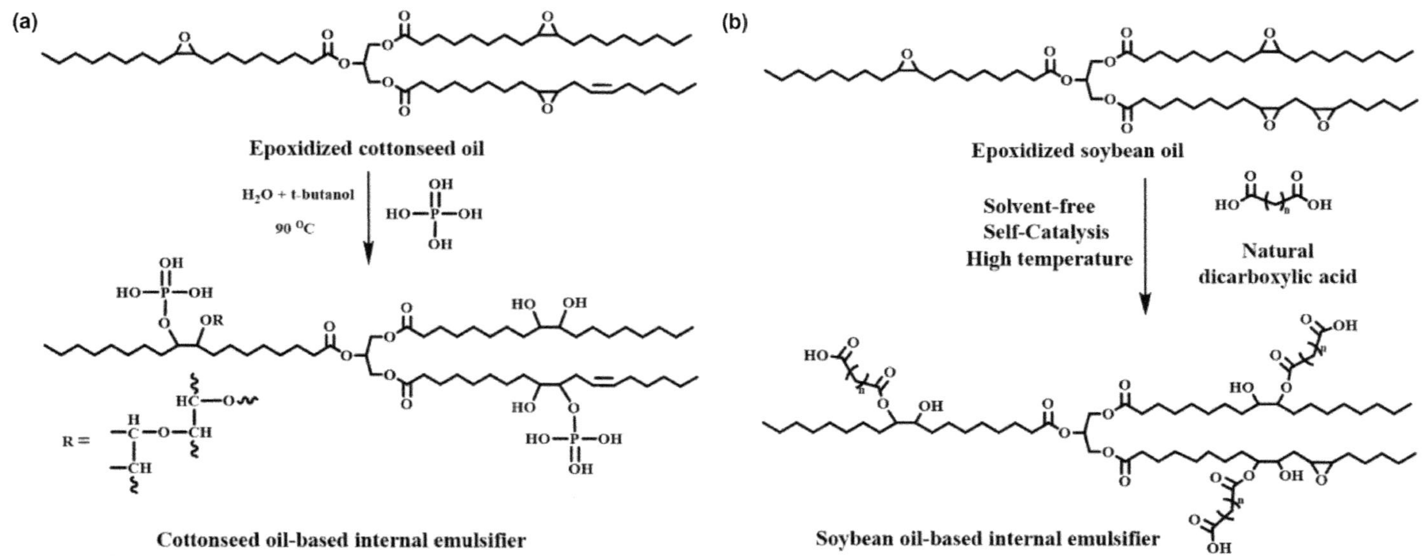

Figure 4.9 Synthesis route of the bio-based internal emulsifier from cottonseed oil (a)[15] and soybean oil (b).[17] (a) Adapted from ref. 15 with permission from American Chemical Society, Copyright 2017. (b) Adapted from ref. 17 with permission from American Chemical Society, Copyright 2021.

4.1.4 Thiol–Ene Photo-click Method

By applying the advantageous characteristics of the thiol–ene photo-click reaction, hydrophilic carboxylic acid groups could be directly grafted to the castor oil molecule in one step and the naturally-occurring hydroxyl groups of castor oil offered the necessary hydroxyl functional groups (Figure 4.10). For example, Chen *et al.* prepared a castor oil-based internal emulsifier from castor oil and 3-mercaptopropionic acid *via* a thiol–ene photo-click reaction under the action of UV1173 as a photoinitiator.[18] The internal emulsifier was obtained with a yield of 92.1% after 6 hours of UV light irradiation (1700 W cm^{-2}, 365 nm) at room temperature, and the obtained internal emulsifier could be used to replace 2,2-dimethylolpropionic acid in the preparation of fully bio-based waterborne polyurethanes with excellent performance with castor oil. Undoubtedly, the direct grafting of a hydrophilic carboxylic acid onto castor oil *via* a thiol–ene photo-click reaction is a simple and convenient strategy for the preparation of bio-based internal emulsifiers. However, the conversion of plant oils into bio-based internal emulsifiers under photochemical conditions has the problems of long reaction time and low reaction efficiency. The following reasons could be attributed to the low reaction efficiency of plant oils during the thiol–ene reaction: (1) the carbon–carbon double bonds which are located in the middle of fatty acid chains and usually are non-conjugated, thus exhibiting low reactivity, and (2) the steric hindrance due to the three long fatty acid chains of plant oils. Therefore, the development of a method for rapidly converting plant oils into bio-based internal emulsifiers has already become a hot research topic. Different from the normal reaction setup, an automated

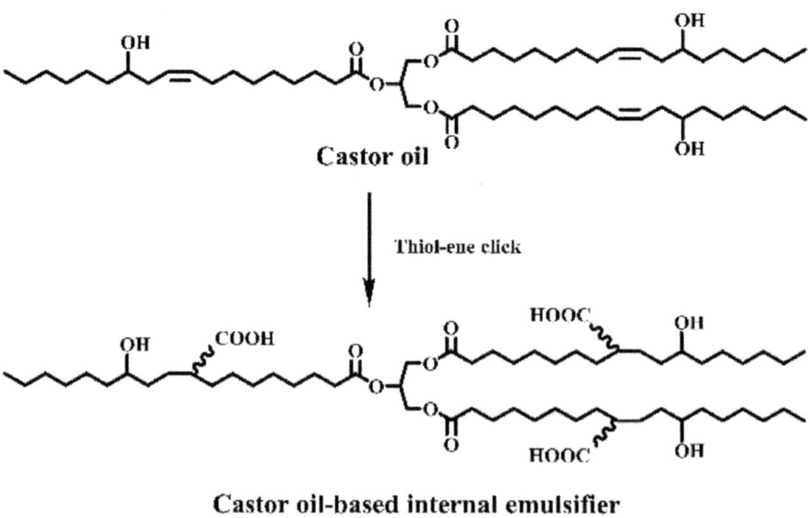

Figure 4.10 Preparation of a castor oil-based internal emulsifier using a thiol–ene photo-click method.

continuous-flow reactor has attracted much attention for green and rapid synthesis of new bio-based chemicals and polymers due to its significant advantages of cost savings, energy efficiency, and increased safety. For example, Zhang *et al.* proposed a novel photo-click, thiol–ene method to rapidly synthesize a fully bio-based internal emulsifier *via* flow chemistry using castor oil and cysteine derivative as raw thiol materials.[19] With a custom flow chemistry setup, the conversion rates of carbon–carbon double bonds were up to 92.0% and the productivity of the resulting internal emulsifiers could reach 360.0 $g\,h^{-1}$. The obtained liquid fully bio-based internal emulsifier exhibited good compatibility with various typical polyols when it was used to synthesize the waterborne polyurethane. Moreover, the resulting waterborne polyurethane film exhibits superior thermo-physical and mechanical properties compared to solvent-based polyurethane and petroleum-based waterborne polyurethane films.

4.1.5 Chain Extenders from Plant Oils

Low molecular weight short-chain diols or diamines are usually used as chain extenders in the synthesis of polyurethane materials. The structure of the primary hydroxyl group or primary amino group makes them usually have higher reactivity, which can effectively increase the molecular weight of the polyurethane and increase the content of hard segments in the polyurethane during polymerization, thereby improving the performance of the obtained polyurethanes.

When the adjacent hydroxyl groups of the plant oil-based polyol are linked by a low number of carbon atoms, the polyols are also considered and applied as chain extenders. Such plant oil-based chain extenders can be synthesized through simple and efficient methods and further used to prepare polyurethanes and tailor their properties. For example, Liu *et al.* developed a method for preparing palm oil-based chain extenders through amidation of palm oil and 2-(2-hydroxyethylamino)ethanol.[20] Palm oil and 2-(2-hydroxyethylamino)ethanol underwent an amidation reaction under sodium methylate as a catalyst at 120 °C for 4 hours to obtain a dark yellow palm oil-based chain extender with a yield of 92.3% and a hydroxyl value of 422.2 mg KOH per g (Figure 4.11). This chain extender has two primary

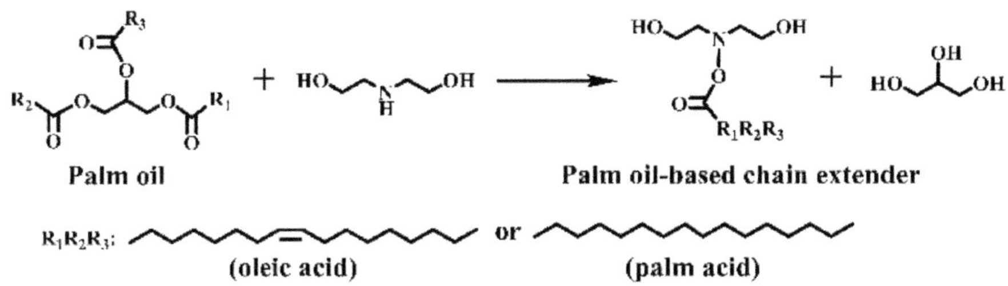

Figure 4.11 Synthesis route of the palm oil-based chain extender.[20] Adapted from ref. 20 with permission from American Chemical Society, Copyright 2022.

hydroxyl groups, the hydroxyl groups are linked by one N atom and four C atoms, and the palm oil fatty acid chain is connected with the N atom through an amide bond as the suspension chain of the chain extender. Subsequently, this chain extender was used to prepare a thermoplastic polyurethane elastomer with superior stretchability (up to a strain of 831%). In addition, other types of plant oils such as niger seed oil and tung oil can also be prepared using the amidation reaction with 2-(2-hydroxyethylamino)ethanol to prepare plant oil-based chain extenders for the preparation of polyurethane materials.[21,22]

Monoglycerides synthesized by transesterification of plant oils and glycerol are another typical plant oil-based chain extender. For example, Hatakeyama *et al.* reported a method for the preparation of palm oil-based chain extenders with a yield of 70% by transesterification of palm oil and glycerol at 90 °C for 2 hours under the action of a basic catalyst.[23] Palmitoleic acid was grafted onto glycerol as a hanging chain, and the remaining two hydroxyl groups connected by two or three carbon atoms on glycerol could react with isocyanate for polymer chain growth. Therefore, this monopalmitolein was further used in the preparation of soft polyurethane foams. In addition to monoglycerides obtained by the transesterification of glycerol with plant oils, the sorbitan monoester obtained by the transesterification of sorbitan and plant oils can also be considered another plant oil-based chain extender. Depending on the type of fatty acid chain, it can be named sorbitan monolaurate, sorbitan monopalmitate, sorbitan monostearate, and sorbitan monooleate. Although they are well known as commercial surfactants, this does not affect their role as chain extenders in the preparation of polyurethane materials. For example, Zhang *et al.* reported a study on the use of sorbitan monooleate to modulate the properties of castor oil-based waterborne polyurethanes.[24] They found that sorbitan monooleate has UV absorption ability, and the resulting waterborne polyurethane could be used for anti-ultraviolet coatings, while the highly active and dense primary hydroxyl sites of sorbitan increased the crosslinking density of the material and could endow waterborne polyurethane with excellent mechanical properties (tensile strength: 26.32 MPa).

The preparation of bio-based chain extenders from plant oil fatty acid methyl esters is another simple approach. For example, Sendijarevic *et al.* developed a new bio-based chain extender using ethylene glycol and an epoxidized palm oil methyl ester *via* a hydroxyl ring-opening reaction.[25] The palmitoleic acid methyl ester-based chain extender needs to be synthesized in two steps. First, the palmitoleic acid methyl ester was epoxidized under the action of formic acid and hydrogen peroxide, and then the epoxidized methyl palm oleate was ring-opened by ethylene glycol under the action of boron trifluoride etherate to obtain methyl palmitoleate diol with a yield of 88.7%. The methyl palmitoleate diol has a primary hydroxyl group derived from ethylene glycol and a secondary hydroxyl group obtained by the ring opening of epoxy, which can react with isocyanate to increase the degree of polymerization of the polyurethane. This chain extender was used to prepare flexible polyurethane elastomers with a hardness of 1–4 Shore A and with a T-peel adhesion of 2.27–1.98 N/25 mm to a polypropylene substrate.

In addition, plant oil-derived polyamine compounds are considered new promising chain extenders for the synthesis of plant oil-based non-isocyanate polyurethanes. The thiol–ene reaction route has been chosen as a choice for the development of a variety of plant oil-based chemicals due to its convenience. It is also a good choice to use the thiol–ene reaction route to develop plant oil-based polyamines. For example, Robin *et al.* proposed a new scheme to develop new plant oil-based polyamines from grape seed oil.[26] Grape seed oil-based polyamines with a yield of 87% were obtained using a thiol–ene photo-click reaction for 8–96 hours between grape seed oil and cysteamine hydrochloride under the action of 2,2-dimethoxy-2-phenyl-acetophenone. Although the preparation of plant oil-based polyamines by the thiol–ene photo-click reaction is convenient, it has the disadvantages of low reaction rate and long reaction time. Meanwhile, Dubois *et al.* reported a method for the preparation of fatty acid diamines by an amide reaction using dimer fatty acid Pripol 1013 and 1,2-diaminoethane.[27] The dimerized fatty acid Pripol 1013 and 1,2-diaminoethane were reacted at 160 °C for one hour with a molar ratio of $[NH_2]$ to $[CO_2H]$ of 1.2 and then heated to 190 °C for 6 hours to obtain this yellow solid bio-based diamine (molecular weight: 4880 $g\,mol^{-1}$; amine content: 0.41 $mol\,kg^{-1}$). Subsequently, the obtained bio-based diamine was used to cure soybean oil-based cyclocarbonate to prepare bio-based non-isocyanate polyurethane with excellent thermal stability ($T_{95wt\%} > 350$ °C). At the same time, some fatty acid diamines have been commercialized, such as Priamine® 1075, Priamine® 1071, acyclic dimerdiamine, cyclic dimerdiamine, aromatic dimer diamine, *etc.*[28,29]

4.1.6 Isocyanates from Plant Oils

Typically, almost all the isocyanates used in the development of polyurethane materials are synthesized using the method associated with the use of toxic phosgene. Scientists from academia and industry have tentatively explored a new routine to prepare bio-based isocyanates for a more sustainable and greener society. Plant oils are undoubtedly a promising option. For example, Kü sefoğlu *et al.* reported a simple and efficient two-step method for the synthesis of soybean oil-based isocyanates. They used *N*-bromosuccinimide to brominate the allyl group on the unsaturated fatty acid chain of soybean oil through a bromination reaction and then replaced the bromine on the unsaturated fatty acid chain of soybean oil with AgNCO to obtain soybean oil-based isocyanate with a yield of 60–70%, and the yield could be increased by increasing the amount of AgNCO.[30] Subsequently, they proposed a simpler one-step method to archive the same purpose: directly reacting carbon–carbon double bonds of soybean oil with iodine isocyanate to obtain a soybean oil-based isocyanate.[31]

However, the bio-based isocyanate prepared directly from the plant oil had a large steric hindrance effect, because the isocyanate groups were located in the middle of the fatty acid chain of plant oil, which leads to its low reactivity and is not suitable for replacing the most commonly used diisocyanates in the industry, such as 2,4-toluene diisocyanate and 4,4′-diphenylmethane diisocyanate. Therefore, much research work is focused on the development of bio-based isocyanates with terminal NCO groups, and the Curtius rearrangement reaction has become the choice of many researchers. For example, Burkart *et al.* proposed a method to prepare isocyanates from fatty acids derived from algal biomass. They used the flow chemistry technology to acyl azidize a fatty acid with hydrazide under the action of methanol and HCl, and then the fatty acid azide was subjected to a Curtius rearrangement reaction under heating at 85 °C to produce a bio-based isocyanate at a rate of 500 mg h^{-1} with a yield of 80%.[32] In addition, the castor oil derivative undecylenic acid can also undergo Curtius rearrangement after hydrazide azide to generate terminal bio-based diisocyanates.[33]

Another simple and efficient synthetic route to the synthesis of terminal bio-based diisocyanates involves the conversion of plant oil fatty acids into terminal dicarboxylic acids followed by Curtius rearrangement. For example, Narine *et al.* proposed a new scheme for the synthesis of novel long-chain linear unsaturated terminal diisocyanates using oleic acid.[34] They first converted oleic acid into terminal dicarboxylic acids with a yield of 64% under the action of Grubbs catalyst, followed by further conversion into yellow oily terminal diisocyanates with a yield of 56% by Curtius rearrangement. Subsequently, the obtained long-chain linear unsaturated terminal diisocyanate was reacted with rapeseed oil polyol to prepare a fully bio-based polyurethane with excellent tensile properties.

4.2 Solvent-based Plant Oil-based Polyurethanes

Compared with waterborne polyurethanes using water as the dispersion medium, the preparation and application of traditional polyurethanes usually involve a certain amount of solvent as the assistance dispersion medium. Using plant oil-based monomers, such as polyols, chain extenders, or isocyanates, solvent-based partially or fully bio-based polyurethanes could be prepared using a similar method to the traditional process.

4.2.1 Partially Bio-based Polyurethanes

Most bio-based polyurethanes still rely on a certain amount of petroleum-based monomers and partially monomers are derived from bio-based resources. For example, Zhang *et al.* systematically synthesized a series of smart polymers with self-toughening and multi-stimulus responsiveness by incorporating scalable rhodamine 6G derivatives into a polymer backbone using castor oil as a cross-linker and soft segment. The new polyurethane elastomers exhibited excellent reversible multicolor behavior in response to light, pH, and mechanics.[35] In another work, Zhao *et al.* synthesized a series of UV-curable polyurethane adhesives from acrylated epoxidized soybean oil for potential application in pigment printing processes. These bio-based polyurethane binders demonstrated excellent color yield, rub resistance, and wash resistance.[36] By replacing 20 wt% of petrochemical polyols with palm oil-based polyols, Krzysztof *et al.* prepared a series of modified open-cell polyurethane foams. It was found that the modified polyurethane foams showed thermal conductivities between 39 mW m^{-1} K^{-1} and 42 mW m^{-1} K^{-1} and apparent densities between 13 and 15 kg m^{-3}, meeting the requirements of commercial products.[2]

4.2.2 Fully Bio-based Polyurethanes

If all the raw materials and monomers are derived from bio-based resources, the resulting polyurethanes should be fully bio-based polyurethanes, which is of great significance for achieving "carbon neutrality". For instance, Roberto *et al.* developed a series of thermosetting elastomer polymers using L-lysine ethyl ester diisocyanate as the isocyanates and blends of castor oil and polypropanediol as the polyols. The resulting polyurethane elastomers showed glass transition temperatures ranging from −41 to 21 °C and high thermal stabilities up to 300 °C.[37] In another work, Swarnalata *et al.* prepared novel fully bio-based polyurethane adhesives by reacting castor oil-based polyols with bio-based poly-isocyanates in the presence of dibutyltin dilaurate as a catalyst. The effect of stoichiometric ratios between the polyols and isocyanates on the bond strength of the resulting bio-based polyurethane adhesives was systematically studied using lap shear strength tests on wood-to-wood substrates.[38]

4.3 Plant Oil-based Waterborne Polyurethane Dispersions

Waterborne polyurethane is an environmentally friendly polyurethane developed on the basis of solvent-based polyurethanes. Replacing petroleum-based chemicals with plant oil-derived polyols, internal emulsifiers, and isocyanates can further alleviate the pressure brought about by the global energy crisis and environmental pollution. Hydrophilic groups distributed on the surface of plant oil-based polyurethane microspheres make them dispersed in the water phase and can be stored stably. The particle size of water-based polyurethane microspheres is generally within a few hundred nanometers, and the charged groups or hydrophilic groups on the surface of the microspheres form a stable electric double layer or interact with water to form a stable colloid. According to the type of hydrophilic group carried by the plant oil-based waterborne polyurethane, it can be classified into anionic waterborne polyurethane, cationic waterborne polyurethane, or nonionic waterborne polyurethane.

4.3.1 Cationic Polyurethane Dispersions

Cationic waterborne polyurethane generally refers to the polyurethane dispersion in which the tertiary amine functional group in the polyurethane segment is neutralized by an organic acid to form a quaternary ammonium salt ion so that the surface of the polyurethane microsphere is positively charged. Cationic waterborne polyurethane has its own advantages. The positive charge on its surface has excellent adhesion to fabrics and so on and shows good flexibility and hand feeling. This also makes cationic waterborne polyurethane have great application prospects in the field of leather finishing agents. Due to the charge interaction, cationic waterborne polyurethane also exhibits strong adhesion to substrates with negatively charged surfaces such as wood.

What is more worth mentioning is that the electropositive surface of a cationic waterborne polyurethane can be combined into the negatively charged bilayer on the surface of bacteria to destroy the structure of bacteria, and thus it can play a role in sterilization and antibacterial activity. The performance of cationic waterborne polyurethane is also affected by its different types of neutralizers. For example, Zhang *et al.* used amino acids including glutamic acid and aspartic acid as neutralizers to prepare a series of cationic plant oil-based waterborne polyurethanes with many functions such as absorbing ultraviolet light, antiseptic activity, and long-lasting antibacterial activity.[39] It is found that the tensile strength and crosslinking density of a waterborne polyurethane prepared using amino acids as neutralizing agents were 5 times and 3 times those of waterborne polyurethanes prepared using known acetic acid as a neutralizing agent, respectively. Furthermore, Zhang *et al.* also used a variety of natural phenolic acids as neutralizing agents to prepare castor oil-based cationic waterborne polyurethanes.[40] The effects of these neutralizing agents on the microstructure (phase separation and hydrogen bonding) and properties (mechanical properties, antibacterial activity, *etc.*) of polyurethanes were further systematically studied and compared with other commonly used neutralizing agents. The results showed that, in addition to being used as a neutralizing agent, phenolic acids also acted as anti-plasticizers for polyurethane films by increasing the rigidity of the hard segment and the number of hydrogen bonds. In addition, the hydrophobic, rigid, and conjugated structures of phenolic acids endowed the resulting polyurethane films with excellent UV protection and water resistance in practical application. The introduction of phenolic acid into the resulting polyurethane films made them possess good antibacterial performance against Gram-positive and Gram-negative bacteria. Additionally, they utilized acrylic acid as a neutralizing agent and further prepared a cationic castor oil-based waterborne polyurethane with an interpenetrating polymer network structure by thermally induced free radical emulsion polymerization.[41] The effect of the resulting waterborne polyurethane–polyacrylic acid dispersion and film properties was systematically investigated. The results showed that with the increase in acrylic acid content, the crosslinking density of the interpenetrating network polymer increased, and the interlocking structure of the soft and hard phase dislocation of the double network molecular segment was tailored, thereby improving the mechanical properties and water resistance of the waterborne polyurethane film. The 96 h water absorption of the film showing the best water resistance was only 3.27%.

The most commonly used organic acid in the preparation of cationic waterborne polyurethanes is acetic acid. For example, Zhang *et al.* prepared a series of cationic waterborne polyurethanes using acetic acid as a neutralizing agent for different plant oil-based polyols.[42] With the increase of the hydroxyl value of the plant oil-based polyols, the tensile strength of the waterborne polyurethane film increased from 1 MPa to 11 MPa, and the Young's modulus increased from 10 MPa to 395 MPa. Of course, theoretically, other water-soluble small molecular acids such as hydrochloric acid can also be used to prepare cationic plant oil-based waterborne polyurethanes.

4.3.2 Anionic Polyurethane Dispersions

Anionic waterborne polyurethane refers to a waterborne polyurethane dispersion that neutralizes carboxylate groups with triethylamine to form a salt or has a sulfonate hydrophilic group. The hydrophilic chain extender includes 2,2-dimethylolbutyric acid, 2,2-dimethylolpropionic acid, sodium ethylenediaminosulfonate, ethylenedihydroxyethanesulfonate, 1,2-diamino-β-propanesulfonate, 1,3-diamino-β-propanesulfonate, *etc.*

Among them, 2,2-dimethylolpropionic acid has become the most commonly used internal emulsifier for the preparation of anionic waterborne polyurethanes due to its low price and good reactivity. Using 2,2-dimethylolpropionic acid as an internal emulsifier to prepare a waterborne polyurethane has the advantage of mature technology, and at the same time, the obtained emulsion is stable and exhibits good coating performance. For example, Shen *et al.* used castor oil as a polyol, 2,2-dimethylolpropionic acid as an internal emulsifier, and a small molecule diol as a chain extender to synthesize a castor oil-based waterborne polyurethane with corrosion resistance.[43] Due to the high density of hydrogen bonds inside the film, the tensile strength of the resulting polyurethane was as high as 29.56 MPa, while the toughness remained 20.12 $MJ\,m^{-3}$. However, 2,2-dimethylol propionic acid reacted heterogeneously with isocyanates during the polymerization process of waterborne polyurethane due to its high melting point. A large amount of organic solvent was usually required in the reaction system to solve the above problem, but the addition of the solvent would reduce the reaction rate. Compared with 2,2-dimethylolpropionic acid, 2,2-dimethylolbutyric acid is a new type of carboxylic acid type internal emulsifier and has the advantages of low melting point and high solubility in solvents. For example, Liang *et al.* used 2,2-dimethylolpropionic acid and 2,2-dimethylolbutyric acid as internal emulsifiers to synthesize two castor oil-based aqueous polyurethane dispersions.[44] The effects of the two emulsifiers and their dosage on the polyurethane dispersion and its film-forming properties were systematically studied. The results showed that aqueous polyurethane dispersions from 2,2-dimethylolbutyric acid exhibited smaller particle sizes than those from 2,2-dimethylolpropionic acid, and films from the former showed a higher tensile strength and a higher elongation at break.

The sulfonate-type hydrophilic chain extender used in the preparation of waterborne polyurethanes is very popular in industrial production because it has a sulfonate group in itself and does not need addition of a neutralizer before emulsification with water. Compared with carboxylic acid groups, the ionic groups of sulfonate have a stronger degree of ionization in the water phase, leading to a thicker electric double layer and higher satiability of the resulting dispersion. However, the preparation process of the waterborne polyurethane with sulfonate-type hydrophilic chain extenders was more complicated than those with carboxylic acid-type hydrophilic chain extenders.

4.3.3 Non-ionic Polyurethane Dispersions

Non-ionic waterborne polyurethanes have characteristics that are significantly different from ionic waterborne polyurethanes. Non-ionic waterborne polyurethanes do not increase the hydrophilicity of the polymer through ionic groups but make it stably dispersed into the water phase through the interaction between the pendant hydrophilic polyether side chain and water. Compared with ionic waterborne polyurethanes, non-ionic waterborne polyurethanes are not affected by the pH value and have the characteristics of a cloud point, that is, non-ionic waterborne polyurethanes lose stability at a certain temperature. This feature makes non-ionic water-based polyurethanes widely used in fine processing, cosmetics, and printing and dyeing fields.

4.4 Non-isocyanate Polyurethanes

Although the concept of polyurethane chemistry and its application have been proposed since the early 1930s, the main process for the development of the most commercially available polyurethanes and synthetic mechanisms so far still rely on the polyaddition reaction between polyisocyanates and polyols. However, the isocyanates were reported to be a toxic volatile compound and can easily cause asthma and dermatitis. More seriously, phosgene used in the production of isocyanates was considered to be a highly toxic compound that can be a serious hazard to the environment and human health.[45,46] In addition, the moisture sensibility of isocyanates poses problems in practical application, such as storage difficulties, short life span, and increased costs. Recently, the European Union has announced the REACH regulation to start a new restriction on the use of products containing more than 0.1% of monomeric diisocyanates.[47] With the increasing awareness of environmental issues and public health safety, the new pathway to prepare polyurethanes without the use of isocyanate has attracted much attention.

In order to avoid the use of isocyanates, various strategies have been proposed to prepare non-isocyanate polyurethanes (NIPUs) with tunable properties compared with the traditional isocyanate-based polyurethanes: (i) a reaction between aziridines and CO_2,[48] (ii) polymerization of cyclic carbamates,[49] (iii) polycondensation between bis(dialkylcarbonate)s and diamines or bis(dialkylcarbamate)s and diols,[50,51] and (iv) polyaddition between poly(cyclic carbonate)s and polyamines.[52,53] Among them, the last method has been widely investigated and holds great promise for large-scale production in the polyurethane industry. In fact, cyclic carbonates can be obtained by cycloaddition of epoxy compounds with CO_2, a 100% atomic economy method that effectively reduces carbon emissions. At the same time, epoxy compounds can be sourced from renewable raw materials, replacing fossil resources with renewable energy sources. One of the most representative renewable epoxides is epoxidized plant oils.[54] Plant oil-based non-isocyanate polyurethanes avoid the use of toxic isocyanate as raw materials, have the advantages of being non-toxic, harmless, and renewable, and are a promising route in the current polyurethane industry.

4.4.1 Carbonated Plant Oils

Carbonated plant oils have the advantages of being non-toxic and non-hazardous, stable storage, biodegradability, and having a high boiling point. They could be produced by the addition reaction of epoxidized plant oils and CO_2 (Figure 4.12). Due to the site-blocking effect of the epoxy functional group of epoxidized plant oils, the carbonylation of epoxidized plant oils usually needs to be catalyzed by an organ catalyst (tetrabutylammonium halide) at supercritical high temperatures and CO_2 pressures (100–140 °C and 3–10 MPa). The harsh conditions for the conversion of carbonated plant oils and the high energy consumption are of great challenge that limit the industrial production of plant oil-based NIPUs.

4.4.2 Solvent-based NIPUs Prepared from Plant Oils

Plant oil-based NIPUs could be prepared by curing carbonated plant oils with aliphatic amines (hexamethylenediamine, diethyltriamine ethylenediamine, tris(2-aminoethyl)amine, *etc.*) or aromatic amines (isophorone diamine, *m*-xylenediamine, *p*-xylenediamine, *etc.*) or polyamines (polyoxypropylene diamines, polyoxypropylene triamine, *etc.*) (Figure 4.8). By selecting different carbonated plant oils, diamines, and their molar ratios, there are many possibilities to tailor the properties of the final polymer NIPUs. NIPUs differ from conventional polyurethanes by the presence of hydroxyl groups in the polymer chains which enhances intramolecular and intermolecular hydrogen bonding resulting in increased intermolecular

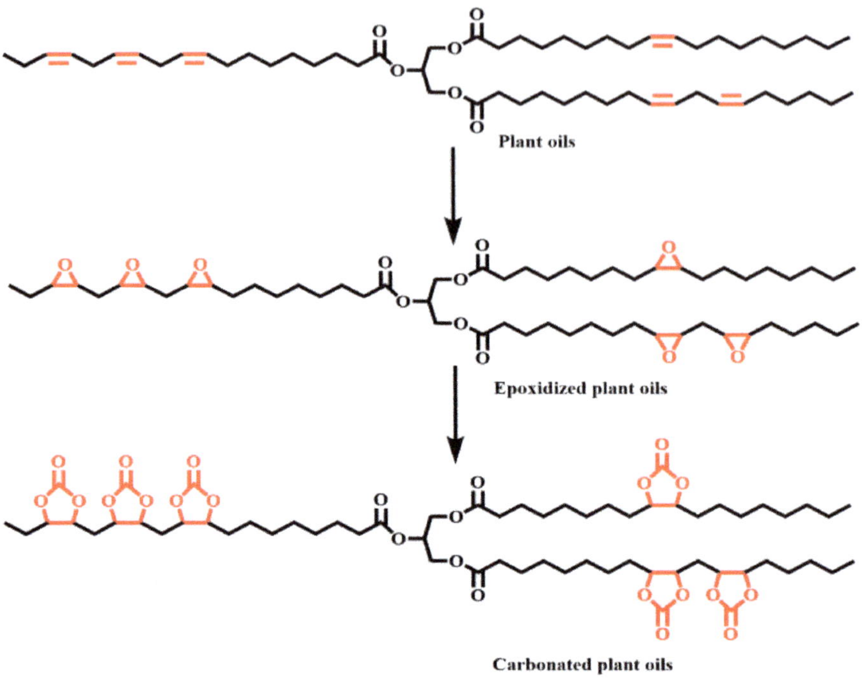

Figure 4.12 Schematic of synthesis procedures of carbonated plant oils.

forces and improved mechanical properties. However, these hydroxyl groups also increase the hydrophilicity of the NIPUs, resulting in increased water absorption. The introduction of carbonated plant oils as hydrophobic chain segments into NIPUs reduces the problem of water absorption caused by hydroxyl groups to a certain extent.

The curing of carbonated plant oils usually needs the assistance of organic solvents with their high viscosity. As early as 2004, Wilkes *et al.* prepared carbonated soybean oil and constructed a bio-based NIPU network for the first time.[55] Later on, this method was extended to carbonated linseed oil, carbonated sunflower oil, and carbonated cottonseed oil. At the same time, new strategies have been developed to improve the performance of solvent-based NIPUs. For example, Poussard *et al.* synthesized extremely tough plant oil-based NIPUs by polymerization of carbonated soybean oil with bio-based short diamines or fatty acid-based amino-telechelic oligoamides. The tensile strength of the resulting NIPUs was increased to 16 MPa and the elongation at break increased to 400%.[27] Yang *et al.* prepared soybean oil-based NIPU vitrimers with 4,4′-diaminodiphenyl methane or 4,4′-diaminodiphenyldisulfide as the curing agents. The introduction of dynamic disulfide bonds into the polymer networks led to the resulting NIPUs' excellent reprocessability and self-healing performance under mild conditions (heat at 50 °C or UV irradiation at room temperature).[56] Gholami *et al.* prepared novel plant oil-based NIPUs *via* a two-step method consisting of polymerization of carbonated soybean oil with tetraethylenepentamine followed by modification with epichlorohydrin. The presence of the cross-linked polymer networks containing cationic aziridinium groups endowed the resulting NIPUs with 100% antibacterial performance against *Staphylococcus aureus* and *Escherichia coli*.[57]

4.4.3 Waterborne NIPUs Prepared from Plant Oils

The newly formed hydroxyl groups on the polymer chains of the NIPUs provide some hydrophilicity, but this is not sufficient to allow the NIPUs to be dispersed directly in water to make waterborne NIPUs. Generally, external or internal emulsifiers were added to the system for obtaining environmentally friendly waterborne plant oil-based NIPUs, avoiding the use of organic solvents. For example, Cramail *et al.* synthesized a series of NIPU latexes by mini-emulsion polymerization with external emulsification from fatty acid dimeric diamine and fatty acid-based bis-carbonates.[58] Different surfactants and hydrophobic additives (Tween® 80, sodium dodecylsulfate, *etc.*) were physically added to assist the dispersion of NIPUs into water. These NIPU emulsions demonstrated particle sizes below 200 nm, which exhibited excellent stability. Using the internal emulsification method, Zhang *et al.* prepared waterborne plant oil-based NIPUs by incorporating ionic segments or a hydrophilic moiety (3,3′-diamino-*N*-methyldipropylamine, amine-polyethylene glycol-monomethyl ether, *etc.*) into the polymer backbone followed by dispersion and self-assembly into water.[59] Cationic, anionic, and non-ionic waterborne plant oil-based NIPUs could be prepared by controlling the types of hydrophilic groups (Figure 4.13). The particle sizes of the resulting NIPU emulsions were concentrated in the range of 100 nm to 350 nm, showed good stability, and exhibited tailored properties.

4.4.4 NIPU Foams Prepared from Plant Oils

Polyurethane foams account for more than 50% of the polyurethane products, with the worldwide production expected to reach 12.7 million tons by the end of 2024. Traditional polyurethane foams usually self-blow by *in situ* formation of CO_2 through the hydrolysis of isocyanates. For NIPU foams, such a method could not be reproduced.

Generally, external blowing agents were widely applied to synthesize NIPU foams in which petroleum-based monomers were initially used. For example, Mülhaupt *et al.* used the phase transition (usually from a liquid to a gas) behavior of hydrofluorocarbon during polymerization and applied it as a physical blowing agent for flexible NIPU foams.[60] Cornille *et al.* reported the development of flexible NIPU foams by using poly(methylhydrogenosiloxane) as an exogenous chemical blowing agent which reacted with amines to release H_2 (a highly flammable gas) and expanded the NIPU thermoset matrix during the polymerization.[61] In addition, Xi *et al.* reported the development of rigid NIPU foams in the presence of $NaHCO_3$ as a chemical blowing agent

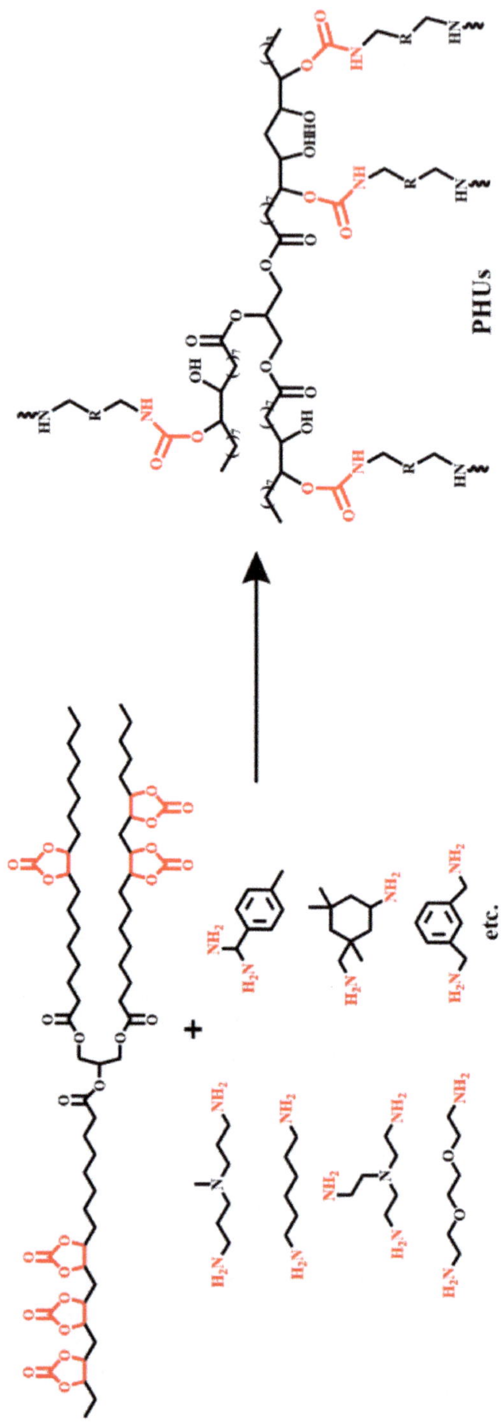

Figure 4.13 Schematic of the preparation procedures of waterborne NIPUs prepared from plant oils.

Table 4.1 Literature reports of blowing agents and characteristics.

Blowing agent	Reaction conditions	Foaming mechanism	Ref.
Citric acid and glutaraldehyde	Stirring for 10–15 s at room temperature	Citric acid carboxylic group reacted with diamine, leading to foaming and glutaraldehyde as a crosslinker increased the viscosity of NIPU resin to prevent foam collapse	63–65
Maleic acid and glutaraldehyde	Stirring at room temperature	The same as above	66
Poly(methyl-hydrosiloxane)	Stir for 1–10 min at room temperature	Reacted with the diamine curing agents to release H_2	67–70

at high temperatures (>200 °C).[62] Other foaming agents such as malic acid, maleic acid, citric acid, and aconitic acid have also been applied to prepare NIPU foams.[63–70] The reaction conditions of various foaming agents and their characteristics are summarized in Table 4.1. Similar methods can also be used to prepare plant oil-based NIPU foams. Dong *et al.* showed a practical and cost-effective approach to producing linseed oil-based NIPU foams by using citric acid and ammonium bicarbonate as the blowing reagents. The results showed that the glass transition temperature of the obtained foam was 42 °C, which was very similar to that of linseed-derived resin samples.[71]

Instead of traditional foaming agents, other strategies have been developed to synthesize environmentally friendly plant oil-based NIPU foams. For example, North *et al.* developed sorbitan bis-carbonate monomer (a plant oil-based derivative), which underwent an intramolecular side reaction in the process of thermally driven ammonolysis, triggering partial decarboxylation of the monomer and spontaneously releasing CO_2 to form rigid NIPU foams.[72] Detrembleur *et al.* reported the synthesis of soybean oil-based NIPU microcellular foams by a supercritical CO_2 foaming technology. CO_2-blown microcellular NIPU foams were prepared with a close porosity, small pores (1–20 μm), low density (110–176 $kg\,m^{-3}$), and a rather low thermal conductivity ($\lambda = 50$ $mW\,m^{-1}\,K^{-1}$).[58] However, these strategies were strongly limited in practical application due to the lack of available monomers whose synthesis was complicated and cost consuming.

Recently, novel strategies have been developed to synthesize environmentally friendly NIPU foams but they have not been applied to plant-oil-based products until now. For example, Anitha *et al.* reported that the cyclic carbonate of resorcinol diglycidyl ether could be reacted with amine-terminated oligomeric phenyl hydroxyl amine, leading to the formation of CO_2 and the resulting poly(hydroxycarbamate) foams, in which the parallel reaction of hydroxyl and amino groups with cyclic carbonate groups competed during the polymerization process.[73] Detrembleur *et al.* reported robust and universal self-blowing flexible or rigid NIPU foams by adding (di)thiol(s) to the NIPU formulations in the presence of DBU as the catalyst. In addition to generating NIPU through aminolysis of cyclic carbonate, cyclic carbonate underwent *S*-thioalkylation to form a thioether bond and released CO_2 as the foaming agent.[74,75] More recently, Detrembleur *et al.* reported the effective hydrolysis of 5-membered cyclic carbonates and the release of CO_2 at 80–100 °C by water added with a base catalyst (*e.g.* DBU or KOH) to obtain self-blowing rigid or flexible NIPU foams.[76] Moreover, Caillol *et al.* synthesized the dithiocarbonate, whose ring opening yielded the *in situ* thiol moiety that could react with usual cyclic carbonate, and prepared self-blowing non-isocyanate polythiocarbamate foams through the aminolysis of cyclic carbonate and dithiocarbonate.[77] Furthermore, Detrembleur *et al.* used rapid aminolysis of thiolactone at 80 °C to create reactive thiol intermediates and form CO_2 for foaming during decarboxylation *S*-alkylation of cyclic carbonates.[78] The above self-foaming methods are expected to be applied to the preparation of plant oil-based NIPU foams in the future.

Despite that much progress has been made in the development of NIPU foams, the versatility and simplicity of the preparation process and the properties of the resulting NIPU foam could not compete with traditional isocyanate-based formulations. It is still a challenge to produce self-foaming NIPU foams by imitating the universal and cost-effective process.

4.5 Application of Plant Oil-based Polyurethanes

Due to their unique chemical structures and specific properties, plant oil-based polyurethanes can have promising applications in a variety of fields, such as protective coatings, drug delivery, slow-release fertilizers, and other fields.

4.5.1 Protective Coatings

Polyurethane coatings have excellent abrasion resistance, water resistance, and chemical resistance, which enhance the appearance of products and extend their service life. Therefore, polyurethane coatings have been widely studied by scholars because of their excellent performance in various aspects (anti-corrosion coatings, wood coatings, UV-curable coatings, *etc.*). For example, Tang *et al.* introduced sustainable diols, isosorbide, and L-tyrosine-derived cyclic dipeptides into the anionic castor oil-based waterborne polyurethane backbone and investigated the stability of the dispersions as well as the cast film properties. The results showed that the incorporation of sustainable diols into the waterborne polyurethane improved the corrosion resistance against 45# steel, which resulted from the strong hydrophobic properties of waterborne polyurethane films.[79] Plant oil-based polyurethane materials are also promising to be used in the field of wood coatings. Bharatkumar *et al.* reported the development of peanut oil-based polyurethane coatings for wood. By testing the properties of the prepared coatings, such as water resistance, mechanical properties, antibacterial properties, and chemical resistance, the performance of the peanut oil-based polyurethane coatings could be compared with commercially available polyurethanes.[80] Zhang *et al.* prepared a series of castor oil-based cationic waterborne polyurethanes with natural acids as the neutral agents. The results showed that the polyurethane films ranging from flexible elastomers to tough plastics were obtained and imparted with excellent UV absorption properties, corrosion resistance, and long-term antimicrobial properties.[81]

4.5.2 Adhesives

Plant oil-based polyurethane adhesives could be potentially used in the field of packaging, construction, automotive, shoe making, and other industries due to their excellent shear strength, impact resistance, and excellent flexibility. Nayak *et al.* synthesized a series of plant oil-based polyurethane–clay nanocomposite adhesives using an *in situ* polymerization process of bio-based polyurethanes with different weights of organically modified nano-clay (1–5 wt%) as the nano-fillers. The resulting nano-composite adhesives exhibited excellent adhesion properties, especially a high bond strength at 70 °C.[82] Christoph *et al.* reported an interesting work about the preparation of a novel soybean oil-based supramolecular polymer binder. It is reported that when the temperature is above the melting temperature of an isophthalic acid-rich structural domain, the viscosity of the polymer binder could drop sharply to 8 Pa s at 120 °C. However, once cooled to room temperature, the material properties were fully restored due to the recombination of the supramolecular network.[83] Christoph *et al.* proposed a novel strategy for the release of the diamine cadaverine using an *o*-nitrobenzyl-protected precursor as a phototrigger for the development of novel castor oil-based polyurethane adhesives. These polyurethane adhesives showed fast curing behavior when UV light is applied. Most impressively, the lap shear strength values of the resulting adhesives were up to 4600 kPa.[84]

4.5.3 Foams

Polyurethane foams exhibited excellent thermal insulation, efficient insulation, sound insulation, waterproof and moisture-proof characteristics after curing, shock and pressure resistance, and other properties. In a recent work, Sylvain *et al.* for the first time used arylated soybean oil and bio-based amines by aza-Michael reactions to obtain highly efficient flexible foams with glass transition temperatures of −20 to −7 °C. These resulting soft foams showed excellent recovery behavior in about a few seconds, which compared favourably with those of commercial petroleum-based polyurethane foams.[85] Angelica *et al.* prepared a series of water-blown polyurethane foams by reacting epoxidized soybean oil and epoxidized linseed oil as polyols with bio-based aliphatic isocyanates. It was shown that the fully bio-based flexible polyurethane foams demonstrated an open cell structure and showed Young's moduli of 14–18 kPa and compression deflection values of 4.6–5.5 kPa.[86] Varaporn *et al.* developed novel bio-based polyurethane foams from palm oil-based polyols and natural rubber. It was shown that the bio-based polyurethane foams showed a polyhedral semi-closed cell structure with hardness in the range of soft foams for potential application in the area of thermal insulation.[87]

4.5.4 Drug Delivery

Waterborne polyurethane dispersion systems have special internal hydrophobic and external hydrophilic structures, which can encapsulate hydrophobic drug molecules, solving the problems of poor compatibility, low circulation capacity, and toxic side effects of traditional hydrophobic drugs in the process of *in vivo* circulation. In a recent work, Luo *et al.* developed a polyurethane micelle with dual stimulatory response properties and the ability to control the pH and the position of the reduction-responsive groups by incorporating tertiary amines in the hydrophobic core of the polyurethane backbone. The rapid and complete redox and pH-stimulated release behavior of the resulting nanocarrier could provide sufficient drug concentration to kill cancer cells and is believed to be a promising anticancer drug delivery system.[88] Similarly, Kan *et al.* synthesized biodegradable castor oil-based polyurethanes using an emulsion solvent evaporation method and used it as a carrier for the nano-emulsion drug delivery system. The avermectin loading capacity could reach 42.3 wt% and the encapsulation efficiency was up to 85%.[89] Recently, controllable pesticide delivery systems have attracted increasing interest in the improvement of pesticide utilization efficiency and environmental safety. For example, Zhang *et al.* developed an eco-friendly aqueous polyurethane delivery system for encapsulation of biopesticides in environmentally friendly modern agricultural applications. These castor oil-based aqueous delivery systems enhanced the deposition, rain tolerance, retention, protection, and sustained release as an efficient pesticide delivery system.[90]

4.5.5 Slow-release Fertilizers

Due to the advantages of plant oil-based polyurethane materials such as their environmental friendliness and renewable, degradable, and hydrophobic characteristics, using them is considered one of the approaches to replace traditional coating materials for biodegradable slow-release fertilizers. For instance, Zhang *et al.* prepared castor oil-based polyurethanes as a slow-release fertilizer encapsulation coating material by adjusting the molar ratio of castor oil hydroxyl group to isocyanate to control the physicochemical properties and microstructure of polyurethanes. The N release longevity of castor oil-based polyurethane slow-release fertilizers can be adjusted from 20 to 140 days at only a coating content of 5%.[91] Guo *et al.* prepared interpenetrated castor oil-based polyurethanes for slow-release fertilizers by a one-pot method. The N release longevity of the polyurethane-coated slow-release fertilizers could reach 210 days, leading to the excellent promotion effects on plant growth.[92] Yang *et al.* synthesized plant oil-based polyurethane acrylates modified with polydimethylsiloxane and thiol-grafted nano-silica to prepare superhydrophobic bio-based polyurethane acrylate slow-release fertilizers. The study showed that the N release longevity of the bio-based polyurethane acrylate slow-release fertilizers was close to 120 days and required only 4% of the coating contents.[93]

4.5.6 Others

With the rapid development of plant oil-based polyurethanes, these new types of bio-based polymers could find specific applications in various fields such as food packaging, defense and military industry, aerospace, *etc.* For instance, electrolyte films could be prepared by adding perchlorate ions into jatropha oil-based polyurethanes using a solution casting technique.[94] Xu *et al.* reported that plant oil-based waterborne polyurethane/tea saponin composites exhibited excellent inhibition of the growth of rice blast mycelia, even after rainfall washing.[40]

References

1. I. Bramhecha and J. Sheikh, *Ind. Eng. Chem. Res.*, 2019, **58**, 21252–21261.
2. K. Polaczek, M. Kurańska, M. Auguścik-Królikowska, A. Prociak and J. Ryszkowska, *J. Cleaner Prod.*, 2021, **290**, 125875.
3. G. d. Feng, Y. Ma, M. Zhang, P. y. Jia, L. h. Hu, C. g. Liu and Y. h. Zhou, *Prog. Org. Coat.*, 2019, **133**, 267–275.
4. C. Zhang, Y. Li, R. Chen and M. R. Kessler, *ACS Sustainable Chem. Eng.*, 2014, **2**, 2465–2476.
5. J. Hong, D. Radojčić, X. Q. Yang, X. Wan and Z. S. Petrović, *J. Appl. Polym. Sci.*, 2019, **137**, 48509.
6. C. Becquet, F. Berche, H. Bricout, E. Monflier and S. Tilloy, *ACS Sustainable Chem. Eng.*, 2021, **9**, 9444–9454.
7. P. Alagi, Y. J. Choi, J. Seog and S. C. Hong, *Ind. Crops Prod.*, 2016, **87**, 78–88.
8. H. Liang, Y. Feng, J. Lu, L. Liu, Z. Yang, Y. Luo, Y. Zhang and C. Zhang, *Ind. Crops Prod.*, 2018, **122**, 448–455.
9. M.-J. Dumont, E. Kharraz and H. Qi, *Ind. Crops Prod.*, 2013, **49**, 830–836.
10. A. Fridrihsone, F. Romagnoli, V. Kirsanovs and U. Cabulis, *J. Cleaner Prod.*, 2020, **266**, 121403.
11. S. Jayavani, S. Sunanda, T. O. Varghese and S. K. Nayak, *J. Cleaner Prod.*, 2017, **162**, 795–805.
12. H. A. Hamid, R. Yunus, U. Rashid, T. S. Y. Choong and A. A. H. Al-Muhtaseb, *Chem. Eng. J.*, 2012, **200–202**, 532–540.
13. S. K. Gaddam and A. Palanisamy, *Colloid Polym. Sci.*, 2016, **294**, 347–355.
14. H. K. Shendi, I. Omrani, A. Ahmadi, A. Farhadian, N. Babanejad and M. R. Nabid, *Prog. Org. Coat.*, 2017, **105**, 303–309.
15. S. K. Gaddam, S. N. R. Kutcherlapati and A. Palanisamy, *ACS Sustainable Chem. Eng.*, 2017, **5**, 6447–6455.

16. L. Liu, J. Lu, Y. Zhang, H. Liang, D. Liang, J. Jiang, Q. Lu, R. L. Quirino and C. Zhang, *Green Chem.*, 2019, **21**, 526–537.
17. L. Liu, H. Deng, W. Zhang, S. A. Madbouly, Z. He, J. Wang and C. Zhang, *ACS Sustainable Chem. Eng.*, 2020, **9**, 147–157.
18. C. Fu, Z. Zheng, Z. Yang, Y. Chen and L. Shen, *Prog. Org. Coat.*, 2014, **77**, 53–60.
19. X. Wang, H. Liang, J. Jiang, Q. Wang, Y. Luo, P. Feng and C. Zhang, *Green Chem.*, 2020, **22**, 5722–5729.
20. Y. Zeng, Y. Chen, D. Sha, Y. Wu, R. Qiu and W. Liu, *ACS Sustainable Chem. Eng.*, 2022, **10**, 11524–11532.
21. L. Man, Y. Feng, Y. Hu, T. Yuan and Z. Yang, *J. Cleaner Prod.*, 2019, **241**, 118341.
22. R. B. Doke, P. M. Paraskar, Y. N. Rajput and R. D. Kulkarni, *Eur. J. Lipid Sci. Technol.*, 2021, **124**, 2100171.
23. R. Tanaka, S. Hirose and H. Hatakeyama, *Bioresour. Technol.*, 2008, **99**, 3810–3816.
24. H. Deng, F. Xie, H. Shi, Y. Li, S. Liu and C. Zhang, *Chem. Eng. J.*, 2022, **446**, 137124.
25. S. Mohd Norhisham, T. I. Tuan Noor Maznee, H. Nurul Ain, P. P. Kosheela Devi, A. Srihanum, M. N. Norhayati, S. K. Yeong, A. H. Hazimah, C. M. Schiffman, A. Sendijarevic, V. Sendijarevic and I. Sendijarevic, *Int. J. Adhes. Adhes.*, 2017, **73**, 38–44.
26. M. Stemmelen, F. Pessel, V. Lapinte, S. Caillol, J. P. Habas and J. J. Robin, *J. Polym. Sci., Part A: Polym. Chem.*, 2011, **49**, 2434–2444.
27. L. Poussard, J. Mariage, B. Grignard, C. Detrembleur, C. Jérôme, C. Calberg, B. Heinrichs, J. De Winter, P. Gerbaux, J. M. Raquez, L. Bonnaud and P. Dubois, *Macromolecules*, 2016, **49**, 2162–2171.
28. E. Rix, E. Grau, G. Chollet and H. Cramail, *Eur. Polym. J.*, 2016, **84**, 863–872.
29. M. Włoch, J. Datta and K. Błażek, *J. Polym. Environ.*, 2017, **26**, 1592–1604.
30. G. Çaylı and S. Küsefoğlu, *J. Appl. Polym. Sci.*, 2008, **109**, 2948–2955.
31. G. Çaylı and S. Küsefoğlu, *J. Appl. Polym. Sci.*, 2010, 2433–2440.
32. T. A. Phung Hai, L. J. S. D. Backer, N. D. P. Cosford and M. D. Burkart, *Org. Process Res. Dev.*, 2020, **24**, 2342–2346.
33. A. S. More, T. Lebarbé, L. Maisonneuve, B. Gadenne, C. Alfos and H. Cramail, *Eur. Polym. J.*, 2013, **49**, 823–833.
34. M. O. Saed, A. Gablier and E. M. Terentejv, *Adv. Funct. Mater.*, 2020, **30**, 1906458.
35. F. Xie, H. Deng, W. Zhang, H. Shi, X. Wang and C. Zhang, *ACS Appl. Mater. Interfaces*, 2022, 50090–50100.
36. C. Li, H. Xiao, X. Wang and T. Zhao, *J. Cleaner Prod.*, 2018, **180**, 272–279.
37. R. Morales-Cerrada, R. Tavernier and S. Caillol, *Polymers*, 2021, **13**(8), 1255.
38. S. Sahoo, H. Kalita, S. Mohanty and S. K. Nayak, *Macromol. Res.*, 2017, **25**, 772–778.

39. H. Liang, Q. Lu, M. Liu, R. Ou, Q. Wang, R. L. Quirino, Y. Luo and C. Zhang, *Chem. Eng. J.*, 2020, 127774.
40. F. Zheng, Y. Li, Z. Zhang, J. Jia, P. Hu, C. Zhang and H. Xu, *J. Cleaner Prod.*, 2020, **264**, 121685.
41. H. Deng, J. Lu, D. Liang, X. Wang, T. Wang, W. Zhang, J. Wang and C. Zhang, *J. Renewable Mater.*, 2023, **11**, 1209–1222.
42. H. Liang, L. Liu, J. Lu, M. Chen and C. Zhang, *Ind. Crops Prod.*, 2018, **117**, 169–178.
43. R. Shen, M. Long, C. Lei, L. Dong, G. Yu and J. Tang, *Chem. Eng. J.*, 2022, **433**, 134470.
44. H. Liang, S. Wang, H. He, M. Wang, L. Liu, J. Lu, Y. Zhang and C. Zhang, *Ind. Crops Prod.*, 2018, **122**, 182–189.
45. D. Bello, C. A. Herrick, T. J. Smith, S. R. Woskie, R. P. Streicher, M. R. Cullen, Y. Liu and C. A. Redlich, *Environ. Health Perspect.*, 2007, **115**, 328–335.
46. A. D. LaMontagne, *Occup. Environ. Med.*, 2005, **62**, 741–742.
47. A. Gomez-Lopez, S. Panchireddy, B. Grignard, I. Calvo, C. Jerome, C. Detrembleur and H. Sardon, *ACS Sustainable Chem. Eng.*, 2021, **9**, 9541–9562.
48. Q.-W. Song, Y.-N. Zhao, L.-N. He, J. Gao and Z.-Z. Yang, *Curr. Catalysise*, 2012, **1**, 107–124.
49. H. Long, Z. Wu, Q. Dong, Y. Shen, W. Zhou, Y. Luo, C. Zhang and X. Dong, *J. Appl. Polym. Sci.*, 2019, **136**, 47709.
50. M. Ghasemlou, F. Daver, E. P. Ivanova and B. Adhikari, *Eur. Polym. J.*, 2019, **118**, 668–684.
51. G. Rokicki, P. G. Parzuchowski and M. Mazurek, *Polym. Adv. Technol.*, 2015, **26**, 707–761.
52. K. Błażek and J. Datta, *Crit. Rev. Environ. Sci. Technol.*, 2019, **49**, 173–211.
53. J. Datta and M. Włoch, *Polym. Bull.*, 2015, **73**, 1459–1496.
54. R. H. Lambeth and T. J. Henderson, *Polymer*, 2013, **54**, 5568–5573.
55. B. Tamami, S. Sohn and G. L. Wilkes, *J. Appl. Polym. Sci.*, 2004, **92**, 883–891.
56. X. Yang, S. Wang, X. Liu, Z. Huang, X. Huang, X. Xu, H. Liu, D. Wang and S. Shang, *Green Chem.*, 2021, **23**, 6349–6355.
57. H. Gholami and H. Yeganeh, *Eur. Polym. J.*, 2021, **142**, 110142.
58. B. Grignard, J. M. Thomassin, S. Gennen, L. Poussard, L. Bonnaud, J. M. Raquez, P. Dubois, M. P. Tran, C. B. Park, C. Jerome and C. Detrembleur, *Green Chem.*, 2016, **18**, 2206–2215.
59. W. Zhang, T. Wang, Z. Zheng, R. L. Quirino, F. Xie, Y. Li and C. Zhang, *Chem. Eng. J.*, 2023, **452**, 138965.
60. H. Blattmann, M. Lauth and R. Mülhaupt, *Macromol. Mater. Eng.*, 2016, **301**, 944–952.
61. A. Cornille, C. Guillet, S. Benyahya, C. Negrell, B. Boutevin and S. Caillol, *Eur. Polym. J.*, 2016, **84**, 873–888.
62. X. Xi, A. Pizzi, C. Gerardin and G. Du, *J. Renewable Mater.*, 2019, **7**, 301–312.

63. X. Y. Chen, X. D. Xi, A. Pizzi, E. Fredon, X. J. Zhou, J. X. Li, C. Gerardin and G. B. Du, *Polymers*, 2020, **12**(4), 750.
64. X. Y. Chen, J. X. Li, X. D. Xi, A. Pizzi, X. J. Zhou, E. Fredon, G. B. Du and C. Gerardin, *Polym. Degrad. Stab.*, 2020, **175**, 109121.
65. P. Singh and R. Kaur, *J. Polym. Environ.*, 2023, **31**, 743–753.
66. X. D. Xi, A. Pizzi, C. Gerardin, H. Lei, X. Y. Chen and S. Amirou, *Polymers*, 2019, **11**(11), 1802.
67. J. Sternberg and S. Pilla, *Green Chem.*, 2020, **22**, 6922–6935.
68. G. Coste, M. Denis, R. Sonnier, S. Caillol and C. Negrell, *Polym. Degrad. Stab.*, 2022, **202**, 110031.
69. G. Coste, D. Berne, V. Ladmiral, C. Negrell and S. Caillol, *Eur. Polym. J.*, 2022, **176**, 111392.
70. V. Valette, N. Kebir, F. B. Tiavarison, F. Burel and L. Lecamp, *React. Funct. Polym.*, 2022, **181**, 105416.
71. T. Dong, E. Dheressa, M. Wiatrowski, A. P. Pereira, A. Zeller, L. M. L. Laurens and P. T. Pienkos, *ACS Sustainable Chem. Eng.*, 2021, **9**, 12858–12869.
72. J. H. Clark, T. J. Farmer, I. D. V. Ingram, Y. Lie and M. North, *Eur. J. Org. Chem.*, 2018, **2018**, 4265–4271.
73. S. Anitha, G. Unnikrishnan and K. S. S. Kumar, *Mater. Lett.: X*, 2022, **14**, 100142.
74. F. Monie, B. Grignard, J. M. Thomassin, R. Mereau, T. Tassaing, C. Jerome and C. Detrembleur, *Angew. Chem., Int. Ed.*, 2020, **59**, 17033–17041.
75. F. Monie, T. Vidil, B. Grignard, H. Cramail and C. Detrembleur, *Mater. Sci. Eng., R*, 2021, **145**, 100628.
76. M. Bourguignon, B. Grignard and C. Detrembleur, *Angew. Chem., Int. Ed.*, 2022, e202213422.
77. G. Coste, C. Negrell and S. Caillol, *Macromol. Rapid Commun.*, 2022, **43**, e2100833.
78. F. Monie, B. Grignard and C. Detrembleur, *ACS Macro Lett.*, 2022, **11**, 236–242.
79. R. Shen, M. Long, C. Lei, L. Dong, G. Yu and J. Tang, *Chem. Eng. J.*, 2022, **433**, 134470.
80. A. J. Raychura, S. Jauhari, V. S. Prajapati and B. Z. Dholakiya, *Bioresour. Technol. Rep.*, 2018, **3**, 88–94.
81. H. Liang, Q. Lu, M. Liu, R. Ou, Q. Wang, R. L. Quirino, Y. Luo and C. Zhang, *Chem. Eng. J.*, 2021, **421**, 127774.
82. S. Sahoo, H. Kalita, S. Mohanty and S. K. Nayak, *J. Renewable Mater.*, 2018, **6**, 117–125.
83. A. del Prado, D. K. Hohl, S. Balog, L. M. de Espinosa and C. Weder, *ACS Appl. Polym. Mater.*, 2019, **1**, 1399–1409.
84. A. M. Borrero-López, D. B. Guzmán, J. A. González-Delgado, J. F. Arteaga, C. Valencia, U. Pischel and J. M. Franco, *ACS Sustainable Chem. Eng.*, 2021, **9**, 11032–11040.
85. G. Coste, C. Negrell, L. Averous and S. Caillol, *ACS Sustainable Chem. Eng.*, 2022, **10**, 8549–8558.

86. A. Cifarelli, L. Boggioni, A. Vignali, I. Tritto, F. Bertini and S. Losio, *Polymers*, 2021, **13**(4), 612.
87. R. Jaratrotkamjorn and V. Tanrattanakul, *J. Appl. Polym. Sci.*, 2020, **137**, 49310.
88. Y. Guan, Y. Su, L. Zhao, F. Meng, Q. Wang, Y. Yao and J. Luo, *Mater. Sci. Eng., C*, 2017, **75**, 1221–1230.
89. H. Zhang, H. Qin, L. Li, X. Zhou, W. Wang and C. Kan, *J. Agric. Food Chem.*, 2018, **66**, 6552–6560.
90. Y. Zhang, B. Liu, K. Huang, S. Wang, R. L. Quirino, Z. X. Zhang and C. Zhang, *ACS Appl. Mater. Interfaces*, 2020, **12**, 37607–37618.
91. D. Liang, Q. Zhang, W. Zhang, L. Liu, H. Liang, R. L. Quirino, J. Chen, M. Liu, Q. Lu and C. Zhang, *J. Cleaner Prod.*, 2019, **210**, 1207–1215.
92. X. Zhao, X. Qi, Q. Chen, X. Ao and Y. Guo, *ACS Sustainable Chem. Eng.*, 2020, **8**, 18044–18053.
93. M. Yang, S. Li, S. Zhang, B. Gao, Z. Tong, D. Cheng, D. Chen, R. Huang and Y. Yang, *J. Mater. Chem. A*, 2022, **10**, 18834–18844.
94. S. R. Mustapa, M. M. Aung, A. Ahmad, A. Mansor and L. TianKhoon, *Electrochim. Acta*, 2016, **222**, 293–302.

CHAPTER 5

Non-isocyanate Polyurethane from Vegetable and Microalgal Oils

TAO DONG,* SKYLAR SCHUTTER, CHENG ZHANG AND JACOB KRUGER

National Renewable Energy Laboratory, Renewable Resources and Enabling Sciences Center, 15013 Denver West Parkway, Golden, CO 80401, USA
*Email: tao.dong@nrel.gov

5.1 Introduction

Polyurethanes (PU) have broadly been used in our daily lives for foam, coating, adhesive, sealant and elastomer applications. PU were ranked the 6th most-produced polymer in 2016 and the global market for PU production will reach 22 million tons by 2023.[1] Currently, most commercial PU are produced through the polymerization of isocyanates with polyols. However, isocyanates, particularly of relatively low molecular weight compounds (*e.g.*, TDI and HDI), are sensitizers that can result in an allergic reaction of hives and asthma.[2] Consequently, the regulatory intention has increased to restrict the use of isocyanates. As a result, research on non-isocyanate polyurethane (NIPU) has attracted attention from industry and academia, as an alternative chemistry to produce PU.[3]

Among the several pathways to produce isocyanate-free polyurethanes, the aminolysis approach in which cyclic carbonates (CCs) react with amines has been recognized as the most attractive method to produce NIPU.[2]

Green Chemistry Series No. 83
Green Chemistry and Green Materials from Plant Oils and Natural Acids
Edited by Zengshe Liu and George Kraus

Published by the Royal Society of Chemistry, www.rsc.org

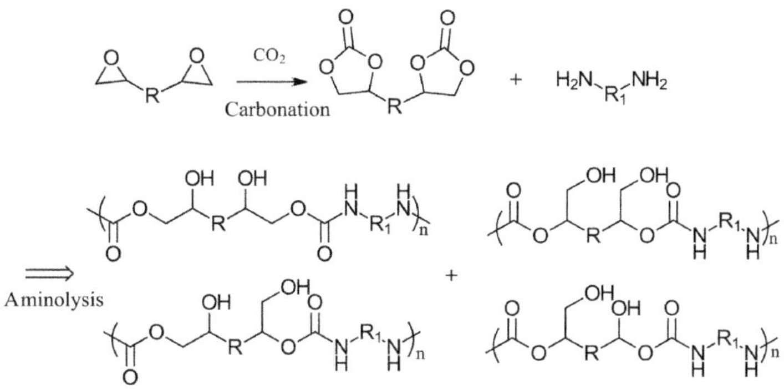

Scheme 5.1 NIPU production *via* the aminolysis pathway.

Notably, 5-membered cyclic carbonates can be synthesized from CO_2 (15–30 wt% of cyclic carbonate) with 100% atom efficiency. The NIPU synthesis approach *via* aminolysis is shown in Scheme 5.1.

There is an abundance of natural fatty acids that could be converted into CC precursors for NIPU production. For example, epoxidized soybean and linseed oil could be used to produce CC precursors for NIPU production.[4] One limitation of these terrestrial fatty acids is that their degree of unsaturation is usually not higher than 3 and, in some applications, higher functionality is preferred to produce highly crosslinked thermoset materials. In general, it is desired to produce CC monomers with high functionality.

In this regard, polyunsaturated fatty acids (PUFA) can be a promising feedstock for NIPU production. Microalgae have the unique capability to produce highly unsaturated fatty acids such as eicosapentaenoic acid (EPA) and docosahexaenoic acid (DHA), which have 5 and 6 double bonds per fatty acid chain, respectively.[5] According to the literature, most cyclic carbonates cannot react with diamines at room temperature, which is a major technical hurdle for NIPU application.[1] However, we have found that the DHA-derived carbonate can react with diamines at room temperature, probably due to its special molecular configuration that drastically increases the reactivity of its derivatives. This result indicates that microalgal PUFA might allow for greater tuning of the polymer production process compared to traditional vegetable lipids. Thus, it is worthwhile to explore the unique performance of these algal lipids for the production of multiple polymer types for varied applications. This research can provide baseline data on algal NIPU and provide insights into how to incorporate natural fatty acids into the NIPU platform for different applications.

More broadly, this approach has wide-ranging implications for both renewable fuels and renewable chemicals, as the production of high-value coproducts from non-fuel-precursor fractions of biomass (algal or other) can offset the costs of biofuel production.[6] As one relevant example, incorporation of a PU coproduct into an algal biorefining approach was calculated to reduce the minimum fuel selling price (MFSP) of algal biofuels by almost \$5 per gallon of gasoline equivalent (GGE), saving almost 50%.[7,8] Combined with the ability to directly incorporate captured CO_2 into the NIPU product (discussed below), NIPU produced from renewable lipid feedstocks can contribute significantly to the stated goals of the US Department of Energy to decarbonize transportation and industry while maintaining competitive pricing with petroleum-derived incumbents.

In this chapter, we report on initial efforts toward engineering a process to produce algae-based NIPU, including fractionation and purification of algal lipids for NIPU production, process flexibility arising from the higher degree of unsaturation available in algal lipids and the impact of oil properties and diamine crosslinkers on the processing and properties of different NIPU formulations.

5.2 Materials and Methods

5.2.1 Materials

Soybean oil was purchased from Alfa Aesar (MA). Amberlite IR-120, 30% hydrogen peroxide, tetrabutylammonium bromide (TBAB), butane diamine (BDA), pentane diamine (PDA) and *m*-chloroperbenzoic acid (*m*-CPBA) were purchased from Sigma-Aldrich (MO). Heterotrophic algal oil (algae-DHA) was purchased from Algarithm (Canada). Phototrophic algal oil Qualitas Almega20 oil (from *Nannochloropsis*) was kindly provided by Qualitas.

5.2.2 Short Path Distillation of *Nannochloropsis* Oil

Nannochloropsis crude lipid was added to a round bottom flask with a glass apparatus for short path distillation (SPD) to separate fatty acids from other impurities. The flask was heated using a heating mantle at 140–240 °C. The glass apparatus was wrapped using heating tape and kept at 240 °C. The collected fractions were cooled down using an ice bath.

5.2.3 Methylation of *Nannochloropsis* Oil

Nannochloropsis free fatty acids (FFAs) (50 g) were mixed with 200 mL of methanol and 2 g of H_2SO_4 as a catalyst in a round bottom flask. The mixture was refluxed at 70 °C for 6 hours to convert FFAs into fatty acid methyl esters (FAME). After refluxing, 200 mL of water and 100 mL of hexane were added to the mixture. The hexane phase was collected, and the solvent was removed by rotary evaporation to isolate the FAME.

5.2.4 Epoxidation by H_2O_2 (Soybean, Linseed and *Schizochytrium* Oils)

Epoxidation was conducted in the presence of acetic acid using Amberlite IR-120 as a solid catalyst. H_2O_2 was used as an oxidation reagent and the reaction was conducted at 70 °C.

5.2.5 Epoxidation by *m*-CPBA (*Nannochloropsis* FAME)

Epoxidation of FAME was conducted using *m*-CPBA in DCM. Briefly, 50 g of FAME, K_2CO_3 and 200 mL of DCM were sequentially fed into a 500 mL round bottom flask that is placed in an ice bath. Then, *m*-CPBA was slowly added into the flask for over 30 min. The mixture was stirred for 48 hours. The obtained white solid was filtered under vacuum. The DCM solution was sequentially washed with Na_2SO_3 solution (1 M) once, NaOH solution (1 M) thrice, $NaHCO_3$ (1 M) thrice and saturated NaCl solution once. The obtained DCM solution was rotary evaporated to obtain the epoxidized oil.

5.2.6 Methylation of Epoxidized Linseed and *Schizochytrium* Oils

Epoxidized linseed oil or *Schizochytrium* oil was mixed with methanol and acetone based on a previous literature study.[9] NaOH was used as a catalyst for methylation for about 10 min. The reaction was terminated by adding water and the catalyst was removed by water washing. The residual solvent was removed using a rotovap and the produced FAME were obtained.

5.2.7 Carbonation

Carbonation of epoxidized oil was carried out using a classic method with tetrabutylammonium bromide (TBAB) as a catalyst.[10] The reactions were conducted at 110 °C for 20 hours under 500 psi CO_2. After the reaction, the catalyst was washed off using water. The number of carbonate groups per triglyceride was calculated from the initial and final epoxy oxygen content (EOC), assuming no side reactions.[11]

5.2.8 Polymer Precursor Fractionation

The carbonated FAME (linseed, *Nannochloropsis* and *Schizochytrium*) were extracted using 20 mL of pentane for 30 min. Pentane was not miscible with the carbonated oil, and only partially carbonated oil can be extracted using pentane. The pentane phase on the top was decanted, and the solvent was removed under a nitrogen flow to obtain the non-polar fraction. The polar fraction, which was not soluble in pentane, was placed in a vacuum oven at 40 °C to remove any residual solvent.

5.2.9 Aminolysis Kinetics Study

Cyclic carbonated oil samples were added to glass tubes and placed in an oil bath at a given temperature. A stoichiometric amount of diamine was added to start the aminolysis reaction under constant stirring. A small aliquot of the sample was taken from the tube and quenched in liquid nitrogen for FTIR analysis. The cyclic carbonate peak (1800 cm^{-1}) is normalized with the alkane group peak (2847 cm^{-1}) to quantify the aminolysis conversion.

5.2.10 Resin Preparation

The carbonated monomers were weighed into a flask and heated in an oil bath at 70 °C. Diamine was added to the carbonated fraction and the mixture was kept stirring for about 30 seconds until the mixture started to gel. The mixture was then poured into a preheated PTFE mold to cast dog bone shaped specimens (ASTM D638 Type IV). The resin specimens were cured in an oven at 70 °C for 12 hours and 110 °C for 3 hours.

5.2.11 Analytical Methods

The fatty acid composition in the oil samples was tested using the NREL FAME protocol. Fourier Transform Infrared Spectroscopy (FTIR) was conducted using a Nicolet iS50 with Attenuated Total Reflectance (ATR) from 4000 to 650 cm^{-1}. Tensile performance was characterized at room temperature following ASTM D638 using an Instron 4465 Tension Frame. Specimens were subjected to extension at a rate of 5 mm min^{-1} until breaking. The tensile properties are reported as average values of three tests. NMR (300 MHz) was conducted using $CDCl_3$ as the solvent.

5.3 Results and Discussion

5.3.1 Fatty Acid Profile

In this research, two kinds of microalgal oil (heterotrophic and phototrophic) were used (Table 5.1). The Alphamega 400 (algae-DHA) oil was produced from *Schizochytrium via* a heterotrophic approach, in which organic carbon was used as the carbon source for cultivation of algae and oil accumulation. Advantages of the heterotrophic approach include high productivity, high culture density and applications in more controlled environments. The Qualitas algal oil was produced using *Nannochloropsis via* a phototrophic approach, in which CO_2 was used as the carbon source, and energy was provided through photosynthesis. Though the productivity of this approach is lower than that of the heterotrophic approach, it does not need organic carbon and can directly utilize sunlight as an energy source.

Of the plant oils, the fatty acid with the highest unsaturated degree is linolenic acid (C18:3), which has three double bonds for modification. In contrast, microalgae could accumulate highly unsaturated fatty acids such as EPA (C20:5) and DHA (C22:6). These highly unsaturated fatty acids allow more functionality of the derived cyclic carbonates and potentially provide more space for polymer engineering using these precursors.

Table 5.1 Fatty acid profiles in plant and algal oils. Adapted from ref. 5 with permission from American Chemical Society, Copyright 2021.

	Plant		Heterotrophic algae	Phototrophic algae
Fatty acid	Soybean	Linseed	Algae-DHA (*Schizochytrium*)	Algae-EPA (*Nannochloropsis*)
C14:0	—	—	8.6	3.2
C16:0	10.8	5.5	20.1	11.8
C16:1n7	—	—	3.8	17.3
C18:0	3.9	4.0	0.7	
C18:1n9	23.4	20.7	1.5	3.5
C18:1n7	1.4	0.7	5.5	
C18:2n6	51.9	15.2	0.6	2.7
C18:3n3	7.0	52.8	—	
C20:1	—	—	—	
C20:4n6				3.9
C20:5n3	—	—	1.8	20.2
C22:5	—	—	8.1	
C22:6n3	—	—	45.1	
Double bond per acyl chain	1.5	2.1	3.3	

5.3.2 Short Path Distillation

There was about 30% of non-FAME impurities in the *Nannochloropsis* oil, and cleanup was needed to enrich fatty acids for NIPU production. Distillation is a scalable approach to purify and fractionate fatty acids for polymer production,[12] but because the lighter fractions tend to have a high proportion of saturated fatty acids that are more suitable for fuel production, distillation also helps in removing impurities in the lipid stream that can contribute to catalyst deactivation during hydroprocessing operations.[13,14] By distillation we have remarkably increased the purity of fatty acids in the light fractions as shown in Table 5.2. The purity of fatty acids in fractions 100–180 °C was higher than 90%, indicating that these fractions were acceptable for subsequent epoxidation and carbonation. In addition, distillation can fractionate fatty acids based on their chain length. At lower temperatures (80–100 °C), most of the short chain fatty acids, such as C14–C16 fatty acids, were removed from the oil *via* distillation. As the temperature increased to 120 °C, longer chain fatty acids were evaporated out. These results indicate that fatty acids in the oil can be fractionated by this method. The distillation was not further optimized, as it is not the focus of this research. Therefore, we combined the fractions 80–180 °C for the methylation and epoxidation reaction that will yield an epoxy product.

5.3.3 Epoxidation of Soybean, Linseed and *Schizochytrium* Triglyceride Oils by H_2O_2

Epoxidation by H_2O_2 with Amberlite IR-120 resin as a catalyst is a classical and effective method to convert unsaturated fatty acids into their

Table 5.2 Short path distillation to fractionate *Nannochloropsis* oil.

Fatty acid recovery in each fraction (%)	Fraction at different temperatures (°C)									
	80	100	120	140	160	180	200	220	240	Leftover
C14:0	40.2	42.5	4.7	0.4	0.4	0.3	0.4	0.2	0.3	10.2
C16:0	15.4	37.5	31.4	2.2	1.1	0.9	1.6	0.2	0.2	10.5
C16:1n11	30.2	19.7	7.5	1.8	2.0	1.7	2.1	0.7	1.1	10.9
C16:1n7	18.0	35.3	29.3	1.7	0.8	0.7	1.2	0.2	0.2	10.4
C18:1n9	4.2	19.6	28.7	14.3	8.0	4.8	4.2	0.9	1.0	12.5
C18:2n6	4.4	19.9	28.7	13.9	7.6	4.5	4.0	0.8	0.9	11.9
C20:4n6	1.4	9.6	16.9	11.3	10.1	9.8	6.0	3.1	3.8	17.9
C20:5n3	1.3	9.2	16.5	10.9	9.9	9.7	5.9	3.1	3.8	17.5
Fatty acid content	75.9	94.5	98.1	94.9	91.9	91.2	88.9	84.2	83.5	24.0
Fatty acid recovery	11.7	24.4	23.5	6.5	5.0	4.5	3.3	1.3	1.7	13.6

corresponding epoxy products. In the reaction, acetic acid and H_2O_2 react on the surface of Amberlite and form acetic peroxide, which can react with unsaturated double bonds to produce epoxy. Post initial epoxidation, acetic peroxide is converted into water and acetic acid and as such the overall reaction only consumes H_2O_2. The epoxidation reaction needs to be terminated by separating epoxy products, because the prolonged reaction will lead to hydrolysis of epoxy groups, producing undesired polyols. Therefore, the epoxidation reaction was closely monitored by analyzing epoxy conversion over the reaction. Once the reaction is finished, the organic phase containing the epoxy product was separated from the aqueous phase (acetic acid and H_2O_2). The organic phase was washed with water to remove acetic acid to stop undesired hydrolysis. To evaluate the effectiveness of our epoxidation of plant and algal oils, we compared our results to commercial epoxidation rates of soybean and linseed oils (Table 5.3). Production of epoxidized soybean and linseed oil has been commercialized *via* a similar approach. In this research, we started with plant oil and algal oil for epoxidation to get a better comparison. For soybean and linseed oil, 95% and 87% of double bonds were converted into epoxy groups, respectively. For algae-DHA oil, we monitored the epoxidation content over the reaction, with an aim to produce epoxy products with different epoxidation degrees (Figure 5.1). By terminating the epoxidation at different time points, we could easily control the epoxidation degree. In this research, we have produced epoxidized algal oil with the epoxidation degree (unsaturated double bond conversion) ranging from 27% to 96%, and those epoxy products are named algae-DHA (27%), algae-DHA (46%), algae-DHA (67%), algae-DHA (83%) and algae-DHA (96%), respectively.

Epoxidation of collected *Nannochloropsis* FAME was first conducted using the classic H_2O_2 method. In this epoxidation, there are two phases generated (the organic phase on top and the aqueous phase at the bottom). In the experiment, it was noticed that partial epoxidized FAME transferred from the organic phase to the aqueous phase, likely due to the increased hydrophilicity of the FAME with multiple epoxy groups. It is very difficult to

Table 5.3 Epoxy contents in different precursors. Adapted from ref. 5 with permission from American Chemical Society, Copyright 2021.

	EOC (%)	IV in epoxy	Epoxy conversion (%)
Soybean	6.5	4	97
Linseed	8.5	21	87
Algae-DHA (27%)	3.2	150	27
Algae-DHA (46%)	5.5	74	46
Algae-DHA (67%)	8.6	43	67
Algae-DHA (83%)	9.9	22	83
Algae-DHA (96%)	11.5	10	96
Nannochloropsis	4.7	4	95

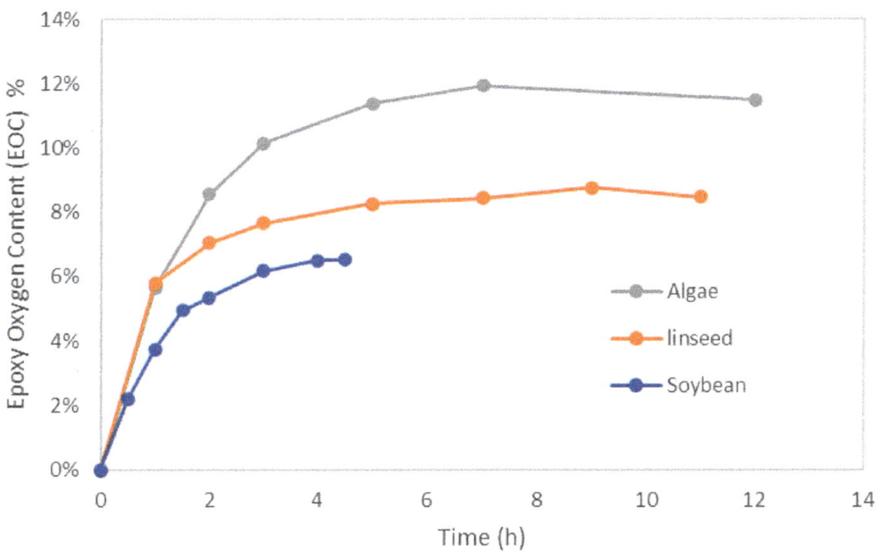

Figure 5.1 Epoxidation of triglyceride oil samples (soybean, linseed and algae-DHA oils).

recover the lost epoxy from the aqueous phase, and the dissolved epoxy might be labile to hydrolysis to produce polyols. Therefore, we adopted a *m*-CPBA epoxidation method in this study to convert *Nannochlorophsis* FAME feedstock.

5.3.4 Epoxidation of *Nannochloropsis* Oil by *m*-CPBA

As shown in Figure 5.2, the epoxidation using *m*-CPBA proceeded very fast at the early stage of the experiment (within 10 min). However, the epoxidation conversion slowed remarkably after 10 min. A conversion of 90% was achieved after 24 h, but it was very difficult to achieve higher than 95% of conversion, probably due to the steric hindrance of the fatty acids and bulky

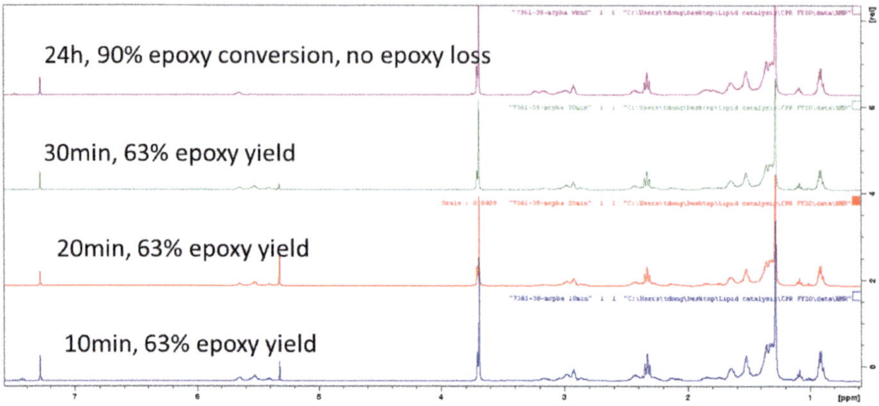

Figure 5.2 NMR analysis of the epoxidized *Nannochloropsis* oil.

structure of *m*-CPBA. After the epoxidation, the DCM phase was washed with an alkaline solution to remove *m*-CPBA and the chlorobenzoic acid by-product. The NMR results showed that there was still a little amount (~5%) of chlorobenzoic acid in the epoxy oil, even after washing more than 5 times with the alkaline solution. The epoxy product was stored in a refrigerator at 4 °C for 3 days, during which time a white precipitate formed. The NMR analysis showed that the chlorobenzoic acid content in the epoxy oil reduced to around 2.5%, indicating that the white solid was crystallized chlorobenzoic acid. Then, the epoxy oil was used for the carbonation reaction.

5.3.5 Methylation of Epoxidized Algae-DHA Triglyceride Oil

The epoxidized algae-DHA triglyceride oil was converted into epoxidized FAME using a simple methylation reaction. The reaction was finished within 15 min. The produced epoxidized algae-DHA FAME stream was used for carbonation to produce FAME with six carbonated groups (CC6 below).

5.3.6 Carbonation

More than 95% of the epoxide groups in most oils were converted to carbonates. From a carbon sequestration perspective, carbonation conversion is an effective approach to capture CO_2. This carbon sequestration during the production of renewable polyurethanes is advantageous to this method as it provides a significant contribution to the sustainability of the process, in addition to the carbon capture inherent in the photosynthetic production of the oil feedstocks.

Regarding *Nannochloropsis* oil, the residual chlorobenzoic acid did not affect the carbonation reaction. Based on the NMR results (Figure 5.3), up to 95% of epoxy disappeared (2.8–3.2 ppm), and a large amount of carbonate groups were formed (4.2–5.0 ppm). The carbonated oil had a dark brown color.

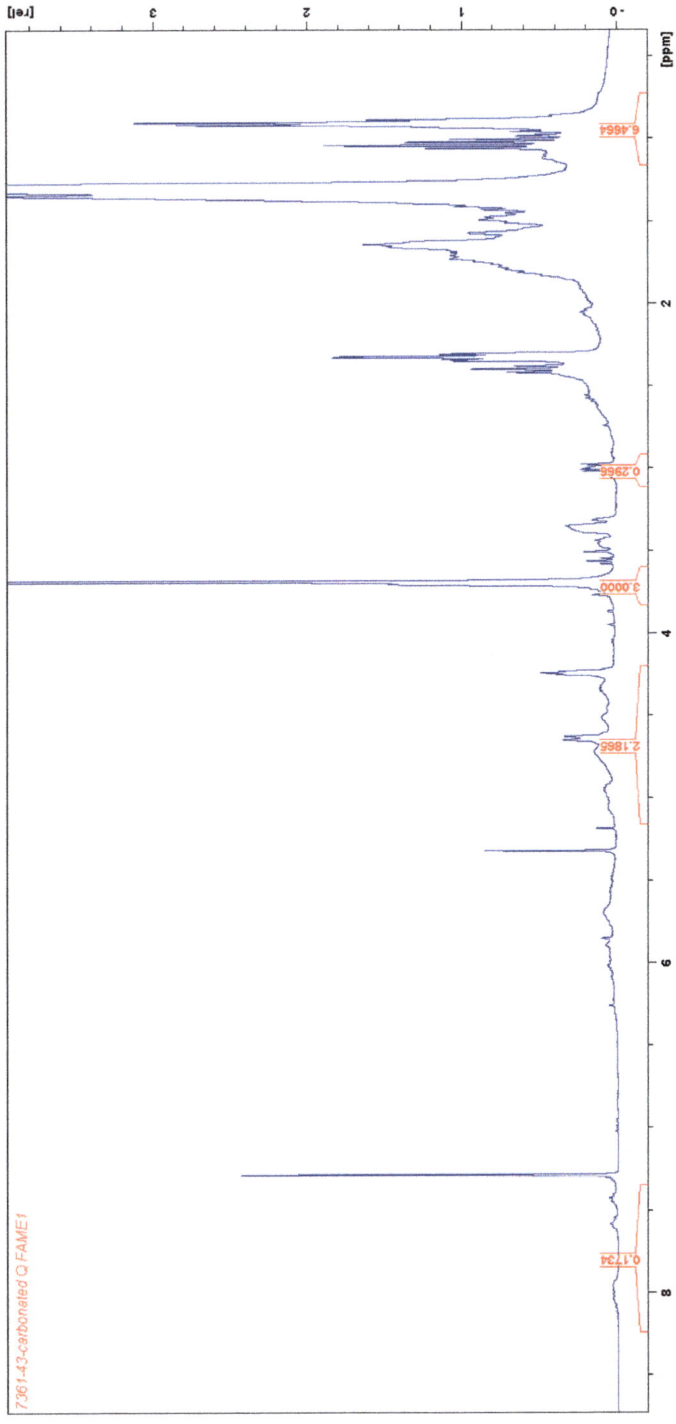

Figure 5.3 NMR of carbonated *Nannochloropsis* oil.

5.3.7 Polymer Precursor Fractionation

We observed that the hydrophilicity of the carbonated lipid increased with carbonate groups. The more the carbonated groups on the FAME, the more hydrophilic the carbonate monomer. In this research, we took advantage of the different hydrophilicity properties of the carbonated FAME and successfully enriched the FAME with more carbonated groups in the polar fraction by using pentane extraction. For carbonated *Nannochloropsis* oil, the yields for the nonpolar and polar fractions were 36.6% and 63.4%, respectively. As shown in Figure 5.4, the NMR spectra clearly showed that more carbonate groups were enriched in the polar fraction, while there were barely any carbonate groups in the non-polar fraction. Thus, we used this practical method to enrich the EPA-derived cyclic carbonate (CC5) for NIPU production.

We also conducted FTIR analysis on the two fractions. In the FTIR analysis (Figure 5.5), inert methyl and methylene groups were used as an internal standard to normalize other peaks for comparison. It is very clear that the carbonate peak in the polar fraction is remarkably stronger than that in the nonpolar fraction, indicating that the solvent fractionation was successful to enrich the FAME with more carbonated groups into the polar fraction (Figure 5.6). The nonpolar fraction mainly consisted of FAME with no or very low content of carbonated groups, which could not serve as good precursors for NIPU production, but can be used for biofuel production. A similar approach was used to enrich linolenic acid and DHA-derived carbonates and they are denoted as CC3 (from linolenic acid) and CC6 (from DHA).

5.3.8 Reaction Kinetics Study

We used cyclic carbonate monomers derived from soybean oil (TAG), linseed oil (TAG), *Schizochytrium* triglyceride (TAG), linolenic acid methyl ester, eicosapentaenoic acid (EPA) methyl ester and docosahexaenoic acid (DHA) methyl ester. Linolenic acid, EPA and DHA fatty acid methyl ester (FAME) derived cyclic carbonates are named CC3, CC5 and CC6, respectively. *Schizochytrium* triglyceride (TAG) derived cyclic carbonate has a melting temperature of 90 °C, and it is too viscous to be processed. In contrast, FAME-derived cyclic carbonates have a much lower viscosity and can be easily processed.

First, we used carbonated triglycerides (soybean, linseed and algae-DHA) for the first phase of the reaction kinetics study (Figure 5.7A). The carbonated soybean oil reacted with butane diamine (BDA) at 70 °C slowly. After 90 min, there were more than 30% of unreacted carbonate groups left. This reaction rate is too slow compared to traditional isocyanate reactions. In contrast, the carbonated linseed and algal oil reacted much faster. Within 30 min, >95% of the carbonated groups underwent aminolysis. These are very promising results, indicating that the reactivity of cyclic carbonate monomers can be adjusted by using different precursors. We surmise that the enriched cyclic carbonate groups on the same alkyl chain increased the electron withdrawing power, making amine attack easier.

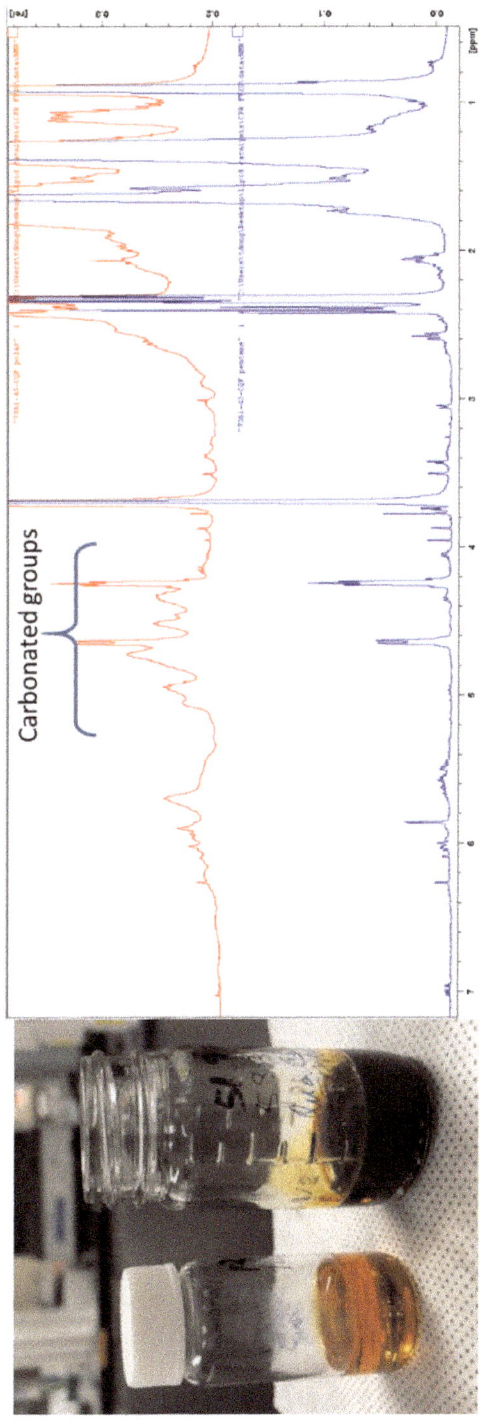

Figure 5.4 Enriched polymer precursor using solvent extraction (orange: non-polar fraction; dark brown: polar fraction). NMR (blue: nonpolar fraction; red: polar fraction).

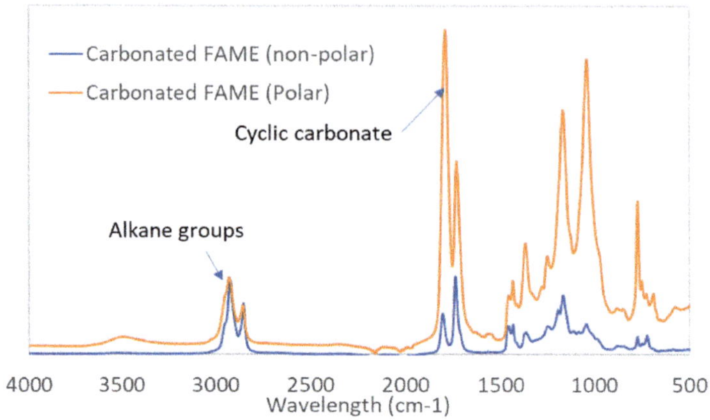

Figure 5.5 FTIR of non-polar and polar carbonated fractions.

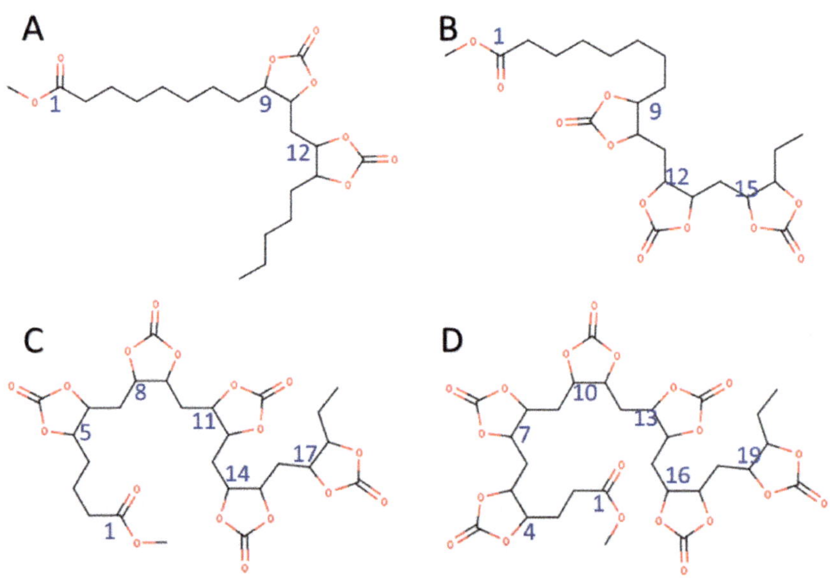

Figure 5.6 Structures of carbonated fractions.

Next, we used the cyclic carbonated fatty acid methyl esters (CC2, CC3, CC5 and CC6) for the second phase of the reaction kinetics study (Figure 5.7B). The results showed a similar trend that the reactivity of monomers with a higher density of cyclic carbonate groups is enhanced. This result further confirmed our hypothesis that the reactivity of cyclic carbonated fatty acids is directly related to the density of the carbonated groups. Higher reactivity can be achieved by using highly functionalized cyclic carbonated monomers, such as CC3, CC5 and CC6. Therefore, microalgal oil with PUFA content can be a preferred feedstock for NIPU production.

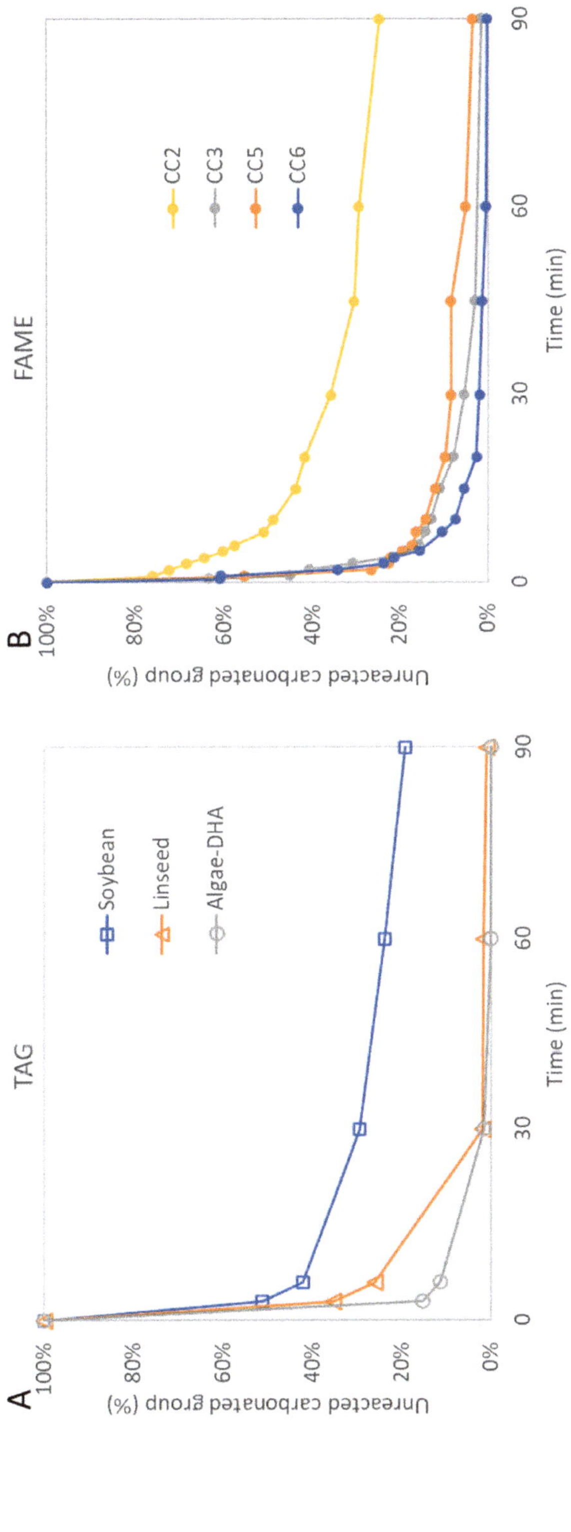

Figure 5.7 Reaction kinetics study of carbonated triglycerides (A) and FAME (B).

5.3.8.1 Characterization of NIPU Polymers from Plant Algal Oil Feedstocks

The produced cyclic carbonate monomers were used to react with butane diamine (BDA), norspermidine (NSD) and EDR-148 to produce NIPU for characterization, and the tensile performance is summarized in Table 5.4.

Triglyceride-derived cyclic carbonates generally have a higher viscosity than FAME-derived cyclic carbonates. Soybean oil gave a weak polymer with a lower tensile strength of less than 3 MPa. Linseed oil generated NIPU with a much higher tensile strength (18 MPa), due to its higher crosslinking degree. Similar trends were observed in the algae-DHA oil with different degrees of carbonation. The higher degree of carbonation in the algae-DHA oil led to better tensile strength (Table 5.4 exp 5–9). Most of the samples were produced by using aliphatic diamines such as BDA, PDA and NSD, and these diamines gave polymer samples with high strength, but low elongation at break. We also tried the commercial EDR-148 diamine, which has ether groups in the backbone. As shown in exp 11 and 14 (Table 5.4), reacting with EDR-148, CC3 and CC5 respectively generated NIPU samples with much higher elongation. The tensile curves for FAME CC3 and CC5-derived NIPU are shown in Figure 5.8. Both of them demonstrate typical elastomer performance, especially the CC5-derived NIPU. This result indicates that CC3 and CC5 can be promising precursors to produce high value NIPU elastomers. FAME-derived cyclic carbonates, such as CC3 and CC5, show better performance than TAG derived monomers in terms of modulus and tensile strength. Considering the much lower viscosity of CC3 and CC5, FAME-derived cyclic carbonates are preferred precursors for NIPU production.

Table 5.4 Polymer performance. Adapted from ref. 5 with permission from American Chemical Society, Copyright 2021.

	Carbonated oil	Source	Diamine	Young's modulus (MPa)	Tensile strength (MPa)	Elongation at break (%)
1	Soybean (TAG)	Plant	BDA	2.7 ± 0.2	1.1 ± 0.1	46.7 ± 3.8
2			PDA	2.1 ± 0.0	1.0 ± 0.1	53.7 ± 3.7
3	Linseed (TAG)	Plant	BDA	590.1 ± 99.2	16.1 ± 2.1	26.0 ± 6.5
4			PDA	335.3 ± 23.3	18.8 ± 1.3	53.2 ± 2.0
5	Algae-DHA-46% (TAG)	Algae	BDA	1.3 ± 0.0	0.5 ± 0.0	50.9 ± 6.6
6			PDA	0.8 ± 0.1	0.3 ± 0.0	46.5 ± 4.2
7	Algae-DHA-67% (TAG)	Algae	BDA	570.4 ± 41.2	16.0 ± 2.0	7.3 ± 1.0
8			PDA	464.3 ± 9.5	13.3 ± 0.4	15.6 ± 1.3
9	Algae-DHA-96% (TAG)	Algae	BDA	1007.5 ± 240.2	14.4 ± 6.9	2.5 ± 1.2
10	CC3 (FAME)	Plant	BDA	1781.7 ± 118.2	29.9 ± 7.4	2.6 ± 0.7
11			EDR 148	757.9 ± 93.8	8.0 ± 0.3	183.0 ± 9.2
12	CC5 (FAME)	Algae	BDA	1289.0 ± 126.8	26.3 ± 3.2	2.7 ± 0.8
13			NSD	1263.7 ± 94.2	31.9 ± 8.0	4.5 ± 2.1
14			EDR 148	398.0 ± 32.0	8.3 ± 0.8	145.2 ± 3.9
15	CC6 (FAME)	Algae	BDA	Too brittle		
16			NSD	Too brittle		
17			EDR 148	0.3 ± 0.0	0.2 ± 0.0	112.9 ± 1.3

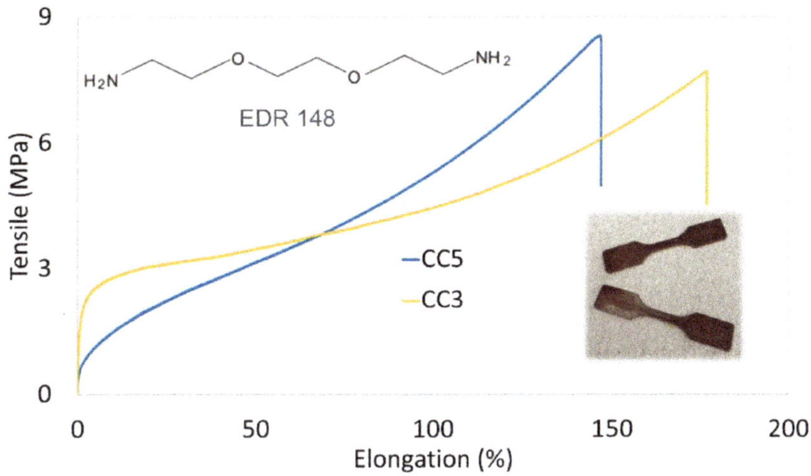

Figure 5.8 Tensile curves for CC3 and CC5 derived NIPU.

5.4 Conclusion

We have demonstrated the use of phototrophic algal oil as the feedstock for NIPU production for the first time. In this research, we have used algae-derived FAME as the precursor for carbonate monomer production. Additionally, the algal carbonate monomers showed a high reactivity in the curing process, indicating their potential for various industrial applications.

Acknowledgements

This work was authored by Alliance for Sustainable Energy, LLC, the manager and operator of the National Renewable Energy Laboratory for the U.S. Department of Energy (DOE) under Contract No. DE-AC36-08-GO28308. Funding was provided by the U.S. Department of Energy, Bioenergy Technologies Office (BETO). The views expressed in the article do not necessarily represent the views of the DOE or the U.S. Government. The U.S. Government retains and the publisher, by accepting the article for publication, acknowledges that the U.S. Government retains a nonexclusive, paid-up, irrevocable, worldwide license to publish or reproduce the published form of this work, or allow others to do so, for U.S. Government purposes.

References

1. A. Cornille, R. Auvergne, O. Figovsky, B. Boutevin and S. Caillol, *Eur. Polym. J.*, 2017, **87**, 535–552.
2. A. Gomez-Lopez, F. Elizalde, I. Calvo and H. Sardon, *Chem. Commun.*, 2021, **57**(92), 12254–12265.
3. L. Maisonneuve, O. Lamarzelle, E. Rix, E. Grau and H. Cramail, *Chem. Rev.*, 2015, **115**(22), 12407–12439.

4. M. Bähr and R. Mülhaupt, *Green Chem.*, 2012, **14**(2), 483–489.
5. T. Dong, E. P. Knoshaug, P. T. Pienkos and L. M. L. Laurens, *Appl. Energy*, 2016, **177**, 879–895.
6. R. Davis, M. Wiatrowski, C. Kinchin and D. Humbird, Conceptual Basis and TechnoEconomic Modeling for Integrated Algal Biorefinery Conversion of Microalgae to Fuels and Products, 2019 NREL TEA Update: Highlighting Paths to Future Cost Goals via a New Pathway for Combined Algal Processing, Technical Report, September 2020, https://www.nrel.gov/docs/fy20osti/75168.pdf.
7. J. S. Kruger, M. Wiatrowski, R. E. Davis, T. Dong, E. P. Knoshaug, N. J. Nagle, L. M. L. Laurens and P. T. Pienkos, *Front. Chem. React. Eng.*, 2022, **3**, 803513.
8. M. Wiatrowski and R. Davis, United States, 2021, p. Medium: ED, Size: 36 p.
9. W. Liu, F. Duan and Y. Bi, *RSC Adv.*, 2018, **8**(23), 13048–13053.
10. T. Dong, E. Dheressa, M. Wiatrowski, A. P. Pereira, A. Zeller, L. M. L. Laurens and P. T. Pienkos, *ACS Sustainable Chem. Eng.*, 2021, **9**(38), 12858–12869.
11. I. Javni, D. P. Hong and Z. S. Petrovič, *J. Appl. Polym. Sci.*, 2013, **128**(1), 566–571.
12. P. F. Martins, V. M. Ito, C. B. Batistella and M. R. W. Maciel, *Sep. Purif. Technol.*, 2006, **48**(1), 78–84.
13. J. S. Kruger, E. P. Knoshaug, T. Dong, T. C. Hull and P. T. Pienkos, *Johnson Matthey Technol. Rev.*, 2021, **65**(2), 227–246.
14. I. Hachemi, K. Jeništová, P. Mäki-Arvela, N. Kumar, K. Eränen, J. Hemming and D. Y. Murzin, *Catal. Sci. Technol.*, 2016, **6**(5), 1476–1487.

CHAPTER 6

Preparation and Application of Renewable Intrinsically Flame-retardant Epoxy Resins from Plant Oils

WENBIN LI, XIAOAN NIE AND JIE CHEN*

Institute of Chemical Industry of Forest Products, Chinese Academy of Forestry, Nanjing 210042, Jiangsu Province, P. R. China
*Email: jiechen@icifp.cn

6.1 Introduction

Epoxy resins with excellent adhesiveness, corrosion resistance, formability, and thermal stability can be used as coatings, adhesives, and molding materials. They are widely used in fields such as electronics, optical machinery, engineering technology, civil construction, and food packaging.[1,2] Traditional epoxy resins are mainly derived from petroleum and its derivatives, but fossil raw materials are non-renewable and the environmental pressure has made the development of bio-based degradable epoxy resins increasingly important.[3,4] In addition, biomass resources have many advantages, such as the wideness of sources, renewability, low cost, and biodegradability. The social attention to sustainable strategies, depletion of fossil raw materials, and environmental issues have greatly promoted the use of biomass resources, and biomass-based polymers can potentially improve or replace traditional petroleum-based polymers.[2,5]

Green Chemistry Series No. 83
Green Chemistry and Green Materials from Plant Oils and Natural Acids
Edited by Zengshe Liu and George Kraus

Published by the Royal Society of Chemistry, www.rsc.org

Plant oils, extracted from plants, are fatty acid glyceride compounds that are ubiquitous in nature.[6] The inherently biodegradable, inexpensive, and stable plant oils can be used in large quantities, making them a very attractive natural resource for synthesizing biomass-based polymers.[2,3,5] In addition, plant oils can be used as chemical raw materials owing to the functional groups in their structure and are often applied as plasticizers and intermediates for bio-polyurethanes or unsaturated polyesters *via* chemical modification of carbon–carbon double bonds, carboxyl groups, hydroxyl groups, and ester groups.[6–8] Common plant oils include soybean oil,[9] peanut oil, rapeseed oil, sunflower oil,[10] palm oil, castor oil,[11] corn oil,[12] tung oil,[2,3,13] and cashew shell oil. The main fatty acids contained in plant oils are shown in Figure 6.1. Triglycerides, the main components of plant oils, are made up of aliphatic chains and thus triglyceride-based materials lack the rigidity and strength required for some applications.[14] Like other organic polymeric materials, the flammability of these materials is a shortcoming in some applications.

Triglycerides contain several reactive positions that are useful in many synthetic transformations, but our research is focused on the reactivity and allylic positions of double bonds.[15] In general, polymers can be prepared from plant oils *via* three main routes (Figure 6.2). The first route is direct polymerization through the double bonds or other reactive functional groups in the fatty acid chain. The second route involves chemical modification of the double bonds and the introduction of functional groups that polymerize more easily. The third route is the chemical transformation of plant oils to produce platform chemicals that can be used to produce monomers for the polymer synthesis and epoxy resins.[15] In fact, plant-oil-based epoxy resins are generally composed of three elements, carbon, hydrogen, and oxygen. They burn easily during use and have a high fire risk, causing economic losses and even endangering human life. Therefore, it is urgent to improve the flame retardancy of epoxy resins through modification.

In essence, flame retardants can delay or even inhibit the material combustion process through technical means, such as adding non-reactive flame retardants to epoxy resins or introducing flame retardant structural units (*e.g.* halogen, phosphorus, nitrogen, and silicon) into the epoxy resin structure.[16] A new flame retardant epoxy resin with the flame retardant effect is designed after retardant structural units are introduced into the main chain or side chain of epoxy resins by chemical bonding and reacted with a curing agent to become a part of a three-dimensional cross-linked network. According to the basic conditions of combustion, flame retardants mainly function through endothermic cooling, dilution of the concentration of flammable gas, and formation of a carbonized thermal insulation layer.[17]

Bio-based epoxy flame retardant materials can be divided into the additive type and the reactive type according to the type of use. Additive flame retardant materials mainly act at the physical level. Blending or coating a flame retardant material with the main material is simple, convenient, and easy for post-processing. However, its low dispersibility, low compatibility,

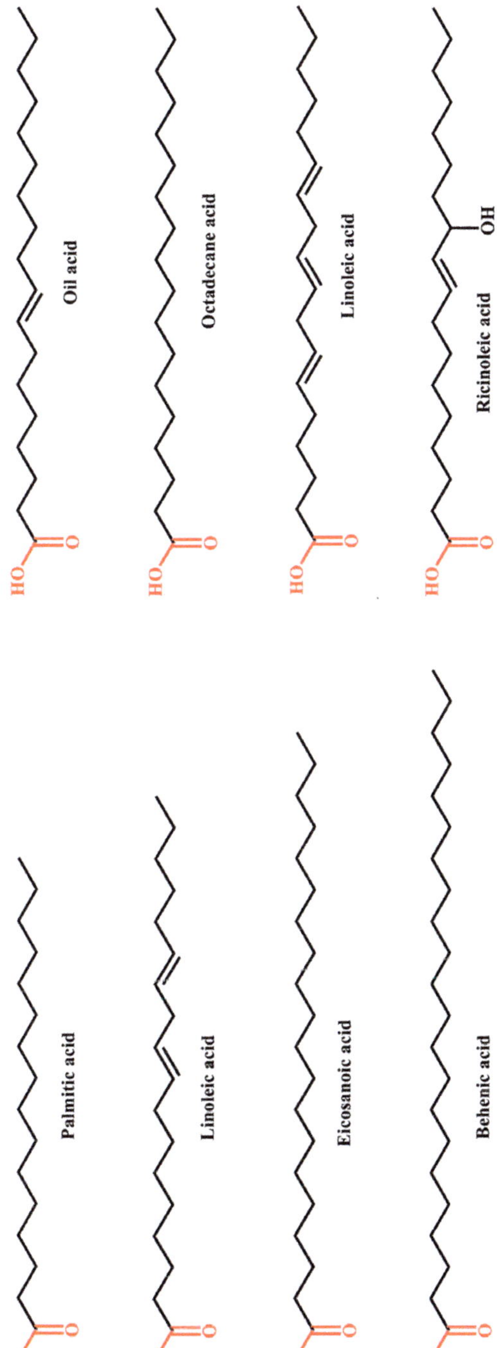

Figure 6.1 Structure of main fatty acids in plant oils.

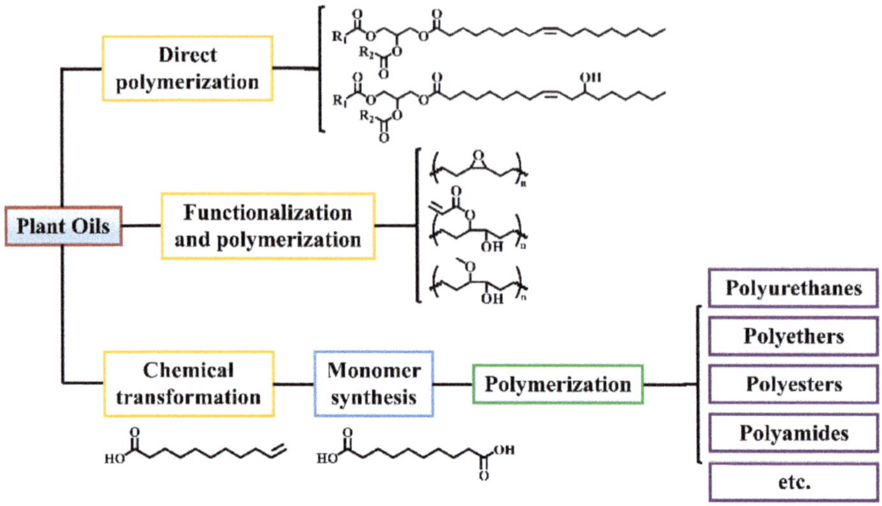

Figure 6.2 General strategies for the synthesis of plant oil-based polymers.

instability, and limited flame retardant effect are defects. Moreover, reactive flame retardant materials mainly act at the chemical level, such as adding flame retardant elements, structures, or materials (grafting/copolymerization) in the polymerization reaction of raw materials or monomers to make them mixed or embedded in the polymerization of the internal structure. Despite their excellent flame retardant performance, these materials are limited by the relatively complex scheme, difficult synthesis, high cost, and difficult industrialization. Apparently, the development of flame retardant technology for plant oil-based epoxy resin materials has always been one of the hot issues and has very important practical significance in today's society.

6.2 Preparation of Epoxy Resins from Plant Oils

Various chemical pathways for functionalizing triglycerides and fatty acids have been studied.[9,18,19] Diverse plant-oil-based epoxy monomers can be divided into two types according to the source of epoxy groups in epoxy resins. (1) In the first type, the carbon–carbon double bond is epoxidized to an epoxy group (Figure 6.3). Plant-oil-based epoxy monomers are usually prepared by oxidizing double bonds into epoxy bonds with strong oxidizing agents such as peroxyacids. For example, Carraro *et al.* used H_2O_2/fluorine polytungstate as the oxidant/catalyst to catalyze the epoxidation of double bonds, which obtained a selectivity greater than 99% and a yield of 98%.[20]

Moreover, epoxidized plant oils including castor oil and soybean oil have been recently used in large quantities for epoxy resins.[9–11,21,22] Badra *et al.* prepared a novel plasticizer of poly(vinyl chloride) (PVC) resins, which was epoxidized sunflower oil, at a mole ratio of C=C : acetic acid : H_2O_2 (30%) of

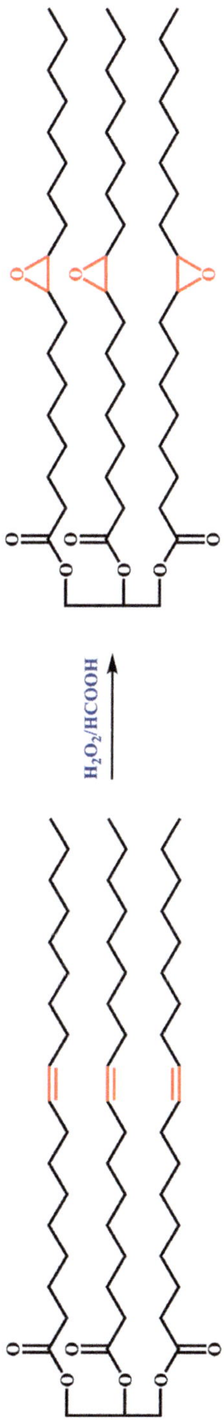

Figure 6.3 Epoxidation of plant oils.

1 : 0.5 : 1.5 at 55–60 °C.[10] Epoxidized sunflower oil was better compatible with PVC and had better mechanical properties than epoxidized soybean oil. Sinadinovic-Fiser *et al.* reported the highest relative epoxy yield (epoxidation of castor oil) of 78.32% when 0.5 mol of acetic acid and 1.5 mol of 30 wt% aqueous hydrogen peroxide per mol of double bond in oil were used in the presence of 15 wt% of Amberlite IR-120.[11] These epoxidized oils can be directly polymerized in the presence of catalysts or can be cross-linked with acid/anhydrides as curing agents. Most noteworthily, epoxidized plant oils are potentially biodegradable because their glyceride bonds can be broken down in the soil.[23]

(2) In the second type, plant oils and their derivatives react with substances with epoxidation. The corresponding alcohols, acids, and phenols in the structure of plant oils can be condensed with epichlorohydrin (ECH) to generate products containing epoxy groups. The most commonly used method is two-step epoxidation: acid, alcohol, or phenol under catalyst conditions undergoes a ring-opening reaction first to form a chlorohydrin intermediate, then an alkaline substance is added to the system, and finally hydrogen chloride is removed. In this method, the ring-opening reaction and the ring-closure reaction are carried out in two separate steps. For instance, Xu *et al.* reported a new plant-oil-based epoxy monomer (epoxy acid: 0.43 mol/100 g) from tung oil, which can undergo methyl esterification, Diels–Alder reaction, and epoxidation reaction (Figure 6.4).[13] This simple and convenient method can be conducted under controllable reaction conditions and yields few by-products, and the prepared epoxy resin has a high epoxy value and a low organic chlorine content.[24]

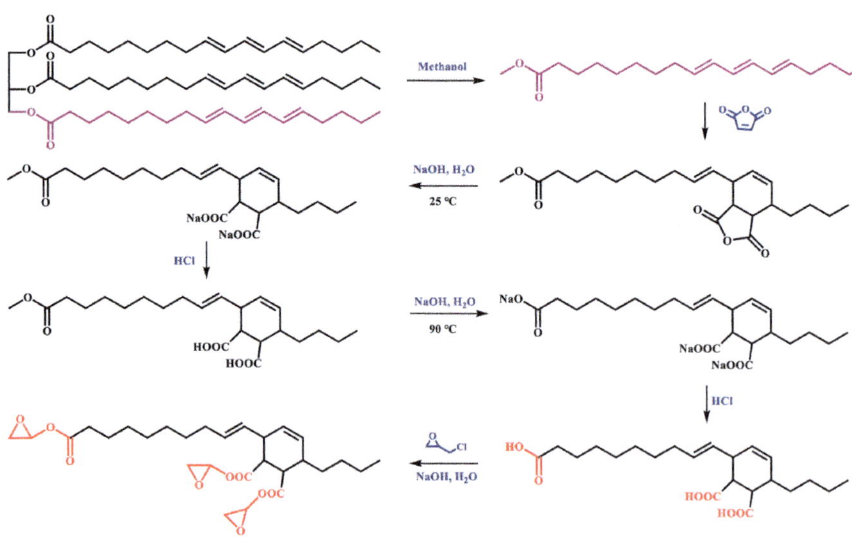

Figure 6.4 Scheme overview of the synthesis route of the tung oil-based epoxy monomer.

Furthermore, enzyme-catalyzed epoxidation of plant oils and fats is a relatively cutting-edge method at present. Because of mild reaction conditions and pollution avoidance, it has a good development prospect. However, because the activity of enzymes in the epoxidation process cannot be guaranteed for a long time, this technology is not applied to industrial production and needs further development.[25] Moreover, natural epoxidized plant oils also exist in iron grass and *Vernonia*, which contain 60–65% epoxy-octadec-9-enoic acid.[26] Noticeably, epoxidized plant oils are limited by very low epoxy group reactivity and the tendency for intramolecular cross-linking. Therefore, epoxidized plant oils can only be used in non-structural applications such as coatings in most industrial applications.

In addition to the preparation of epoxy resins from epoxidation, plant oils can also be chemically modified to obtain a series of polyacids used as curing agents for epoxy resins. Jaillet *et al.* obtained an oil-based polybasic acid/anhydride after the addition of soybean oil and thioglycolic acid (Figure 6.5), which was then cured with diglycidyl ether of bisphenol A (DGEBA).[27] Nevertheless, this epoxy resin has a relatively low crosslinking density, a glass transition temperature of −3 °C, and a storage modulus of only 0.59 MPa. Li *et al.* reported a series of polyfunctionality acids/anhydrides derived from renewable tung oil *via* methyl esterification and Diels–Alder reaction (Figure 6.6).[3] The epoxy vitrimer shows high thermal stability and excellent dynamic mechanical properties with a tensile strength of 59.21 MPa and a glass transition temperature of 76.73 °C.

6.3 Types of Flame-retardant Epoxy Resins from Fatty Acid Derivatives

6.3.1 Epoxy Resins from a Halogen-containing Fatty Acid Derivative

Halogenated flame retardants (HFRs) are a large group of chemicals that are used as additives to increase the fire resistance of products such as televisions, computers, textiles, and furniture upholstery.[28] The halogen-containing flame retardant system is the earliest and most widely-used epoxy resin flame retardant system in the world. In the combustion stage, halogens play the role of chemical interference (•H & OH• capture mechanism) in the gas phase through the free radical chain mechanism. While the total amount of combustible substances remains unchanged, the amount of released combustion heat is reduced and the combustion generation of hydroxyl radicals and hydrogen radicals is eliminated to achieve flame retardance.[29]

In the actual production process, only two types of flame retardants *viz.* chlorine-based and bromine-based flame retardants are widely used.[30] For instance, Howell *et al.* achieved good flame retardancy in the DGEBA epoxy resin by introducing bromine at the double bond.[30] Compared with resin

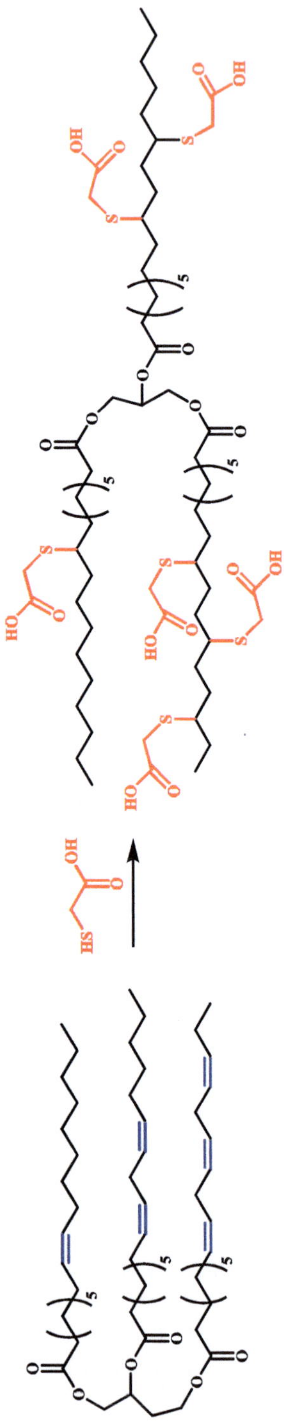

Figure 6.5 Functionalization of soybean oil with thioglycolic acid, idealized structure.

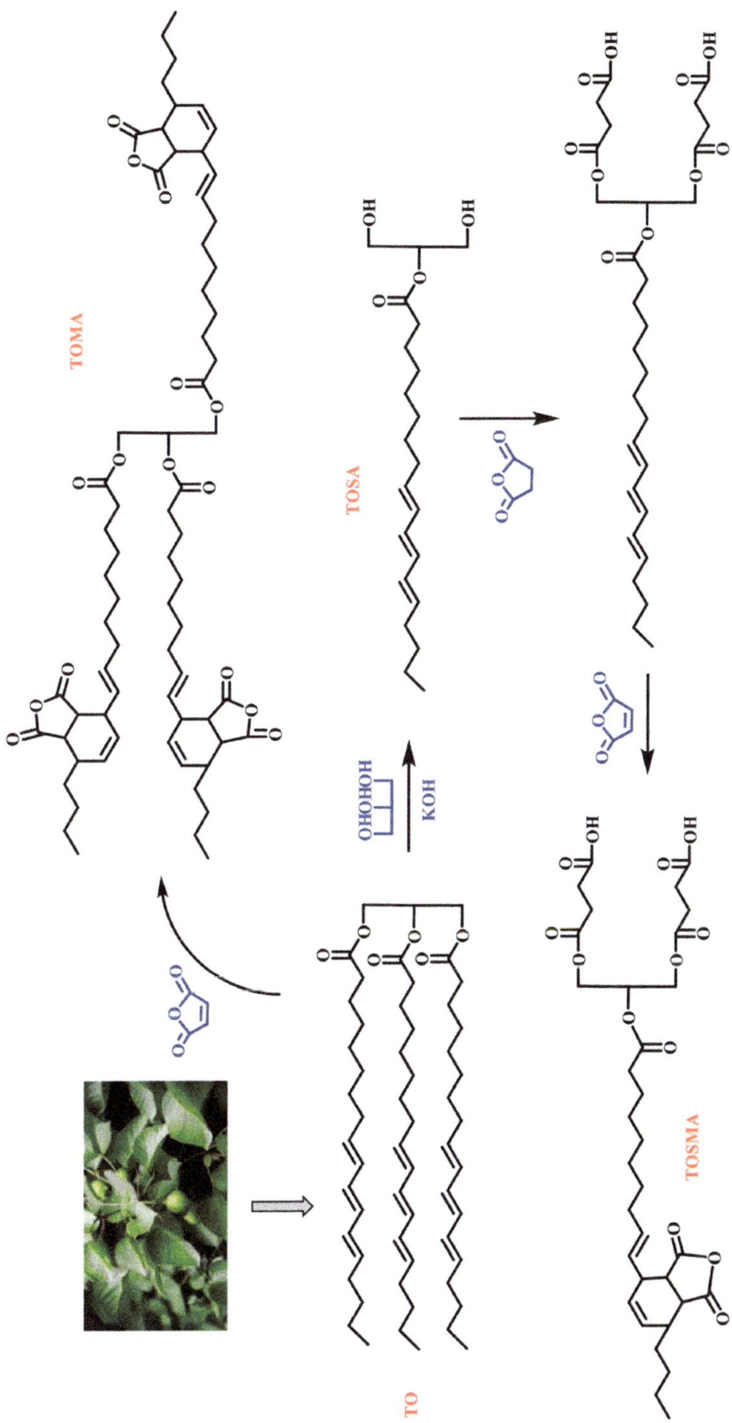

Figure 6.6 Polyfunctionality acids/anhydrides derived from renewable tung oil.

without additives, the polymer containing the brominated 9,10-dihydro-9-oxa-10-phosphaphenanthrene-10-oxide (DOPO) ester at a sufficient level to provide one percent phosphorus (5.1% bromine) displays a peak heat release rate reduction of 50%. Eren *et al.* synthesized and characterized flame-retardant polymers from bromoacrylated plant oil triglycerides through the simultaneous addition of bromine and acrylate to the double bonds of fatty acids in triglycerides.[31]

However, a large number of smoke particles will be released to suffocate the human body in the fire process because of the unique chemical properties of halogen elements. At the same time, hydrogen halide, as the most critical element of HFRs to terminate the combustion reaction, has strong toxicity and corrosiveness, which pose a huge threat to ecological development and human life safety.[32] Today, research and development of HFRs have been greatly challenged with the increasing awareness of environmental protection. Although HFRs have excellent flame retardant effects, finding and designing new green molecules and formulations and developing new halogen-free flame retardants have become the mainstream research directions of flame retardant materials.[33]

6.3.2 Epoxy Resins from a Carbon-containing Fatty Acid Derivative

Biological carbon-based flame retardant materials have become a research hotspot owing to their wide sources, non-toxicity, and environmental friendliness. Starch is a natural flame retardant material that acts as a char-forming agent in flame retardant systems. Notably, starch can release CO_2 and CO to form a carbon layer that inhibits heat and oxygen transmission during combustion. Furthermore, the acid source can promote starch dehydration to form a carbon layer with higher thermal stability when an acid source exists in the system.[34] However, new starch-based flame retardant materials are still restricted by the low stability of composite materials and the difficulty in using epoxy resins from a carbon-containing fatty acid derivative.

For epoxy resins from a carbon-containing fatty acid derivative, the current research is based on the choice of curing agents (Figure 6.7), which mainly include methyl tetrahydrophthalic anhydride (MTHA), methyl hexahydrophthalic anhydride (MHHPA), 4,4′-diaminodiphenyl sulfone (DDS), and 4,4′-diaminodiphenylmethane (DDM).[35,36] For example, the new potential flame retardant thermoset resin from epoxidised corn oil and MTHY reported by Song *et al.* had a significant reduction from 17% to 40% in peak heat rate release (pHRR) and 26 to 30% in total heat release (THR).[12] Notably, such epoxy resins from a carbon-containing fatty acid derivative have poor flame retardancy. Mandal *et al.* reported that copolymerization with triallyl maleopimarate (TMPA) effectively enhanced the mechanical, flame retardancy, and thermal properties of the wood polymer composites derived from the methacrylic anhydride modified epoxidized soybean oil (Figure 6.8). Wang *et al.* synthesized a cardanol-based epoxy monomer (ECFGE, Figure 6.9) with DGEBA

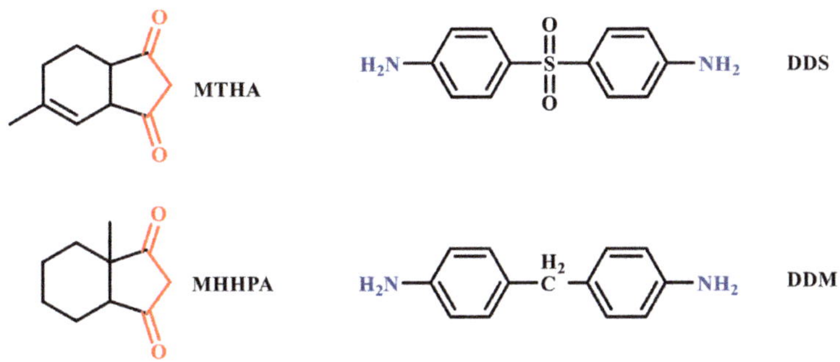

Figure 6.7 Types of carbon-containing fatty acid derivative epoxy curing agents.

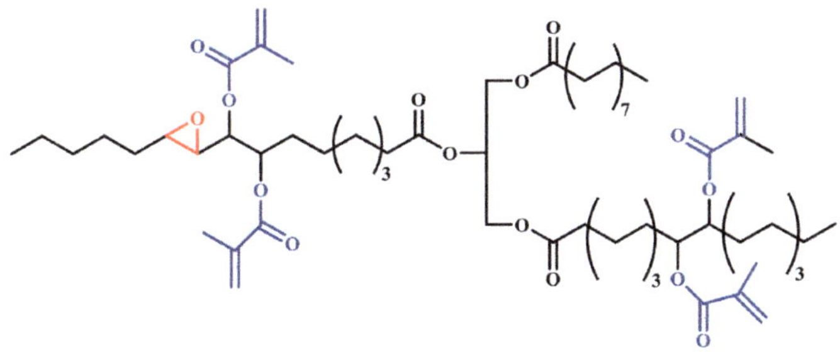

Figure 6.8 Schematic diagram of methacrylic anhydride-modified epoxidized soybean oil.

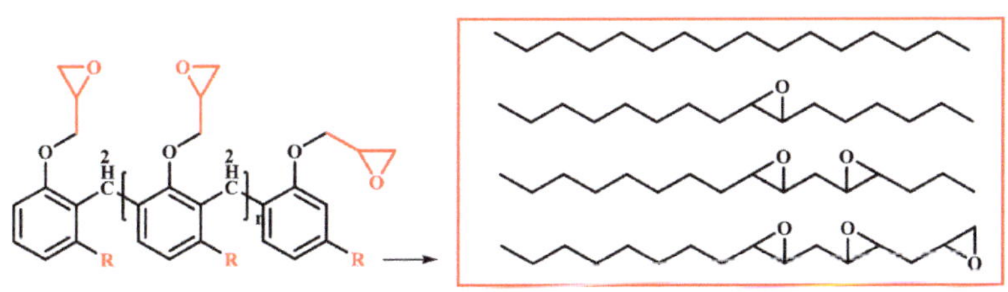

Figure 6.9 Schematic diagram of a cardanol-based epoxy monomer (ECFGE).

as the control.[37] ECFGE/DDM showed comparable thermal stability to DGEBA/DDM, and the PHRR of ECFGE was 46.0% lower than that of DGEBA, but the flame retardancy was not greatly improved. Nevertheless, the glass transition temperature (T_g) and the initial decomposition temperature of the cured ECFGE/DDM can be made less bad after flame-retardant modification.

6.3.3 Epoxy Resins from a Phosphorus-containing Fatty Acid Derivative

Phosphorus-containing polymers are well recognized for their flame-retardant properties, and they become increasingly more popular than their halogen counterparts, as they generally give off nontoxic combustion products.[33] Therefore, incorporation of an organo-phosphorous functionality into the polymeric structure is recognized as one of the most efficient ways to obtain an environmentally friendly flame-retardant system. For example, Sun *et al.* synthesized plant-oil-based flame retardants and plant-oil-based self-retarding functional materials through the polymerization pathways of epoxidized and hydroxylated triglycerides with phosphoric acid (H_3PO_4) with significant phosphate ester-based cross-links.[38] Jia *et al.* reported a chlorinated phosphate ester derived from castor oil for preparing flame-retardant PVC materials and found that the limiting oxygen index (LOI) of plasticized PVC blends reached 36.3%.[39] Wang *et al.* reported novel biobased pressure-sensitive adhesives (PSAs) derived from epoxidized soybean oil and carboxylic acid-terminated polyesters.[40] With an increase in the content of flame-retardant monomers, the flame retardancy of two phosphorus-containing PSAs improved. However, the phosphorus content introduced into plant oils is generally low and usually used in the field with plant-oil-based flame-retardant coatings.

Phosphorus-based flame retardants generally play a flame retardant role in the condensed phase. They will decompose to phosphoric acid with strong dehydration and easy condensation, which will dehydrate and carbonize the polymer matrix at high temperatures to form a carbon layer, thereby protecting the polymer matrix from further decomposition. Therefore, the flame retardant effect of phosphorus-based flame retardants in the condensed phase can be attributed to the reduced release of combustible products, generating non-combustible water vapor and forming a dense and continuous phosphorus-rich carbon layer. In this way, Yang *et al.* synthesized bio-based organophosphorus-containing epoxy curing agents (Figure 6.10) with different rigid or flexible properties by using castor oil and after curing with epoxy resin (E-51).[41] Moreover, the LOIs of the cured epoxy resins reached 22.4–24.5%. The mechanical properties of the epoxy resins were relatively balanced with a high elongation at break of 160.8% and a moderate tensile strength of 9.95 MPa. Moreno *et al.* reported phosphorus-containing thermosetting resins derived from high oleic sunflower oil.[42] These results demonstrate that the flame retardant properties and LOIs of vegetable-oil-based thermosets can be significantly improved by adding covalently-bonded phosphorus to the polymer. Specifically, the LOIs increase from 21.5 in the phosphorus-free material to 38.0 in the final material with 4.2% P.

Besides, phosphorus-based flame retardants also play a flame retardant role in the gas phase as a radical scavenger. Phosphorus-based flame retardants will be thermally decomposed to a large number of phosphorus-containing small-molecular fragments or phosphorus-containing radicals (*e.g.* $PO^•$ and $HPO^•$) during combustion (Figure 6.11), which can effectively capture the highly active $H^•$ and $OH^•$.[33] Thereby, the pyrolysis rate of the polymer is decelerated *via* reducing the concentrations of active free radicals in the combustion zone. In addition, phosphorus-based flame retardants more

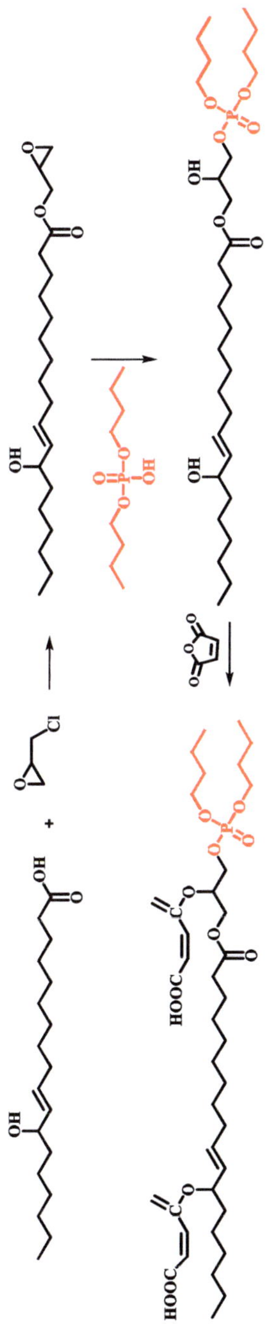

Figure 6.10 The synthesis route of bio-based organophosphorus-containing epoxy curing agents.

$$PO\cdot + H\cdot \longrightarrow HPO$$
$$PO\cdot + OH\cdot \longrightarrow HPO_2$$
$$PO\cdot + OH\cdot + H_2 \longrightarrow HPO_2\cdot + H_2O$$
$$HPO\cdot + H\cdot \longrightarrow PO\cdot + H_2$$
$$HPO_2\cdot + H\cdot \longrightarrow PO\cdot + H_2O$$
$$HPO_2\cdot + OH\cdot \longrightarrow PO_2\cdot + H_2O$$

Figure 6.11 Schematic diagram of the gas phase flame-retardant mechanism of phosphorus-containing flame retardants.

quickly and efficiently capture highly reactive free radicals in the gas phase compared to halogen-based flame retardants. Furthermore, the contact layer between the polymer matrix and oxygen can release a certain amount of water molecules under the promotion of phosphorus-based flame retardants during combustion, taking away a part of the heat generated during combustion and further improving the combustion resistance of the polymer matrix.

Notably, DOPO, a cyclic organic phosphorous compound, and its derivatives are used to synthesize many synthetic resin flame retardants.[43] The highly reactive P–H inherent in its structure can undergo addition reactions with unsaturated bonds, epoxy groups, and carbonyl groups to form various derivatives. In addition, because of a unique cyclic O=P–O pendant phosphorus group, DOPO is chemically and thermally more stable than conventional organophosphates and has a better flame retardant effect.[44] Lligadas *et al.* obtained a new phosphorus-containing fatty acid diepoxide (Figure 6.12) from 10-undecenoyl chloride and 10-(2′,5′-dihydroxyphenyl)-9,10-dihydro-9-oxa-10-phosphaphenanthrene-10-oxide after crosslinking with DDM and bis(*m*-aminophenyl)methylphosphine oxide.[14] LOIs indicate good flame-retardant properties that are related to the formation of a protective phosphorus-rich layer that slows down the degradation and prevents it from being total. As reported, the addition of DOPO-based flame retardants can significantly improve the flame retardant properties of polymers but will reduce the T_g and mechanical properties. Therefore, DOPO is often combined with other flame retardants.[45]

Cardanol, a natural product extracted from cashew nut shell oil, has been one of the research hotspots in the field of bio-based materials in recent years. Its chemical structure has a phenol-benzene ring, a hydroxyl group, and a C15 straight chain with unsaturated double bonds.[37,46] Wang *et al.* synthesized a cardanol-derived phosphate (DPP-CFR, Figure 6.13) to obtain flame-retardant bio-based epoxy materials, as a substitute for diglycidyl ether bisphenol A (DGEBA)-type epoxy, which can increase LOIs from 26.0% to 31.0%, achieving UL-94 V-0 rating.[37] Guo *et al.* reported a novel intrinsically anti-flammable and self-toughened cardanol-derived novolac epoxy monomer (ECF-PO(OPh)$_2$, Figure 6.14).[47] Specifically, the cured ECF-PO(OPh)$_2$/DDM possessed a high LOI of 33.5% and UL-94 V-0 rating. This phosphorylated cardanol-derived novolac epoxy simultaneously possesses fire-safe and toughening properties

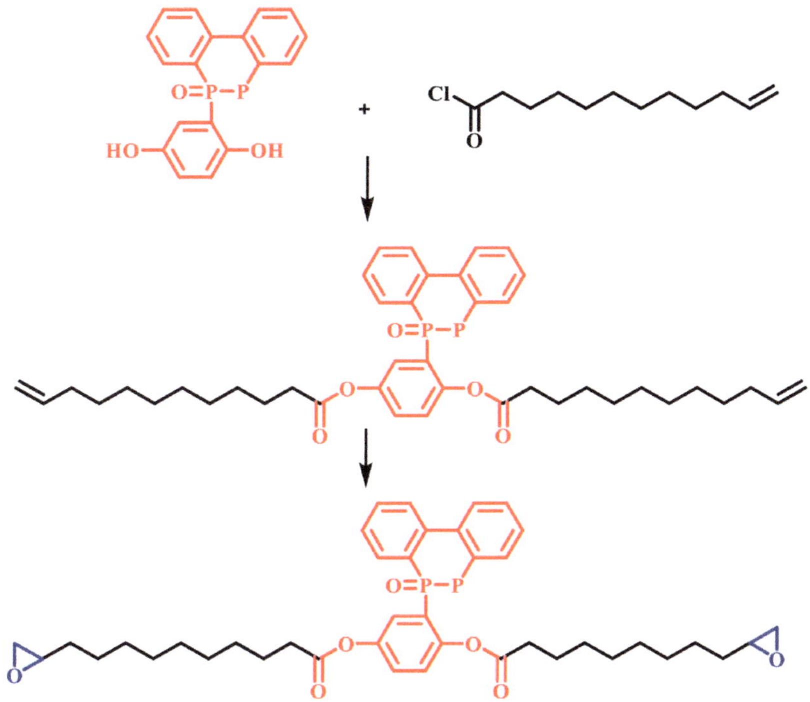

Figure 6.12 Synthesis of epoxy resin containing phosphorus derived from plant oil.

and thus can be a renewable substitute for commercial DGEBA. Besides, Huang *et al.* synthesized two cardanol-based curing agents (Car-DCP-MAH and Car-DPC-MAH, Figure 6.15) as alternatives for petroleum-based DDM.[48] Compared with the traditional DGEBA/DDM thermoset, the cured DGEBA/Car-DCP-MAH and DGEBA/Car-DPC-MAH thermosets both displayed high intrinsic fire resistance in terms of higher LOIs (29% and 28%, respectively, *versus* 24.5%) and V-0 rating in the UL-94 test.

6.3.4 Epoxy Resins from a Nitrogen-containing Fatty Acid Derivative

Nitrogen-based fatty acid derivative flame retardants, with a remarkable flame retardant effect in the gas phase, have been extensively studied in recent years owing to their high heat resistance and excellent flame retardancy. Eganeh *et al.* reported a novel epoxy-terminated polyurethane using renewable castor oil after blending with different weight percents of poly(ethylene glycol)-based epoxy-terminated polyurethane (Figure 6.16).[49] Nitrogen-based flame retardants will generate a large amount of non-flammable gases (*e.g.* N_2, NH_3, NO_2), which can effectively dilute the oxygen

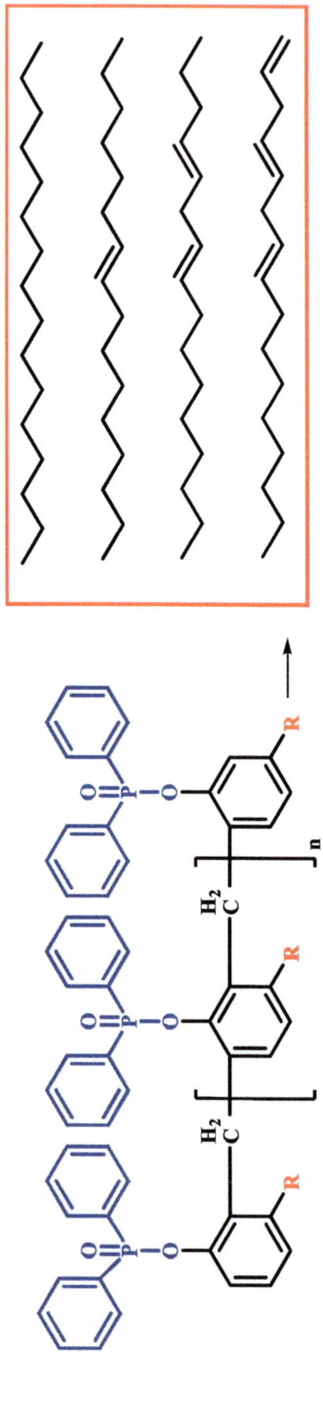

Figure 6.13 Schematic diagram of a cardanol-derived phosphate (DPP-CFR).

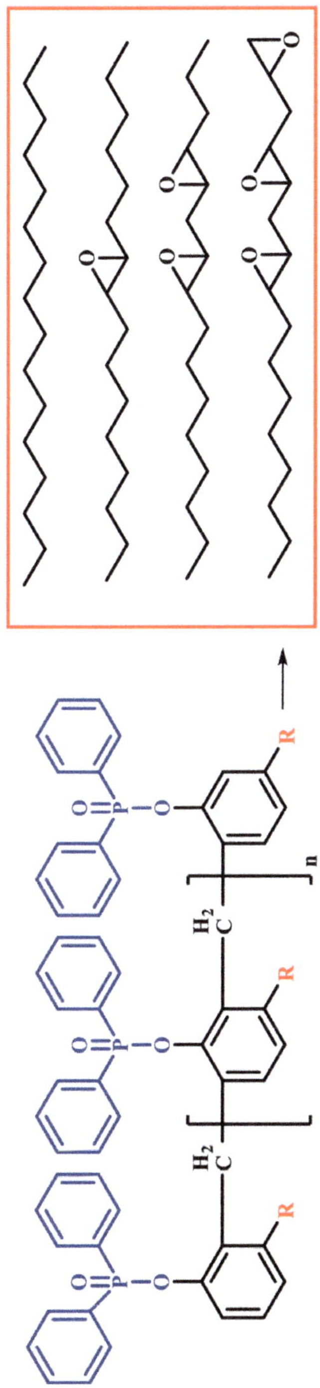

Figure 6.14 Schematic diagram of cardanol-derived novolac epoxy (ECF-PO(OPh)$_2$).

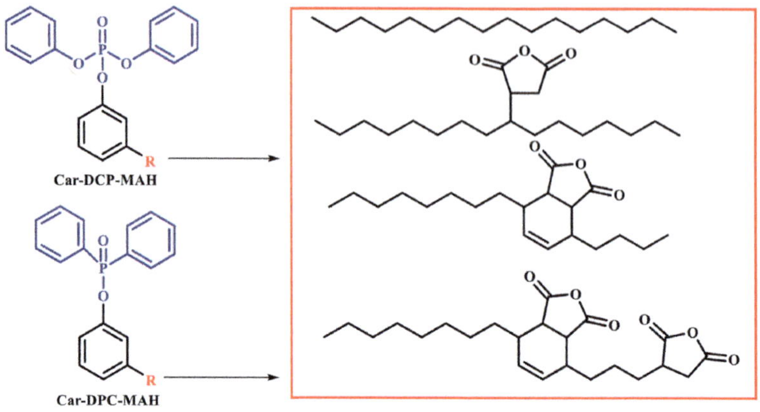

Figure 6.15 Schematic diagram of cardanol-based curing agents (Car-DCP-MAH and Car-DPC-MAH).

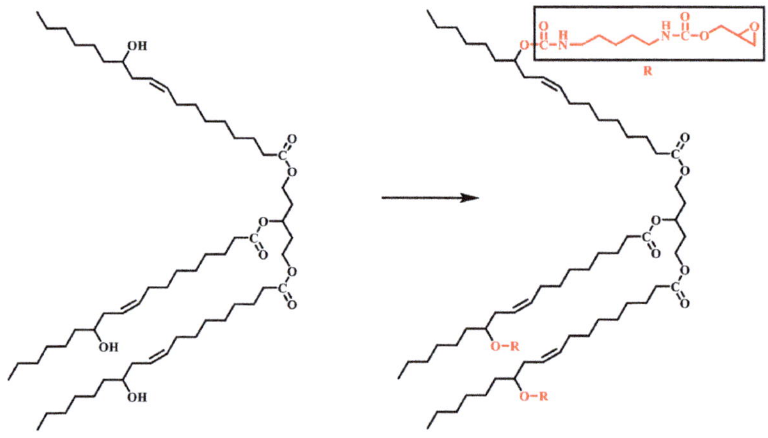

Figure 6.16 Synthetic route for preparation of epoxy-terminated polyurethane prepolymers.

concentration in the gas phase and take away much heat through absorption during combustion. In fact, in the condensed phase, nitrogen-containing compounds will form various cross-linked nitrogen-containing polycondensation products on the surface of the polymer matrix during combustion. These polycondensation products can affect the polymer matrix decomposition and help promote polymer carbonization, as well as improving the structural stability of the carbon layer.

Recent research shows that when nitrogen-based flame retardants are physically blended or chemically bonded with other types of flame retardants, the flame retardant efficiency of the polymer matrix can be further improved.[50] Taking phosphorus-based flame retardants as an example, they can interact with nitrogen-based flame retardants during combustion to generate

P–N intermediates, which can improve the carbonization rate of the polymer matrix. For example, Ding *et al.* successfully synthesized a novel castor-oil-based flame-retardant polyol containing phosphorus and nitrogen (Figure 6.17) and significantly improved the LOI of flame-retardant polyurethane foams to 25.4%.[51]

In addition, nitrogen-based flame retardants can generally promote the oxidation of phosphorus atoms and reduce the loss of phosphorus compounds due to volatilization in the condensed phase. Therefore, research on synergistic flame retardants including nitrogen-based flame retardants will be one important direction in the future. Nevertheless, the issue of eutrophication caused by phosphorylation needs to be addressed, wherein phosphorus enters water and undergoes a phosphorous cycle. The metric denotes compounds containing phosphorus (P), nitrogen (N), and neither of them as of high, medium, and low risk respectively.[52]

6.3.5 Epoxy Resins from a Silicon-containing Fatty Acid Derivative

Silicon-based flame retardants show broad development prospects because of their low costs, excellent mechanical strength, and high thermal decomposition temperature. These retardants do not cause harm to the environment or human health during thermal decomposition and thus are widely used in industrial production.[53] Some researchers investigated the hydrosilylation reaction of terminal unsaturated fatty acid esters with several polyfunctional hydrosilylating agents, including 1,4-bis(dimethylsilyl)benzene (DMSB), tetrakis(dimethylsilyloxy) silane (TKDS), and 2,4,6,8-tetramethylcyclotetrasiloxane (TMCTS).[54] In addition, epoxidized methyl oleate (EMO)-based polyether polyols are used in the synthesis of polyurethanes with specific applications, such as silicon-containing polyurethanes with enhanced flame-retardant properties.[55] Besides, novel silicon-containing biobased polyurethanes also were obtained from an EMO-based poly(ether polyol) derived from unsaturated fatty acid methyl esters (Figure 6.18). Subramani *et al.* reported some highly flexible bio-silica-reinforced poly(benzoxazine-*co*-epoxy) hybrid composites using renewable cardanol-based benzoxazine, bisphenol-F, and functionalized bio-silica.[46] The degradation temperature ($T_{d5\%}$) of poly(benzoxazine-*co*-epoxy) hybrid composites ranges from 268 to 317 °C, which indicates high thermal stability (degradation temperature and char yield).

Silicon-based flame retardants mainly include organic and inorganic types, which are both considered to exhibit a flame retardant effect mainly in the condensed phase. The incorporation of an organosilicon functionality into the polymeric structure is recognized as one of the most efficient ways to obtain an environmentally friendly flame-retardant system. This is because these compounds dilute the more combustible organic components partly because the siliceous residues can form a barrier to an advancing flame.

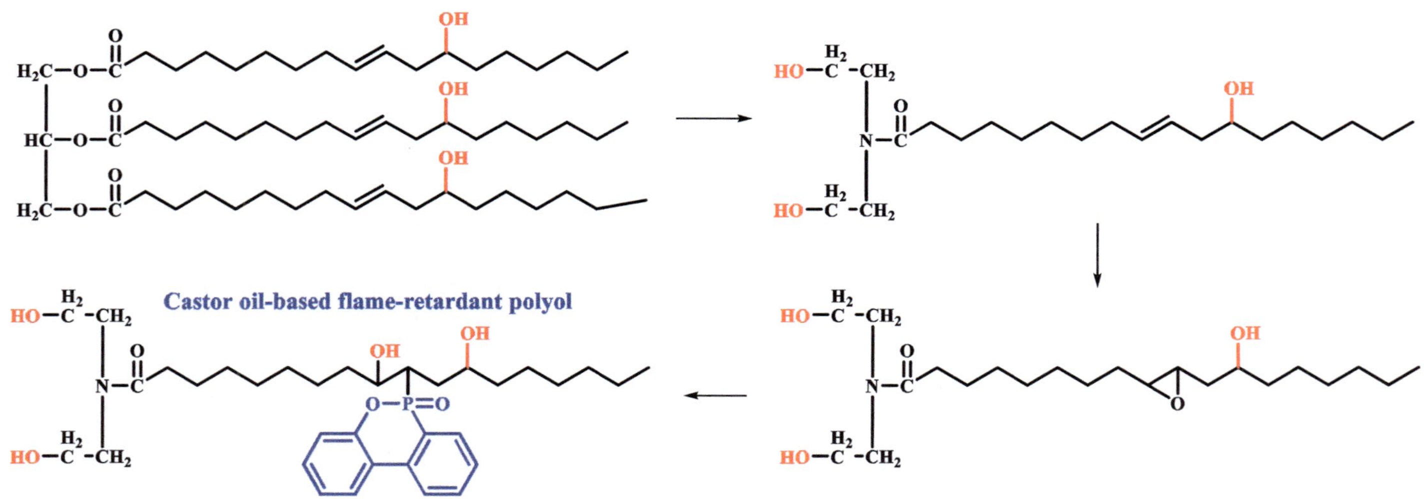

Figure 6.17 Synthesis route and structure of castor oil-based flame-retardant polyol.

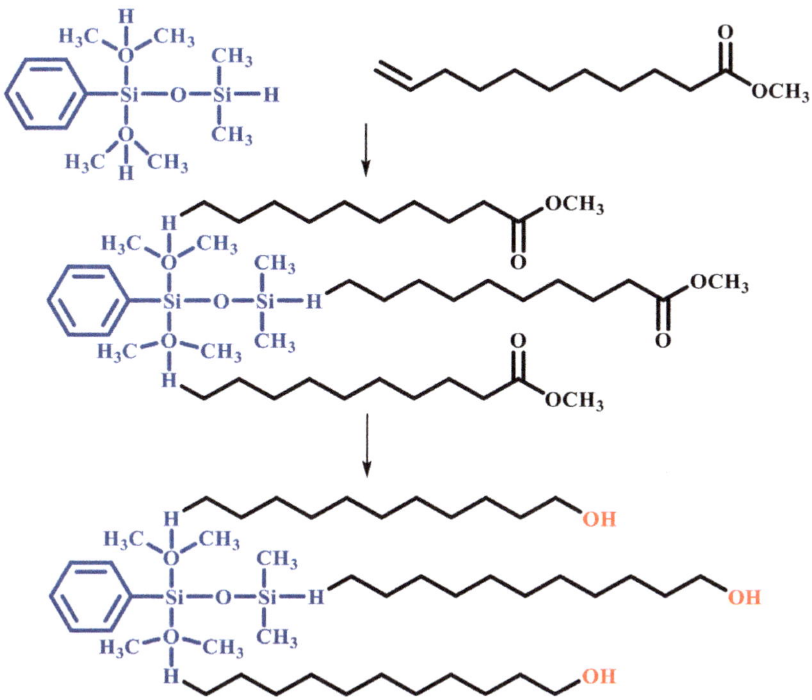

Figure 6.18 Synthesis of a silicon-containing polyol from natural ricinoleic acid.

As one of the most common silicone-based flame retardants, organopolysiloxane can usually be used in combination with several synergistic flame retardants such as pentaerythritol and aluminum hydroxide. Polysiloxane can form interpenetrating polymers during combustion. The network combines with the polymer matrix to form a barrier against the exchange of heat and oxygen. Polysiloxane can also promote the formation of a carbon layer during combustion, improving the LOI of the polymer, and has an obvious smoke suppression effect. During combustion, the organopolysiloxane exhibits a melted morphology with low viscosity at high temperatures and can be well dispersed in the polymer matrix to form a siliceous carbonized layer with a flame retardant effect. In particular, the silicone flame retardant does not generate toxic gas during thermal decomposition and can effectively improve other physical and chemical properties of the polymer.[56]

There are various inorganic silicon-based flame retardants. As one of the most widely-used silicon-based halogen-free flame retardant additives, silica is non-toxic, tasteless, and thermally stable. During combustion, SiO_2 can migrate to the surface of the polymer matrix and interact with it to form a carbonized layer containing Si–O–Si and Si–C, which are unique to silicon-based flame retardants. A silicon-containing cross-linked network formed in the material acts as a protective layer to insulate oxygen and heat and prevent the escape of gaseous products during combustion. In addition, SiO_2 has a large specific surface area and contains a large number of surface silyl hydroxyl groups, which can be used as a template or active sites to physically or chemically modify the surface of SiO_2 to form novel halogen-free flame retardants. However, the addition amount of the silicon-based flame retardant in the polymer matrix shall not be too large, as excessive addition will destroy the mechanical properties of the polymer matrix. Therefore, investigating the dispersion of silicon-based flame retardants in the polymer matrix is very important.

6.3.6 Epoxy Resins from Intumescent Flame Retardants (IFRs) and Nanoscale Flame Retardants

IFRs can improve the flame retardancy of polymers through the synergistic flame retardant effect of three components: acid source, carbon source, and gas source.[57] IFRs use nitrogen and phosphorus as the main elements, which can foam and expand when decomposed by heat and finally form a porous expanded carbon layer. Among them, the acid source generally refers to compounds that can generate acids during thermal decomposition. The acid source can undergo an esterification reaction with the carbon source in the system to remove water molecules and generally has a higher boiling point. The carbon source generally is a class of compounds that are rich in carbon atoms and hydroxyl groups in the molecular structure. It is the most important basic substance that can form a carbon layer in the IFR system. The surface of the polymer produces a foamed carbon layer to effectively protect the polymer matrix. Moreover, the gas source refers to nitrogen-containing compounds that can release abundant non-toxic and non-flammable volatile gases during thermal decomposition, such as melamine, urea, and dicyandiamide.[58]

When the IFR system is heated, the acid source first decomposes to an acid that can strongly dehydrate and esterify polyols. As the temperature increases, the acid interacts with the carbon source to start the esterification reaction. At this time, the IFR system is molten and dehydrates the carbon layer. With the progress of the reaction, the gas source decomposes to generate water vapor and other non-flammable nitrogen-containing gases, so that the system expands under the gas promotion and finally forms a foamed carbon layer, which acts as a physical barrier to effectively protect the polymer matrix.[59]

Chitosan, a carbohydrate containing hydroxyl and amino groups, can be degraded and regenerated and undergoes acylation, esterification, etherification, alkylation, graft copolymerization, and crosslinking to form a series of derivatives for IFR materials. Its flame retardant mechanism lies in the presence of highly reactive amino groups and hydroxyl groups in its molecular structure, which mainly act as carbon sources and foaming agents in the system. In addition, chitosan will release NH_3, which promotes the formation of a swelling system during combustion. However, due to the low nitrogen content, it is generally synergistic with phosphorus-containing compounds in flame retardancy.

Since the IFR system improves the flame retardant properties of the polymer by mutually mixing the three components, the dosages of different components shall be well controlled. A stable expanded foam carbon layer can be formed only when the progress rate and the rate at which the flame retardant system changes from the molten state to the solid state are well controlled, and otherwise, no physical barrier with flame retardant effect can be achieved. Therefore, research on monomeric IFRs is increasingly extensive, and they are a class of flame retardants containing the three components required by the IFR system in the same molecular structure and are more efficient than mixed IFRs. The above advantages have become the mainstream trend of the future development of IFR systems.

Although the above nanomaterials can act as flame retardant fillers to effectively improve the flame retardant properties of polymers, most of the nanoscale flame retardants cannot pass the most important fire performance benchmark tests in the flame retardant industry – the vertical flame test (UL-94). In addition, introduction of nanoscale flame retardants usually leads to an increase in system viscosity, which affects the dispersion of fillers in the polymer matrix and further limits the practical application of nanoscale fillers in the field of flame retardants.

6.4 Conclusions

With increasing awareness about sustainable development and environmental protection, plant oil-based epoxy thermosets made from various renewable products have attracted extensive attention to substitute petroleum-based counterparts over the past few decades. Like most polymer materials, plant oil-based epoxy resins easily burn, which limits their application. Therefore, it is necessary to endow biological epoxy resins with flame retardancy. Until now, intrinsically flame-retardant plant oil-based epoxy thermosets have not been widely reported, suggesting that research in this field is just beginning. Generally, intrinsically flame-retardant bio-based epoxy thermosets can be designed by two approaches: (i) selecting bio-based chemicals with high charring ability and abundant aromatic structures and (ii) incorporating flame retardant elements such as phosphorus, silicon, and boron into the molecular structure of bio-based epoxy monomers or curing agents.

Although some plant oil-based epoxy resins have high flame retardancy, their thermal stability and glass transition temperature are low. In addition, the costs of bio-based materials used to synthesize bio-based epoxy resins are relatively high, which limits their large-scale application. Plant oil-based materials mostly contain hydroxyl and carboxyl groups. Although they are good carbon sources and can promote carbon formation, their low thermal stability and low environmental tolerance cannot meet the application requirements in special fields. Because many plant oil-based materials do not exhibit flame retardancy, preparation of intrinsic flame-retardant bio-based epoxy resins requires introduction of phosphorus, nitrogen, and other flame retardant elements as well as the use of organic solvents and other chemicals. At present, the application of intrinsic flame retardant bio-based epoxy resins still faces some limitations, and their commercialization still has a long way to go.

The future development directions of intrinsic flame retardant bio-based epoxy resins must be low carbon, environmental protection, and greenness. This chapter discusses examples that highlight the potentiality of triglyceride oils in the synthesis of polymeric materials from renewable resources. In particular, the chemical modification of fatty acid derivatives with functional groups that can facilitate subsequent polymerization will be of considerable interest in the future synthesis of new efficient thermosetting materials. Intrinsic flame retardant bio-based epoxy resins need to meet the requirements in special fields, such as electronics, electronics, and aerospace, and solve the problems of low thermal stability and reduced mechanical properties. In addition, increasing the functionality of bio-based epoxy monomers or curing agents and increasing the biomass content and biodegradability are the key factors for developing high-performance bio-based epoxy resins in the future.

Acknowledgements

The authors are grateful for the support and funding from the Fundamental Research Funds for the Central Non-profit Research Institution of CAF (CAFYBB2022XB001) and the National Natural Science Foundation of China for Youth Science Funds (32101471).

References

1. G. M. Roudsari, A. K. Mohanty and M. Misra, *ACS Sustainable Chem. Eng.*, 2017, **5**, 9528–9541.
2. T. Liu, J. Peng, J. Liu, X. Hao, C. Guo, R. Ou, Z. Liu and Q. Wang, *Composites, Part B*, 2021, **224**, 109188.
3. W. B. Li, L. H. Xiao, J. R. Huang, Y. G. Wang, X. A. Nie and J. Chen, *Compos. Sci. Technol.*, 2022, **227**, 109575.
4. L. Xiao, S. Li, Y. Wang, W. Li, J. Chen, J. Huang and X. Nie, *Ind. Crops Prod.*, 2021, **170**, 113723.
5. R. Wang and T. P. Schuman, *eXPRESS Polym. Lett.*, 2013, **7**, 272–292.
6. S. Ataei, S. N. Khorasani and R. E. Neisiany, *Prog. Org. Coat.*, 2019, **129**, 77–95.
7. E. Sharmin, F. Zafar, D. Akram, M. Alam and S. Ahmad, *Ind. Crops Prod.*, 2015, **76**, 215–229.
8. F. Mustata, N. Tudorachi and I. Bicu, *J. Anal. Appl. Pyrolysis*, 2016, **122**, 191–201.
9. W. He, Z. Fang, D. Ji, K. T. Chen, Z. D. Wan, X. Li, H. F. Gan, S. G. Tang, K. Zhang and K. Guo, *Org. Process Res. Dev.*, 2013, **17**, 1137–1141.
10. B. Bouchareb and M. T. Benaniba, *J. Appl. Polym. Sci.*, 2008, **107**, 3442–3450.

11. S. Sinadinovic-Fiser, M. Jankovic and O. Borota, *Chem. Eng. Process.*, 2012, **62**, 106–113.
12. M. Cabo, M. N. Prabhakar and J. I. Song, *Sci. Rep.*, 2021, **11**, 24140.
13. Y. Z. Xu, P. Fu, S. L. Dai, H. B. Zhang, L. W. Bi, J. X. Jiang and Y. X. Chen, *Ind. Crops Prod.*, 2021, **171**, 113978.
14. G. Lligadas, J. C. Ronda, M. Galia and V. Cadiz, *J. Polym. Sci., Part A: Polym. Chem.*, 2006, **44**, 5630–5644.
15. J. C. Ronda, G. Lligadas, M. Galia and V. Cadiz, *Eur. J. Lipid Sci. Technol.*, 2011, **113**, 46–58.
16. W. T. He, P. A. Song, B. Yu, Z. P. Fang and H. Wang, *Prog. Mater. Sci.*, 2020, **114**, 100687.
17. Q. Y. Liu, D. H. Wang, Z. K. Li, Z. F. Li, X. L. Peng, C. B. Liu, Y. Zhang and P. L. Zheng, *Materials*, 2020, **13**, 2145.
18. V. P. Chavan, A. V. Patwardhan and P. R. Gogate, *Chem. Eng. Process.*, 2012, **54**, 22–28.
19. S. La Tegola, C. Annese, M. Suman, I. Tommasi, C. Fusco and L. D'Accolti, *Sci. World J.*, 2013, **2013**, 212156.
20. M. Carraro, M. Gardan, A. Sartorel, C. Maccato and M. Bonchio, *Dalton Trans.*, 2016, **45**, 14544–14548.
21. G. Liang and K. Chandrashekhara, *J. Appl. Polym. Sci.*, 2006, **102**, 3168–3180.
22. M. R. Jankovic, O. M. Govedarica and S. V. Sinadinovic-Fiser, *Ind. Crops Prod.*, 2020, **143**, 111881.
23. H. Uyama, M. Kuwabara, T. Tsujimoto and S. Kobayashi, *Biomacromolecules*, 2003, **4**, 211–215.
24. L. P. Wang, Y. Y. Liang, Q. Yu, S. F. Chen, J. H. Zhang, M. H. Miao and D. H. Zhang, *Polym. Compos.*, 2018, **39**, E2046–E2055.
25. E. Milchert, K. Malarczyk and M. Klos, *Molecules*, 2015, **20**, 21481–21493.
26. L. H. Sperling, J. A. Manson, S. Qureshi and A. M. Fernandez, *Ind. Eng. Chem. Prod. Res. Dev.*, 1981, **20**, 163–166.
27. F. Jaillet, M. Desroches, R. Auvergne, B. Boutevin and S. Caillol, *Eur. J. Lipid Sci. Technol.*, 2013, **115**, 698–708.
28. A. Moller, Z. Y. Xie, M. H. Cai, G. C. Zhong, P. Huang, M. G. Cai, R. Sturm, J. F. He and R. Ebinghaus, *Environ. Sci. Technol.*, 2011, **45**, 6793–6799.
29. Q. An, M. Aamir, S. D. Mao, Y. X. Liu, Y. T. Wang, P. Zheng and W. P. Liu, *Sci. Total Environ.*, 2022, **803**, 150043.
30. B. A. Howell and E. A. Ostrander, *Fire Mater.*, 2020, **44**, 242–249.
31. T. Eren and S. H. Kusefoglu, *J. Appl. Polym. Sci.*, 2004, **91**, 2700–2710.
32. Z. G. Cao, Q. Y. Chen, X. X. Li, Y. C. Zhang, M. H. Ren, L. F. Sun, M. M. Wang, X. T. Liu and G. Yu, *Sci. Total Environ.*, 2019, **646**, 1090–1096.
33. S. Q. Huo, P. A. Song, B. Yu, S. Y. Ran, V. S. Chevali, L. Liu, Z. P. Fang and H. Wang, *Prog. Polym. Sci.*, 2021, **114**, 101366.
34. P. Joseph, D. Bakirtzis, Q. Richard and R. Brulfert, *Mater. Today: Proc.*, 2017, **4**, 5282–5289.
35. Z. Q. Chen, Q. Shi, X. Kuang, H. J. Qi and T. J. Wang, *Composites, Part B*, 2020, **194**, 108054.

36. X. F. Liu, B. W. Liu, X. Luo, D. M. Guo, H. Y. Zhong, L. Chen and Y. Z. Wang, *Chem. Eng. J.*, 2020, **380**, 122471.
37. X. Wang, H. X. Niu, W. W. Guo, L. Song and Y. Hu, *Chem. Eng. J.*, 2021, **421**, 129738.
38. B. J. K. Ahn, S. Kraft and X. S. Sun, *J. Mater. Chem.*, 2011, **21**, 9498–9505.
39. P. Y. Jia, L. H. Hu, M. Zhang and Y. H. Zhou, *J. Therm. Anal. Calorim.*, 2016, **124**, 1331–1339.
40. X. L. Wang, L. Chen, J. N. Wu, T. Fu and Y. Z. Wang, *ACS Sustainable Chem. Eng.*, 2017, **5**, 3353–3361.
41. X. J. Yang, C. P. Wang, J. L. Xia, W. Mao and S. H. Li, *Prog. Org. Coat.*, 2017, **110**, 195–203.
42. M. Moreno, G. Lligadas, J. C. Ronda, M. Galia and V. Cadiz, *J. Polym. Sci., Part A: Polym. Chem.*, 2013, **51**, 1808–1815.
43. X. Wang, Y. Hu, L. Song, W. Y. Xing, H. D. A. Lu, P. Lv and G. X. Jie, *Polymer*, 2010, **51**, 2435–2445.
44. K. A. Salmeia and S. Gaan, *Polym. Degrad. Stab.*, 2015, **113**, 119–134.
45. A. Bifulco, D. Parida, K. A. Salmeia, R. Nazir, S. Lehner, R. Stampfli, H. Markus, G. Malucelli, F. Branda and S. Gaan, *Mater. Des.*, 2020, **193**, 108862.
46. D. Subramani, K. Krishnamoorthy, N. Eeda, S. Salendra and A. K. Achimuthu, *J. Macromol. Sci., Part A: Pure Appl. Chem.*, 2022, **59**, 46–52.
47. W. W. Guo, X. Wang, J. L. Huang, W. Cai, L. Song and Y. Hu, *Ind. Crops Prod.*, 2021, **166**, 113496.
48. J. L. Huang, H. L. Ding, X. Wang, L. Song and Y. Hu, *Chem. Eng. J.*, 2022, **450**, 137906.
49. H. Yeganeh and P. Hojati-Talemi, *Polym. Degrad. Stab.*, 2007, **92**, 480–489.
50. D. Wang, X. W. Mu, W. Cai, L. Song, C. Ma and Y. Hu, *Composites, Part A*, 2018, **109**, 546–554.
51. H. Y. Ding, K. Huang, S. H. Li, L. N. Xu, J. L. Xia and M. Li, *J. Anal. Appl. Pyrolysis*, 2017, **128**, 102–113.
52. P. Jessop, *Green Chem.*, 2020, **22**, 13–15.
53. S. Y. Lu and I. Hamerton, *Prog. Polym. Sci.*, 2002, **27**, 1661–1712.
54. M. Galià, L. M. de Espinosa, J. C. Ronda, G. Lligadas and V. Cádiz, *Eur. J. Lipid Sci. Technol.*, 2010, **112**, 87–96.
55. G. Lligadas, J. C. Ronda, M. Galia, U. Biermann and J. O. Metzger, *J. Polym. Sci., Part A: Polym. Chem.*, 2006, **44**, 634–645.
56. L. J. Qu, Y. L. Sui, C. L. Zhang, P. H. Li, X. Y. Dai, B. S. Xu and D. N. Fang, *Eur. Polym. J.*, 2020, **122**, 109383.
57. S. P. Bhoite, J. Kim, W. Jo, P. H. Bhoite, S. S. Mali, K. H. Park and C. K. Hong, *Polymers*, 2021, **13**, 2662.
58. X. Q. Qiu, Z. W. Li, X. H. Li and Z. J. Zhang, *Chem. Eng. J.*, 2018, **334**, 108–122.
59. K. S. Lim, S. T. Bee, L. T. Sin, T. T. Tee, C. T. Ratnam, D. Hui and A. R. Rahmat, *Composites, Part B*, 2016, **84**, 155–174.

CHAPTER 7

Reactions of Mixed Natural Acids and Esters in Acidic Media

WILLIAM J. BRADLEY, YANG QU, ZACHERY CRANDALL, THERESA L. WINDUS AND GEORGE A. KRAUS*

Department of Chemistry, Iowa State University Ames, IA 50011, USA
*Email: gakraus@iastate.edu

7.1 Introduction

The global market for citric acid is growing at about 4.5% per year and is expected to reach \$3.9 billion dollars by 2024.[1] The global market for malic acid is projected to reach \$257.7 million by 2025.[2] The structures of these diacids are shown in Figure 7.1. Although malic acid is produced to a lesser extent than citric acid, malic acid has industrial applications in foods and pharmaceuticals. Malic acid is used in the food industry as a pH control agent, a flavor enhancer, and a flavoring agent.[3] Malic acid also has many applications in the cosmetic industry.[4] Microbially-produced citric acid is an important organic acid that is widely used in pharmaceutical, food, beverage, detergent, and cosmetic industries.[5] Citric acid is produced by different types of microorganisms. Reactions that transform these feedstock acids into more valuable materials will enhance the overall value of these feedstocks. Many transformations on these feedstocks have already been reported, including selective dehydration,[6] anhydride formation,[7] and anion formation.[8]

There are many fruits that have high concentrations of citric acid and malic acid.[9] Separating both acids from carbohydrates and polyphenols is

Green Chemistry Series No. 83
Green Chemistry and Green Materials from Plant Oils and Natural Acids
Edited by Zengshe Liu and George Kraus

Published by the Royal Society of Chemistry, www.rsc.org

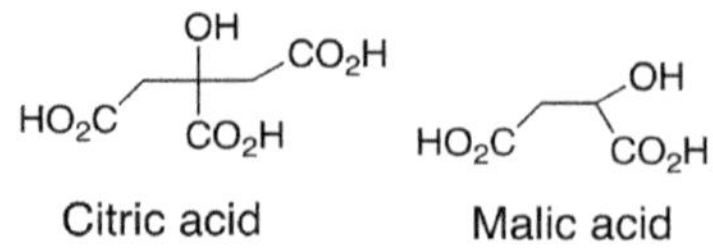

Figure 7.1 Structures of citric acid and malic acid.

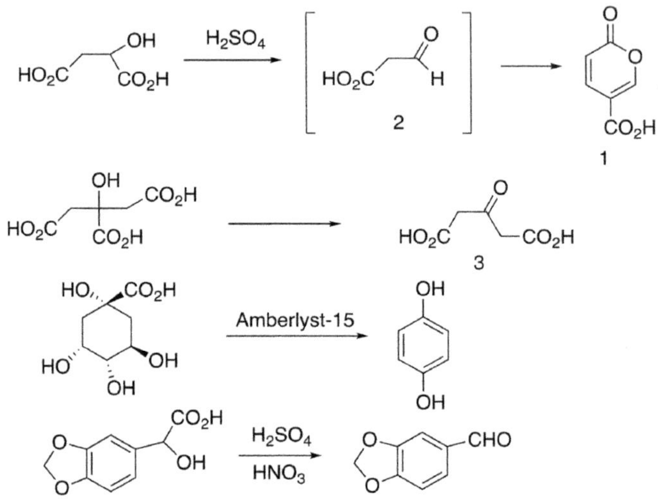

Scheme 7.1 Transformations of hydroxy acids in strongly acidic media.

straightforward. However, separating these acids from one another is difficult, due to multiple carboxylic acid functional groups in each molecule. Herein, we disclose transformations that can use both acids without separation to develop novel products.[10]

The reaction of hydroxy acids with strong acids such as sulfuric acid has a long history. In 1893, von Pechmann reported that the reaction of malic acid in hot sulfuric acid with sulfur trioxide produced coumalic acid **1**, as shown in Scheme 7.1.[11] In this reaction, an aldehyde acid **2** was generated along with carbon monoxide and water. The aldehyde acid was not stable under the reaction conditions and dimerized to produce coumalic acid. Citric acid, when treated with sulfuric acid at 80 °C, produces acetone dicarboxylic acid **3** in 90% yield.[12] The use of a concentrated acid is essential, since dilute acids produce only dehydration products.[13] Other hydroxy acids such as mandelic acid[14] and quinic acid[15] also undergo a reaction with sulfuric acid. Interestingly, quinic acid produces an intermediate trihydroxy ketone which undergoes two dehydrations to afford hydroquinone.

7.2 Results and Discussion

As part of a project to understand the chemistry of mixed organic acids and esters, we reacted a 1 : 1 mixture of citric acid and malic acid in concentrated

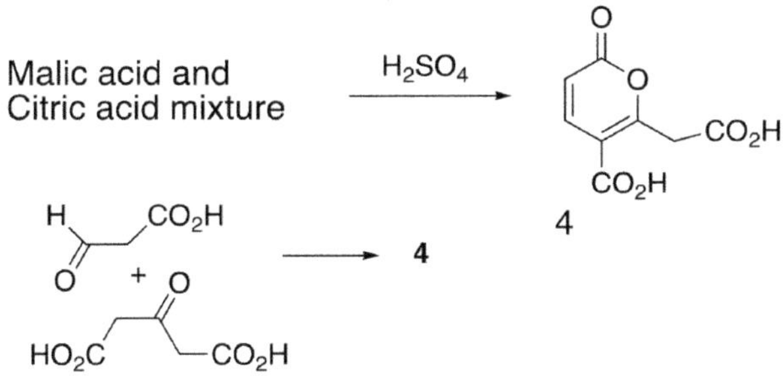

Scheme 7.2 Conversion of citric acid and malic acid into **4**.

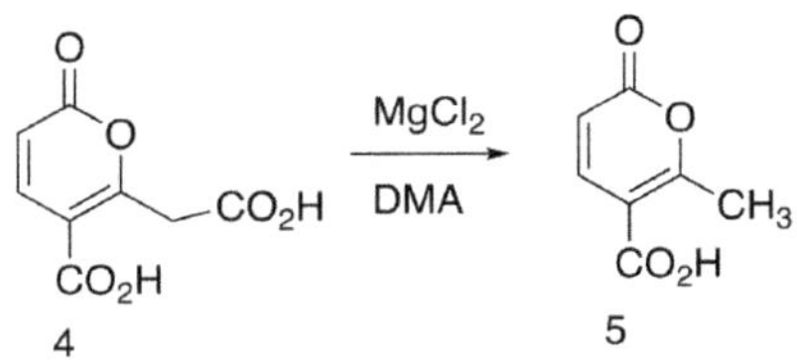

Scheme 7.3 Selective decarboxylation of **4**.

sulfuric acid at 80 °C. To our surprise, the product was not a mixture of **1** and **3** but a new material in 45% yield whose structure we determined to be pyrone diacid **4**. The mechanism for its formation is illustrated in Scheme 7.2.

Diacid **4** was formed by the reaction of the aldehyde acid from malic acid with acetone dicarboxylic acid produced from citric acid. The yield of diacid **4** could be increased to 76% by adding 2 equivalents of citric acid. Diacid **4** is a new pyrone and has functionality useful for making sustainable polymers. With catalytic magnesium chloride in boiling dimethylacetamide (DMA), diacid **4** was regioselectively decarboxylated to pyrone acid **5** in 95% yield as shown in Scheme 7.3.[8]

Although the production of coumalic acid is well precedented, we have not found any precedent for converting a hydroxy ester into an aldehyde. Synthesis of the diester of **4** directly would have a number of advantages. If the reaction was successful, the resulting diester would be much less water soluble than acid **4**. The starting esters such as dimethyl malate or diethyl malate are commercially available. As shown in Scheme 7.4, the reaction of dimethyl malate with sulfuric acid in 1,2-dichloroethane affords methyl coumalate in 84% isolated yield as a white solid. Its proton NMR spectrum matched the literature NMR spectrum of the commercial material.

This reaction could be scaled to produce 10 g of methyl coumalate. Different malate–sulfuric acid ratios were tested. Four equivalents of sulfuric acid per equivalent of malate gave the best result. The reaction did not proceed below 80 °C. The use of an immiscible co-solvent such as 1,2-dichloroethane enhanced the yield, dissolving methyl coumalate and insulating it from further reaction with sulfuric acid. The complete reaction optimization results are shown in Table 7.1.

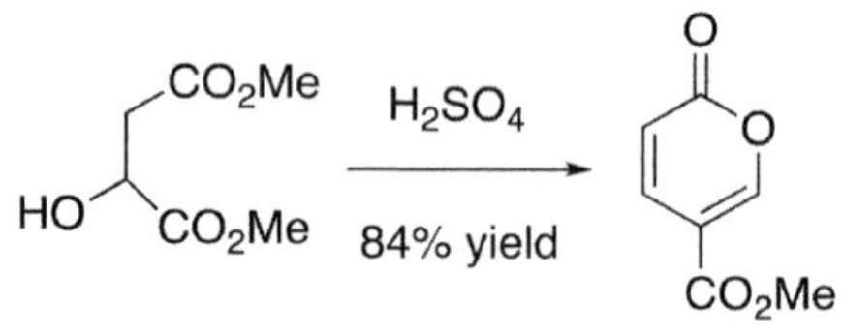

Scheme 7.4 Synthesis of methyl coumalate.

Table 7.1 Reaction optimization of Scheme 7.4.

Entry	Diester (mmol)	H_2SO_4 (mmol)	Temp. (°C)	Co-solvent (10 mL)	Yield
1	4.5	9	80	Neat	43%
2	4.5	13.5	80	Neat	67%
3	4.5	18	80	Neat	82%
4	4.5	22.5	80	Neat	81%
5	4.5	18	80	DCE	84%
6	4.5	18	80	CH_3CN	Mixture
7	4.5	18	80	1,4-Dioxane	NR
8	4.5	18	80	CH_3NO_2	Mixture
9	4.5	18	60	DCE	NR
10	4.5	18	100	DCE	84%

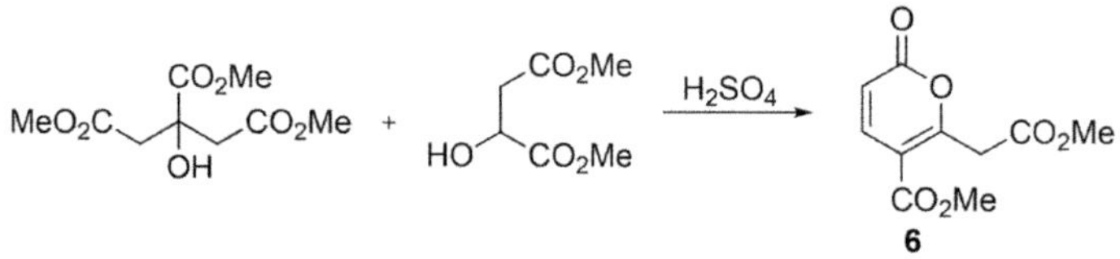

Scheme 7.5 Formation of pyrone diester **6**.

The mixture of dimethyl malate and trimethyl citrate afforded diester **6** in 50% yield as shown in Scheme 7.5. The use of a 1.3 : 1 ratio of citrate to malate afforded **6** in 55% yield. This reaction is scalable.

In these reactions, it was found that more than four equivalents of sulfuric acid were necessary to convert methyl malate to methyl coumalate. The use of 1,2-dichloroethane resulted in a clear product, in comparison with the other solvents used.

A related pyrone, 4,6-pyrone dicarboxylic acid, is produced from lignin *via* biocatalysis. Its structure is shown in Scheme 7.6. This diacid has been converted into polymers.[16]

Pyrones exhibit a range of biological activities, compared to ethyl butyrate which has the same number of carbon atoms. As shown below, pyrones have a larger dipole moment, compared to ethyl butyrate. They also have a much larger polar surface area. Additionally, pyrones often behave as a phenol isostere.

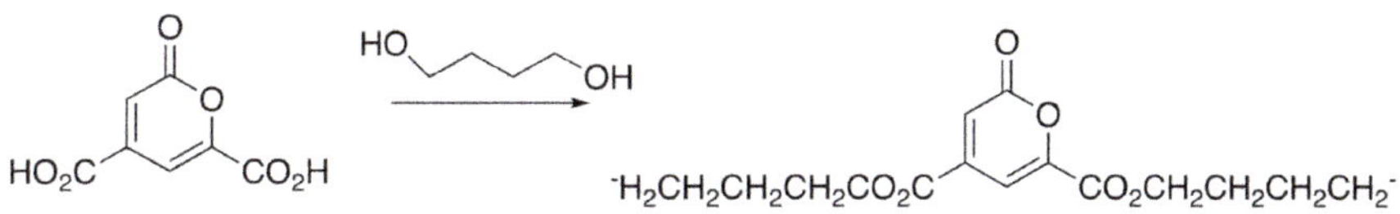

Scheme 7.6 Polymer from pyrone dicarboxylic acid.

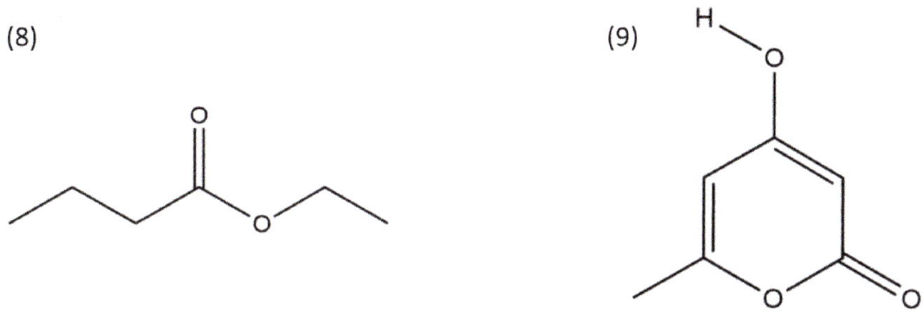

Figure 7.2 Structures for ethyl butanoate (**8**) and 4-hydroxy-6-methyl-2-pyrone (**9**).

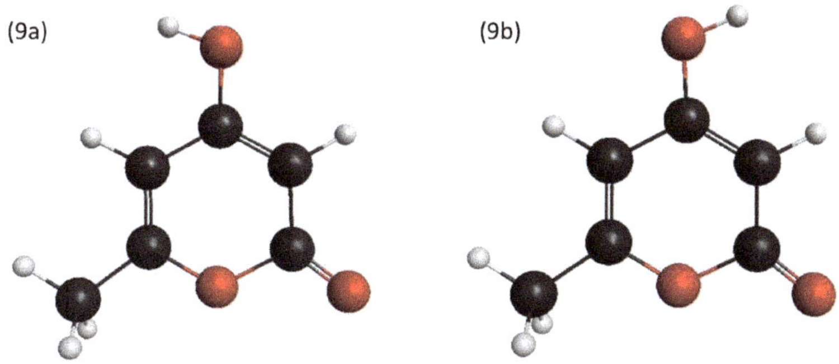

Figure 7.3 Optimized structures for the two conformations (**9a** and **9b**) necessary to describe 4-hydroxy-6-methyl-2-pyrone, formed by flipping the hydrogen position on the hydroxyl group.

7.3 Computational Evaluation

In order to better understand the differences, ethyl butyrate **8** and 4-hydroxy-6-methylpyrone **9** (Figure 7.2) were optimized to find structures with minimum energies. One conformation was sufficient to describe ethyl butyrate (**8**), while two conformations were needed to describe 4-hydroxy-6-methyl-2-pyrone (**9a** and **9b**), corresponding to flipping the hydrogen position on the hydroxyl group (Figure 7.3).

MEP maps are used to visualize charge distributions and charge-related properties of molecules. They map the contours of the total electron density based on color grading (different colors represent and visualize different surface electrostatic potentials) from which the active sites of nucleophiles and electrophiles in a molecular system can be determined. In most MEP maps, the red outline corresponds to the largest negative site (easy to donate electrons), and the blue outline corresponds to the largest positive site (easy to accept electrons), which are the preferred locations for electrophilic and nucleophilic attacks, respectively. Molecular electrostatic potential (MEP) surfaces for **8** (Figure 7.4) and **9a** and **9b** (Figure 7.5) were generated using the optimized structures. The dipole moment and the polarizability were also calculated, as shown in Table 7.2.

While polarizabilities are similar between **8** and **9**, the dipoles of both conformers of **9** are significantly larger than those of **8**. This is likely due to the more symmetric appearance of **8** with the carbon chains dampening the net dipole induced by the carbonyl oxygen, while **9** has a protruding, negatively charged carbonyl oxygen on one side of the molecule. Interestingly, the oxygen of the hydroxy group on **9** is only slightly red (negative), with the carbonyl oxygen bearing most of the negative charge. In addition, the hydrogen of the hydroxy group on **9** is blue, or electrophilic, suggesting that it is open to nucleophilic attack. It is important to note that the charge

Figure 7.4 MEP surface of ethyl butanoate.

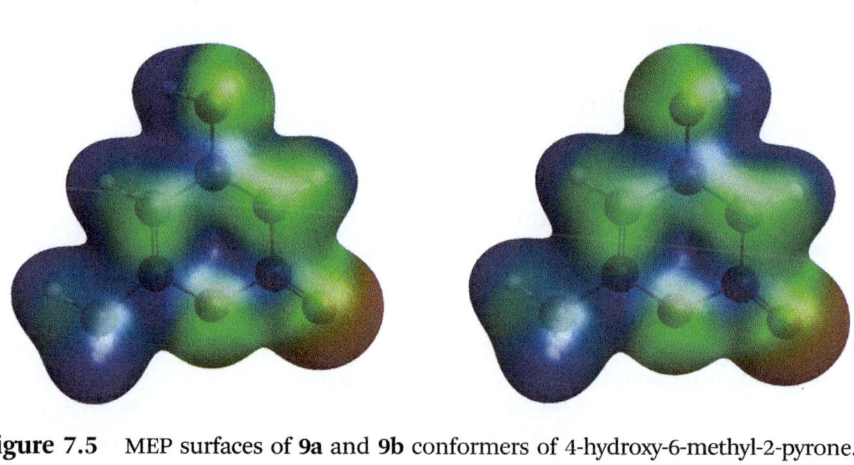

Figure 7.5 MEP surfaces of **9a** and **9b** conformers of 4-hydroxy-6-methyl-2-pyrone.

Table 7.2 Energy difference of conformers for each molecule is shown. The calculated dipole moment of each molecule and conformer are also reported.

Molecule	E_{Diff} (kcal mol^{-1})	Dipole (Debyes)	Pol (A^3)
8	0.0	1.84	12.1
9a	0.0	6.35	12.3
9b	1.4	4.32	12.3

on the hydroxyl hydrogen is about three times larger than that of any of the hydrogens in **8**, explaining the difference between the dipoles in **9a** and **9b**. The larger dipole moment and potential for electrophilic and nucleophilic attack are likely contributors to the larger biological activity of the pyrone over ethyl butyrate.

The reaction of mixed acids or esters provides a rapid route to a novel pyrone. These pyrones can be produced on a gram scale from biorenewable sources. The procedure using diesters[9] offers an improved yield and is an operationally more convenient route to synthesize methyl coumalate. Importantly, the esters are preserved and allow for direct synthesis of pyrones **4** and **6**.

7.4 Materials and Methods

7.4.1 Procedure for the Synthesis of 4

To a flame dried round bottom flask was added malic acid (1 g, 7.46 mmol) and citric acid (2.86 g, 14.9 mmol) followed by the addition of concentrated sulfuric acid (2 mL, 36 mmol). The reaction mixture was heated to 80 °C for 12 hours. After cooling to room temperature, the brown reaction mixture was poured onto 40 g of ice and stirred for 30 min. The mixture was extracted with ethyl acetate (3 × 20 mL). The combined organic extracts were washed with ice-cold water, dried over anhydrous sodium sulfate, concentrated and collected in 76% yield. ^{1}H NMR (400 MHz, DMSO) δ 7.86 (d, J = 9.8 Hz, 1H), δ 6.33 (d, J = 9.8 Hz, 1H), δ 3.98 (s, 2H).

7.4.2 Procedure for the Synthesis of 6

To a flame dried round bottom flask was added dimethyl malate (1 g, 6.17 mmol) and trimethyl citrate (2.86 g, 12.2 mmol) followed by the addition of concentrated sulfuric acid (2 mL, 36 mmol) and 5 mL of dichloroethane. The reaction mixture was heated to 80 °C for 12 hours. After cooling to room temperature, the brown reaction mixture was poured onto 100 g of ice and stirred for 30 min. The mixture was extracted with ethyl acetate (3 × 20 mL) and the combined organic extracts were washed with ice-cold water and dried over anhydrous sodium sulfate. Concentration *in vacuo* followed by flash chromatography gave **5** in 55% yield. ^{1}H NMR (400 MHz, $CDCl_3$) δ 7.84 (d, J = 9.8 Hz, 1H), δ 6.28 (d, J = 9.8 Hz, 1H), δ 4.07 (s, 2H), δ 3.85 (s, 3H), δ 3.74 (s, 3H).

7.4.3 Methyl Coumalate from Dimethyl Malate

To a flame dried round bottom flask was added dimethyl malate (729 mg, 4.5 mmol) followed by concentrated sulfuric acid (1 mL, 18 mmol) and dichloroethane (3 mL). The reaction mixture was heated to 80 °C for 12 hours. After cooling to room temperature, the brown reaction mixture was poured onto 40 g of ice and stirred for 30 min. The mixture was extracted with ethyl acetate (3 × 20 mL) and the combined organic extracts were washed with ice-cold water and dried over anhydrous sodium sulfate. Concentration and flash chromatography gave methyl coumalate in 84% yield. ^{1}H NMR (400 MHz, $CDCl_3$) δ 8.3 (dd, $J = 2.6$, 1.1 Hz, 1H), 7.8 (dd, $J = 9.8$, 2.6 Hz, 1H), 6.3 (dd, $J = 9.8$, 1.1 Hz, 1H), 3.9 (s, 2H).

Acknowledgements

We thank the NSF Engineering Center for Biorenewable Chemicals which awarded the NSF grant EEC-0813570 for support of this research. ZC and TLW thank the U.S. Department of Energy, Office of Science, Basic Energy Sciences, Division of Chemical Sciences, Geosciences, and Biosciences through the Ames Laboratory. The Ames Laboratory is operated for the U.S. Department of Energy by Iowa State University under Contract No. DEAC02-07CH11358. We thank the Department of Chemistry for partial support for WB.

References

1. Citric acid: https://www.prnewswire.com/news-releases/global-citric-acid-market-to-reach-3-9-billion-by-2024–301349523.html.
2. Malic acid: https://www.grandviewresearch.com/press-release/global-malic-acid-market.
3. A. Koevilein, C. Kubisch, L. Cai and K. Ochsenreither, *J. Chem. Technol. Biotechnol.*, 2020, **95**(3), 513–526.
4. V. K. Singh, *Cosmet. Toiletries*, 2017, **132**(2), 64–67.
5. S. Chozhavendhan, M. Aniskumar, C. Santhaseelan, D. Muthu Gayathri, R. Sathya and K. Shanmugapriya, *Int. J. Pharm. Biol. Sci.*, 2021, **11**(2), 79–85.
6. R. R. Umbdenstock and P. F. Bruins, *Ind. Eng. Chem.*, 1945, **37**, 963–967.
7. A. J. Repta and T. Higuchi, *J. Pharm. Sci.*, 1969, **58**(9), 1110–1114.
8. M. Sefkow, A. Kelling and U. Schilde, *Tetrahedron Lett.*, 2001, **42**, 5101–5104.
9. https://www.ehow.com/facts_5744584_fruits-citric-acid_.html.
10. J. Yang, A. R. Webb and G. A. Ameer, *Adv. Mater.*, 2004, **16**, 511–516.
11. (a) H. von Pechmann, *Ann.*, 1891, **264**, 272; (b) R. H. Wiley and N. R. Smith, *Organic Syntheses*, 1963, coll. vol. 4, p. 201.

12. H. Wu, Y. Li, Y. Liu, S. Lyu, C. Wu and T. Li, *ChemBioChem*, 2012, **13**(6), 862–871.
13. G. E. Hilbert, *J. Am. Chem. Soc.*, 1932, **54**, 3413.
14. K. Harada, M. Aoi and M. Shirai, *Jpn. Kokai Tokkyo Koho*, JP07258246A 19951009, 1995.
15. B. Assoah, L. F. Veiros, C. A. M. Afonso and N. R. Candeias, *Eur. J. Org. Chem.*, 2016, 3856–3861.
16. Y. Qian, Y. Otsuka, T. Sonoki, B. Mukhopadhyay, M. Nakamura, J. Jellison and B. Goodell, *BioResources*, 2016, **11**(3), 6097–6109.

CHAPTER 8

Industrial Chemicals Via *Decarboxylation of Natural Carboxylic Acids*†

KENNETH M. DOLL* AND BRYAN R. MOSER

United States Department of Agriculture, Agricultural Research Service, National Center for Agricultural Utilization Research, Bio-Oils Research Unit, 1815 N. University Street, Peoria, IL 61604, USA
*Email: Kenneth.Doll@usda.gov

8.1 Plant Feedstocks for Industrial Use

The use of biomass for food, products, chemicals, and fuels is as ancient as humanity itself.[1–4] For example, humankind has used fatty acids for thousands of years as soap. More recently, lipids have found widespread use as substrates for lubricants[5–8] and biodiesel,[9–12] which consists of methyl esters of fatty acids (FAME). Plant biomass originates from carbon dioxide and water using sunlight as the energy source, producing oxygen as a fortunate byproduct. The main components of plant biomass can be roughly divided into carbohydrates, lignin, and lipids. Some of these are structural while others are for energy storage. Within the carbohydrate fraction, the primary constituents are C_6- and C_5-sugars that form cellulose (polymerized glucose) and hemicellulose (copolymers of glucose and xylose), which provide

† Mention of trade names or commercial products in this publication is solely for the purpose of providing specific information and does not imply recommendation or endorsement by the U.S. Department of Agriculture. USDA is an equal opportunity provider and employer.

Green Chemistry Series No. 83
Green Chemistry and Green Materials from Plant Oils and Natural Acids
Edited by Zengshe Liu and George Kraus

Published by the Royal Society of Chemistry, www.rsc.org

structure to plants. Energy is also stored in carbohydrate-based starches and sugars. Lignin is a highly cross-linked polymer composed of substituted phenols and is part of the mixture that gives strength and structure to plants.[1]

For plants, the primary purpose of lipids is energy storage. However, in addition to the abovementioned soaps and biodiesel, humans have used lipids for the production of a variety of oleochemicals. Products produced from lipids include monomers, polymers, detergents, paints, plasticizers, linoleum, textiles, adhesives, cosmetics, and many others.[13–16] These renewable materials often serve as direct drop-in biobased replacements for existing petrochemically-derived materials. Beyond drop-in replacements is the concept of "bioprivileged molecules," which are those that are biologically derived by minimal conversion steps, with high atom economy, and that cannot be effectively accessed from petrochemical feedstocks.[17,18] These molecules can either be direct replacements or new chemical species with advantageous new product properties.[19,20] Despite several examples of commercial success and numerous promising academic reports, the industry has been hesitant to adopt biobased products in greater volumes. Much of this is due to performance and cost issues, real or perceived, that accompany biobased materials. Natural fats and oils have poor oxidation stability,[21,22] a characteristic that transfers to many of the products derived from them. In addition, in some cases, the performance of renewable materials is lower than that of the conventional products they are intended to replace. Furthermore, issues such as the need for new processing technologies, feedstock variability, price, availability, and the food *versus* fuel debate leave a limited pool of willing industrial adopters, even with favorable life cycle assessments[23] and BioPreferred programs.[24] This chapter will review a few cases where the resulting biobased materials will be attractive to industries willing to incorporate renewable components into their product portfolios, thereby increasing their likelihood of industrial adoption and commercial success. Such an outcome would not only benefit consumers looking to use biobased products, but also the agricultural sector by providing expanded markets for biologically derived materials.

8.1.1 Plant Oil Production and Availability

The emergence of food and industrial applications has resulted in a dramatic increase in worldwide production of fats and oils, which have more than tripled since 1985 when production was only 68.2 MMT.[16] Concomitant with the increase in lipid production and the corresponding industrial uses of fatty acids and their derivatives, especially biodiesel, is a rise in glycerol as a coproduct and its utilization as a building block for renewable chemicals.[15,25–28] Worldwide production of vegetable oils (Figure 8.1, left) from October 2021 to September 2022 was 212 MMT, with a similar or slightly higher amount of 219 MMT projected for 2013. Palm (75.9 MMT), soybean

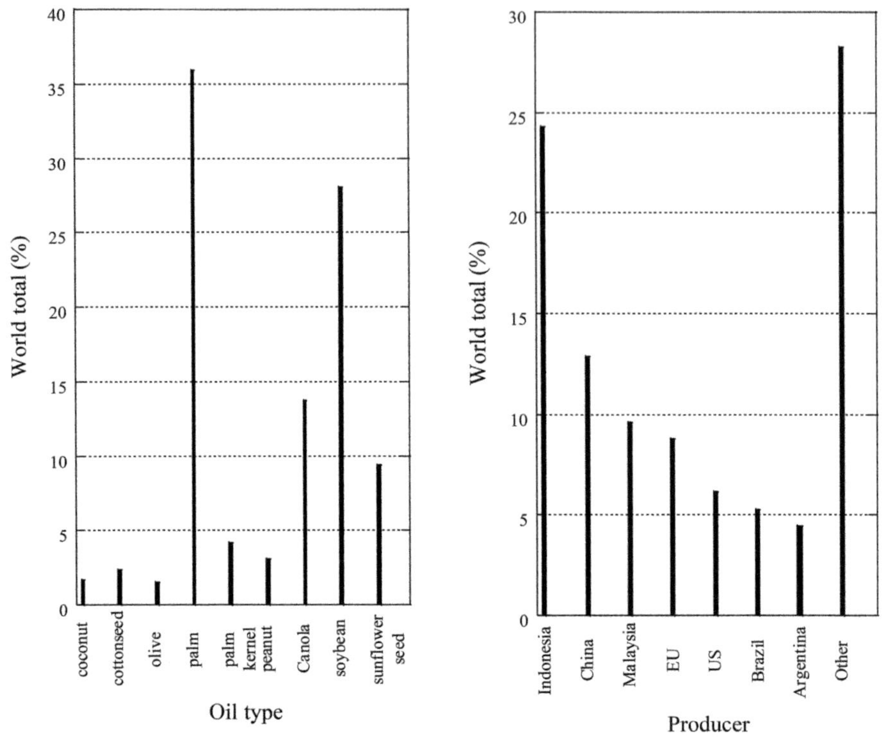

Figure 8.1 Worldwide lipid production in 2022 by percentage of total oil type and by producer.

(59.4 MMT), canola (29.0 MMT), and sunflower (19.9 MMT) oils were the largest by weight, with palm kernel (8.9 MMT), peanut (6.5 MMT), cottonseed (5.0 MMT), coconut (3.6 MMT), and olive (3.3 MMT) oils comprising most of the remaining major commodity vegetable oils (Figure 8.1, right). Indonesia (51.6 MMT), China (27.3 MMT), Malaysia (20.4 MMT), the European Union (EU) (18.7 MMT), the United States (U.S.) (13.1 MMT), Argentina (9.4 MMT), and Brazil (11.5 MMT) are the most significant producers of these oils.[29]

Regional plant oil availability is dictated by geography and climate. Soybean oil is the most significant vegetable oil produced in the U.S. and is also the dominant oil from Brazil and Argentina. Palm and coconut oils are predominantly produced in tropical countries such as Malaysia and Indonesia, whereas canola oil is the major oil produced in the EU and Canada.[29] Sunflower and olive oils represent additional commercially significant lipids produced in the EU.[10] In the majority of lipids, 68% are utilized in food applications, 13% are converted to biofuels, and the remaining 19% serve as feedstocks for chemicals, polymers, and other industrial products.[15]

8.2 Removal of Oxygen from Natural Carboxylic Acids

The three main approaches to preparing hydrocarbons from carboxylic acids that have received significant attention are hydrogenation, decarbonylation, and decarboxylation (Figure 8.2). All are thermodynamically favorable[30,31] and produce the desired hydrocarbon with carbon dioxide and/or carbon monoxide and/or water as elimination byproducts.

Hydrogenation of carboxylic acids and esters (Figure 8.2, top) is a straightforward and well-studied approach.[1,32] It is capable of producing fully saturated hydrocarbons, which can be useful in some applications. However, this method requires significant quantities of hydrogen, and when utilized with larger fatty acids, such as those from soybean oil, its fully saturated products have fairly limited applications.

Decarbonylation (Figure 8.2, middle) produces an unsaturated hydrocarbon as well as carbon monoxide through a mechanism that involves the coordination of a carboxylate moiety to the metal center of a catalyst in an oxidative addition. The first observation of this reaction, catalyzed by ruthenium and osmium, was made approximately 50 years ago.[33] Methodologies using palladium or rhodium catalysts to decarbonylate acids of 8–12 carbons were later developed,[34,35] and extension to biobased fatty acids was finally made.[30,36–44] There are some drawbacks to this technology. First, for efficacy, the mechanism requires either the addition of or the *in situ* production of a hydrogen acceptor such as an anhydride. This disadvantage is best demonstrated by the example of palladium-catalyzed decarbonylation of decanoic acid.[35] Turnovers of up to 12 370 were observed when equimolar acetic anhydride was added, whereas less than five turnovers were noted without anhydride.

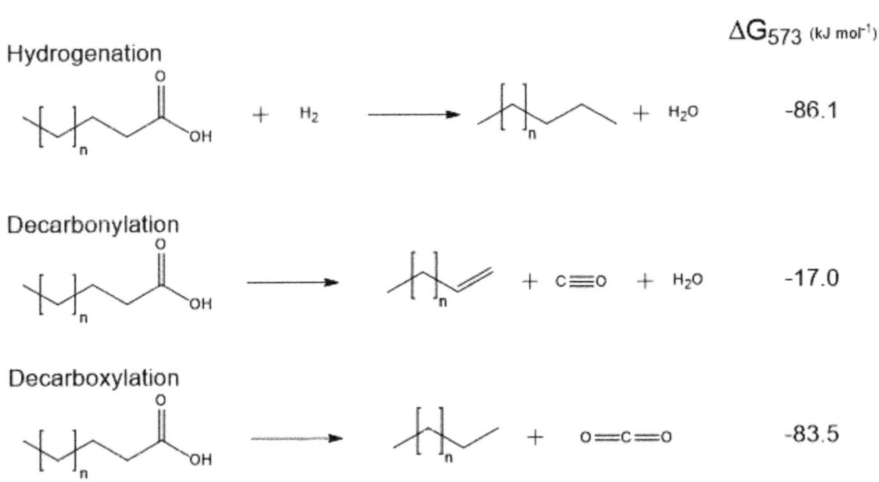

Figure 8.2 Thermodynamic data of hydrogenation, decarboxylation, and decarbonylation.

Decarboxylation (Figure 8.2, bottom) is arguably the most versatile of these three technologies. It may produce either saturated or unsaturated hydrocarbons along with carbon dioxide as a byproduct. Thermal decarboxylation begins with simple oxide materials[45] on various branched[46] and cyclic[47] compounds. Several zeolites are also catalytically active[48–51] at temperatures ranging from 370–440 °C. This technology is effective at producing hydrocarbon fuels.[52–55] Addition of a metal catalyst that works in an oxidative mode is an approach that has gained attention. Silver in its +2 oxidation state initially coordinates to a fatty acid, which then undergoes an electron transfer that leads to a reduced silver species, a proton, and a carbonyl radical. Under most conditions, the carbonyl radical is unstable and will spontaneously decarboxylate to yield a hydrocarbon.[56] A copper co-catalyst is needed if an unsaturated hydrocarbon is desired.[57] Although this reaction works under mild conditions, it leaves silver and/or copper in an inactive oxidative state.

Regeneration of the catalysts, therefore, requires an oxidant such as sodium persulfate that increases cost. In addition, the difficulty with controlling radical reactions is sufficient to limit the industrial appeal of this method. Another method to reduce the reaction barrier to decarboxylation, without the need for sodium persulfate, is manipulation of the substrate's isomeric structure. Specifically, unsaturated carboxylic acids in which the double bond is conjugated or nearly conjugated with the carboxylate moiety exhibit significant enhancement in the rate of decarboxylation relative to other isomers.[46] Therefore, decarboxylation can be conducted at considerably lower temperatures when the carboxylic acid moiety is in proximity to a double bond (Figure 8.3). Catalysts that specifically take advantage of this process will be discussed in Sections 8.4 and 8.5.

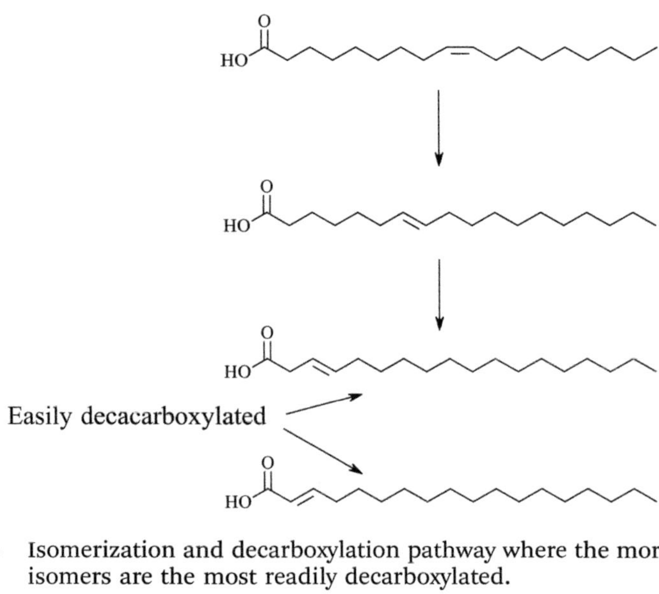

Figure 8.3 Isomerization and decarboxylation pathway where the more proximate isomers are the most readily decarboxylated.

8.3 Chemicals from Decarboxylation of Natural Carboxylic Acids

Biobased alternatives to petroleum-derived products have been an important goal for many years. Although progress toward a biobased economy is evident in a few areas, there is still a lack of successful product examples in the private sector. The newly developed decarboxylation technology may be the key to the conversion of natural carboxylic acids into direct drop-in biobased replacements for nonrenewable commodity chemicals. These include polymer precursors, olefins of varying chain lengths, new lubricants, and renewable aviation fuel that mimics the composition of the corresponding petroleum-derived jet fuel.

Heating plant oils to temperatures above 300 °C results in indiscriminate isomerization,[58] but controlled isomerization is needed to obtain well-defined industrial products *via* the route described in the previous section (Figure 8.3). Iodine,[59] enzymes,[60] and organometallic complexes using complicated phosphine ligands[61–65] isomerize double bonds in plant oil-based materials. Transition metal carbonyls and related complexes are especially effective. In one route, iron carbonyl was used to isomerize fatty acids,[66] but it is difficult to work with due to its volatility and toxicity.

8.4 Ruthenium-based Decarboxylation

In a relatively new development, triruthenium dodecacarbonyl [$Ru_3(CO)_{12}$] will spontaneously form a lower temperature isomerization catalyst when reacted with a carboxylic acid.[67] This precursor facilitates isomerization of alkenes, both with[68] and without[69] added phosphine ligands, and can decarboxylate formic acid.[33,70] Thus, isomerization of a double bond of an unsaturated fatty acid into a position where subsequent decarboxylation is more readily achievable is possible using $Ru_3(CO)_{12}$.

In a series of recent publications,[71,72] $Ru_3(CO)_{12}$-mediated decarboxylation has been demonstrated (Table 8.1) on a variety of substrates. Conversion to alkenes was observed using 9-*cis*-octadecenoic acid (oleic acid), *trans*-cinnamic acid, and 10-undecenoic acid. Catalyst loadings of 0.1 wt% were employed and turnover frequencies of $\sim$100 $h^{-1}\,Ru^{-1}$ were observed.[71] Osmium carbonyl ($Os_3(CO)_{12}$) displayed similar reactivity to $Ru_3(CO)_{12}$, but with a slightly lower reaction rate.[73]

8.4.1 Other Reactions Observed in the Ruthenium System

Follow-up studies[74,75] were conducted, which took a closer look at the minor products of ruthenium-catalyzed decarboxylation. It was observed (Figure 8.4) that, in addition to the expected series of alkene isomers, other minor reactions were occurring. Materials of both more and less unsaturation were observed in approximately equal amounts. Specifically, hydrogenation of some of the alkene products or mono-unsaturated substrates

Table 8.1 The yields and products from the decarboxylation-isomerization of unsaturated substrates and related isomerization only of ester and alkene substrates.

Substrate	Conditions	Conversion to alkene (GC %)	Turnover frequency (h^{-1} Ru^{-1})	Observed product
trans-Cinnamic acid	No precatalyst, 200 °C, 4 hours	1	NA	Styrene
	0.11 wt% $Ru_3(CO)_{12}$, 200 °C, 4 hours	29	93	
	10.0 wt% $Ru_3(CO)_{12}$, 200 °C, 4 hours	74	3	
9-*cis*-Octadecenoic acid	No precatalyst, 250 °C, 24 hours	0	NA	Heptadecene, multiple isomers
	0.10 wt% $Ru_3(CO)_{12}$, 250 °C, 4 hours	20	39	
	0.10 wt% $Ru_3(CO)_{12}$, 250 °C, 24 hours	36	12	
10-Undecenoic acid	0.89 wt% $Ru_3(CO)_{12}$, 250 °C, 4 hours	60 (GC–MS)	20	Decene, Multiple isomers
Methyl 10-undecenoate	0.10 wt% $[Ru(CO)_2(EtCO_2)]_n$, 250 °C, 4 hours	NA	NA	Isomerized ester (apparent steady state of isomers)
	0.14 wt% $[Ru(CO)_2(EtCO_2)]_n$, 150 °C, 4 hours	NA	NA	Starting material (minor amount of isomerization)
Methyl 9-*cis*-octadecenoate	0.13 wt% $[Ru(CO)_2(EtCO_2)]_n$, 250 °C, 4 hours	NA	NA	Isomerized ester (apparent steady state of isomers)
Methyl 9-*cis*-octadecenoate	0.12 wt% $[Ru(CO)_2(EtCO_2)]_n$, 150 °C, 4 hours	NA	NA	Starting material (minor amount of isomerization)
7-*trans*-Tetradecene	0.015 wt% $[Ru(CO)_2(EtCO_2)]_n$, 150 °C, 4 hours	NA	NA	Isomerized alkenes (apparent steady state of isomers)

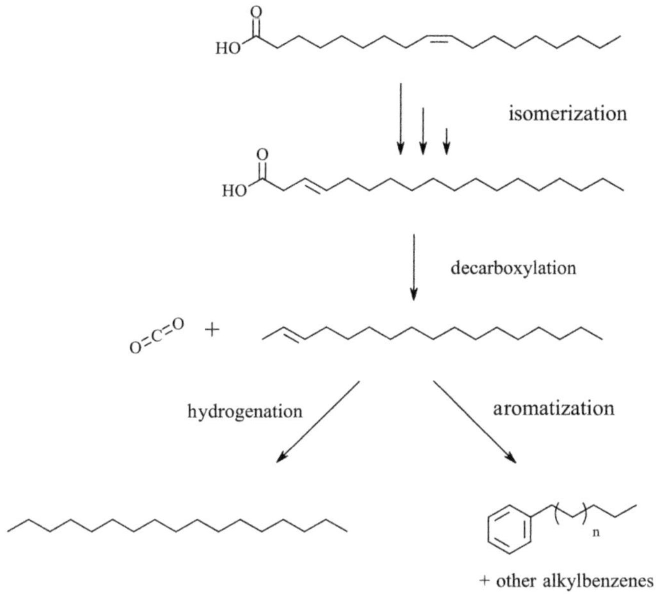

Figure 8.4 Reaction scheme accounting for the observed decarboxylation, hydrogenation, and aromatization products observed in the catalytic decarboxylation of oleic acid.

was occurring, producing saturated materials, concurrent with the production of polyunsaturated products. These polyunsaturated compounds then underwent cyclization/aromatization reactions to produce alkyl-substituted benzenes.

8.5 Iridium-based Decarboxylation and Aromatization Reactions for Aviation Fuel

Production of aromatics during decarboxylation of fatty acids using transition metal catalysts is of interest for renewable aviation fuel applications. The synthesis of biobased hydrocarbons suitable as jet fuel has been an especially high-profile research topic in recent years.[76–78] Although significant progress has been made, aromatics content is an important consideration for potential aviation fuels. Aromatics are beneficial because they alter fuel density and, more importantly, are proven to reduce shrinkage in elastomeric seals and reduce seal failure. Literature reports[79] demonstrate that a minimum aromatic content of 8% by volume is needed in aviation fuels to address density and seal shrinkage issues, with a best-case operability of 14% under some conditions.[79] However, if the content of aromatics is too high, these same seals will over-soften, which may also result in failure. According to ASTM D1655, the American aviation fuel specification, a maximum limit of 25% is specified to maintain other desirable fuel properties, with a further requirement that naphthalenes be restricted to only 3%.[80]

Conventional aviation fuels contain approximately 16% aromatics, which limits the blending ratio of any non-aromatic containing component to 50%.[76–78] This effectively caps the amount of biobased aliphatic fuels made via earlier processes that can be blended with conventional aviation fuels. However, iridium-based decarboxylation of fatty acids produces a mixture of renewable aliphatic and aromatic hydrocarbons, thereby increasing the permissible blend ratio beyond 50%. Consequently, higher volumes of renewable aviation fuel can be incorporated into existing aviation fuel schemes, thereby increasing markets for vegetable oil-based materials and enhancing agricultural utilization while simultaneously reducing consumption of petroleum-derived materials. Oleic acid and other monounsaturated fatty acids must be dehydrogenated to form the desired aromatic species needed for the American aviation fuel specification ASTM D1655 (Table 8.2).

While ruthenium is strongly active in catalyzing decarboxylation and isomerization, it is not considered highly active for hydrogenation or dehydrogenation reactions which are needed for the production of aromatics. However, iridium, combined with appropriate ligands, has been shown to dehydrogenate many carbon-based substrates,[81,82] specifically for the production of polyunsaturated compounds.[83] The key to increasing the frequency of these side

Table 8.2 Selected specifications for jet fuels from ASTM D-1655-12. Specifications related to aromatic content are printed in bold print.

Property		ASTM method	
Composition			
Acidity	mg KOH per g	0.10 max	D3242
Aromatics	**vol%**	**25 max**	**D1319**
Aromatics	**vol%**	**26.5 max**	**D6379**
Sulfur, mercaptan	mass%	0.003	D3227
Sulfur, total	mass%	0.30	D1266, D2622, D4294, or D5453
Volatility			
Distillation	10% °C	205	
Final boiling point	°C	300 max	
Flash point	°C	38	D56 or D3828
Density	15 °C $kg\,m^{-3}$	775–840	D1298 or D4052
Fluidity			
Freezing point	°C	−40 (Jet A) −47 (Jet A-1)	D5972, D7153, D7154, or D2386
Viscosity	−20 °C $mm^2\,s^{-1}$	8.0	D445
Combustion			
Net heat of combustion	$MJ\,kg^{-1}$	42.8	D4529, D3338, or D4809
Naphthalenes[a]	**vol%**	**3.0 max**	**D1840**

[a] Can alternatively meet the higher smoke point specification tested by D1322.

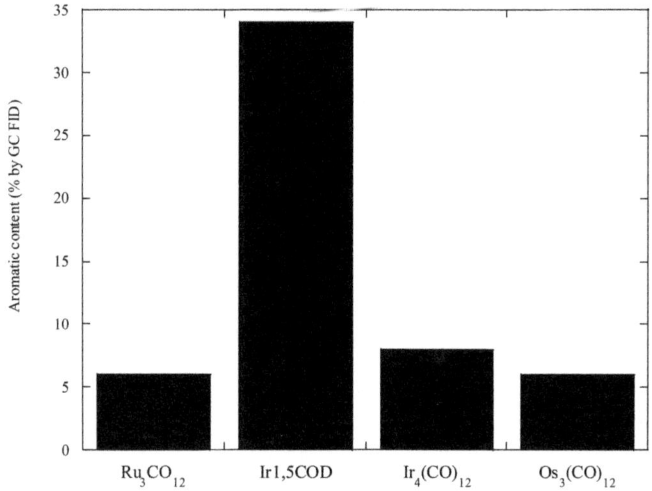

Figure 8.5 A comparison of the observed aromatic content achieved using ruthenium, osmium, and iridium catalysts, in the decarboxylation of oleic acid, and 1 wt% catalysts at 250 °C, and 24 hours. Conversions were high: $Ru_3(CO)_{12}$ - 92%; $(C_8H_{12}IrCl_2)_2$ - 96%; $Ir_4(CO)_{12}$ - 84%; and $Os_3(CO)_{12}$ - 54%.

reactions, while preserving activity, was in switching from ruthenium to iridium (Figure 8.5). A variety of catalysts were studied and quantified for both reactivity and selectivity. While platinum and palladium did not give significant advantages over ruthenium, iridium did. Utilizing the carbonyl compound of iridium, $Ir_4(CO)_{12}$, the aromatic content was increased from ~7 to ~10%. However, an even further increase was observed by also changing the ligand on the catalyst.[84,85] A maximum aromatic content of 34.9% was achieved using catalytic $(C_8H_{12}IrCl_2)_2$.

8.6 Methacrylic Acid from Selective Decarboxylation of Itaconic Acid

The classic route to methacrylic acid begins with the condensation of acetone with hydrocyanic acid to give an acetone cyanohydrin intermediate (Figure 8.6). Dehydration and hydrolysis of the nitrile group with a stoichiometric excess of concentrated sulfuric acid at 140 °C provide methacrylic acid along with an ammonium bisulfate byproduct that is produced in a ratio of 1.5 : 1 relative to methacrylic acid.[86,87] Both of these routes are environmentally unfriendly because they use nonrenewable starting materials, utilize stoichiometric amounts of harmful reagents, generate toxic intermediates, and, in the case of methacrylic acid, produce an excess of low-value ammonium bisulfate byproduct. Consequently, biobased routes to methacrylic acid are of interest to obviate the deficiencies of conventional petrochemical production methods.

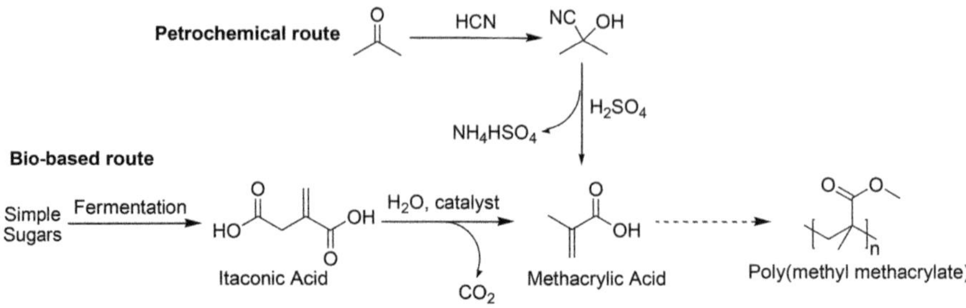

Figure 8.6 Biobased methacrylic acid from itaconic acid along with the conventional petrochemical route from acetone.

Only a few biobased routes to methacrylic acid have been reported. One such approach is liquid-phase dehydration and decarboxylation of citramalic acid at 250–400 °C and 450–4000 psi, in the presence of basic and/or acidic catalysts.[88] Application of these methods on maleic acid yields acrylic acid.[88] Another approach relies on biomass-derived ethene, carbon monoxide, and methanol in a two-stage procedure referred to as the "Alpha Process" to yield methyl methacrylate.[89] The first stage entails catalytic carbonylation of ethene in the presence of carbon monoxide to yield methyl propionate. In the second stage, methyl propionate is converted to methyl methacrylate using formaldehyde.[90] However, such a methodology requires considerable processing and purification of the biomass source to obtain the necessary chemical feedstocks, and the processing steps themselves consume considerable quantities of fossil fuels.

The most promising new method for this process involves conversion of citric acid to itaconic acid, followed by decarboxylation of itaconic acid in the presence of bases. Methacrylic acid is formed under a variety of conditions.[91] However, appreciable accumulation of byproducts, such as 2-hydroxy-butyric and crotonic acids, was noted. In addition to low selectivity, high temperatures >350 °C are required and yields are relatively low (70% or lower). In an optimized supercritical water embodiment, selectivity was improved to over 90% and reaction temperatures were lowered slightly, but relatively high temperatures (245–270 °C) and pressures (450–3000 psi) were still needed. Stoichiometric amounts of base catalysts were used, and byproducts such as crotonic and 2-hydroxybutyric acids were still produced along with propene.[92]

In other studies,[93,94] manganese oxalate was used, along with a triphenylphosphine additive, in order to perform the transformation of itaconic acid into methacrylic acid at 14% yield. In the second part of the study, the well-known ruthenium decarboxylation catalyst discussed in the previous section, $[Ru(CO)_2(CH_3CH_2COO)]_n$, was employed for the same reaction, but without the triphenylphosphine additive. After collection by distillation, isolated yields of up to 33.8% were reported. One of the keys to the reaction was the utilization of 400 psi of nitrogen headspace pressure, utilized in both systems.

Overall, these studies show that it is possible to overcome the difficulties in selectively decarboxylating a natural sugar. Such a process benefits the producers of acrylate and methacrylate polymers as well as itaconic acid manufacturers and farmers who produce the sugars necessary for fermentative production of itaconic acid from citric acid.

8.7 Conclusions

The synthesis of renewable hydrocarbons from biobased resources is an important research priority. By use of decarboxylation chemistry, new synthetic routes are realized for the production of biobased lubricants, feedstocks, aviation fuels, and polymer precursors. Efficient catalysts that enable these reactions from both vegetable oil-based and carbohydrate-based starting materials have been identified.

An effective decarboxylation system utilizing ruthenium was discovered, which can produce a desirable mixture of alkene isomers. The technology is effective, requires no oxidation co-reactants, and has relatively high total turnovers.

By changing the metal used in the catalyst from ruthenium to iridium, an appropriate level of important aromatic molecules can also be added to the product mixtures for aviation fuel applications. These aromatics give fuel of appropriate density and, more importantly, they reduce shrinkage in elastomeric seals and stop seal failure.

The key to using decarboxylation technology on carbohydrates, specifically from natural itaconic acid, has been uncovered. Two different technologies were reviewed here. One of them utilized ruthenium while the other employed the even more common manganese. Utilizing these catalysts, methacrylic acid, suitable for a variety of high-performance polymeric materials, was produced.

All these new technologies have been studied under a variety of conditions and compared well with previously reported methods. This research will benefit the producers of vegetable oil and carbohydrate feedstocks as well as the polymer and renewable fuel industries.

Acknowledgements

This work was supported in part by the U.S. Department of Agriculture, Agricultural Research Service.

References

1. A. Corma, S. Iborra and A. Velty, *Chem. Rev.*, 2007, **107**, 2411.
2. G. W. Huber, S. Iborra and A. Corma, *Chem. Rev.*, 2006, **106**, 4044.
3. P. Gallezot, *Chem. Soc. Rev.*, 2012, **41**, 1538.
4. M. Besson, P. Gallezot and C. Pinel, *Chem. Rev.*, 2014, **114**, 1827.
5. N. J. Fox and G. W. Stachowiak, *Proc. Int.*, 2007, **40**, 1035.
6. P. Nagendramma and S. Kaul, *Renewable Sustainable Energy Rev.*, 2012, **16**, 764.

7. A. Campanella, E. Rustoy, A Baldessari and M. A. Baltanas, *Bioresour. Technol.*, 2010, **101**, 245.
8. T. M. Panchal, A. Patel, D. D. Chauhan, M. Thomas and J. V. A. Patel, *Renewable Sustainable Energy Rev.*, 2017, **70**, 65.
9. T. P. Durrett, C. Benning and J. Ohlrogge, *Plant J.*, 2008, **54**, 593.
10. B. R. Moser, *In Vitro Cell. Dev. Biol.: Plant*, 2009, **45**, 229.
11. T. Issariyakul and A. K. Dalai, *Renewable Sustainable Energy Rev.*, 2014, **31**, 446.
12. M. R. Avhad and J. M. Marchetti, *Renewable Sustainable Energy Rev.*, 2015, **50**, 696.
13. U. Biermann, U. Bornscheuer, M. A. R. Meier, J. O. Metzger, J. O. Metzger and H. J. Schafer, *Angew. Chem., Int. Ed.*, 2011, **50**, 3854.
14. A. Llevot, P. K. Dannecker, M. von Czapiewski, L. C. Over, Z. Soyler and M. A. R. Meier, *Chem. - Eur. J.*, 2016, **22**, 11510.
15. A. Behr and J. P. Gomes, *Eur. J. Lipid Sci. Technol.*, 2010, **112**, 31.
16. H. Baumann, M. Buhler, H. Fochem, F. Hirsinger, H. Zoebelein and J. Falbe, *Angew. Chem., Int. Ed.*, 1988, **27**, 41.
17. B. H. Shanks and P. L. Keeling, *Green Chem.*, 2017, **19**, 3177.
18. N. Hernandez, R. C. Williams and E. W. Cochran, *Org. Biomol. Chem.*, 2014, **12**, 2834.
19. A. LLevot and M. A. R. Meier, *Green Chem.*, 2016, **18**, 4800.
20. P. Anastas and N. Eghbali, *Chem. Soc. Rev.*, 2010, **39**, 301.
21. N. A. Porter, B. A. Weber, H. Weenen and J. A. Khan, *J. Am. Chem. Soc.*, 1980, **102**, 5597.
22. J. P. Cosgrove, D. F. Church and W. A. Pryor, *Lipids*, 1987, **22**, 299.
23. S. A. Miller, A. E. Landis, T. L. Theis and R. L. Reich, *Environ. Sci. Technol.*, 2007, **41**, 4143.
24. Federal Register, 2005, 70, 1792.
25. C. H. Zhou, J. N. Beltramini, Y. X. Fan and G. Q. Lu, *Chem. Soc. Rev.*, 2008, **37**, 527.
26. F. Yang, M. A. Hanna and R. Sun, *Biotechnol. Biofuels*, 2012, **5**, 13.
27. Y. Zheng, X. Chen and Y. Shen, *Chem. Rev.*, 2008, **108**, 5253.
28. M. Pagliaro, R. Ciriminna, H. Kimura, M. Rossi and C. Della Pina, *Eur. J. Lipid Sci. Technol.*, 2009, **111**, 788.
29. United States Department of Agriculture. Foreign Agricultural Service. Oilseeds: World Markets and Trade, https://apps.fas.usda.gov/psdonline/circulars/oilseeds.pdf, October 2022.
30. M. Snåre, I. Kubičková, P. Mäki-Arvela, K. Eränen and D. Y. Murzin, *Ind. Eng. Chem. Res.*, 2006, **45**, 5708.
31. M. Czaun, A. Goeppert, R. May, R. Haiges, G. K. S. Prakash and G. A. Olah, *ChemSusChem*, 2011, **4**, 1241.
32. J. Fu, X. Lu and P. E. Savage, *ChemSusChem*, 2011, **4**, 481.
33. R. H. Prince and K. A. Raspin, *Chem. Commun.*, 1966, 156.
34. D. M. Fenton, *US Pat.*, 3530198, 1970.
35. J. A. Miller, J. A. Nelson and M. P. Byrne, *J. Org. Chem.*, 1993, **58**, 18.
36. T. A. Foglia and P. A. Barr, *J. Am. Oil Chem. Soc.*, 1976, **53**, 737.

37. G. A. Kraus, *US Pat. Appl.*, 2013/0102823, 2013.
38. G. A. Kraus and S. Riley, *Synthesis*, 2012, **44**, 3003.
39. J. G. Immer, M. J. Kelly and H. H. Lamb, *Appl. Catal., A*, 2010, **375**, 134.
40. Y. Liu, K. E. Kim, M. B. Herbert, A. Fedorov, R. H. Grubbs and B. M. Stoltz, *Adv. Synth. Catal.*, 2014, **356**, 130.
41. P. Maki-Arvela, I. Kubickova, M. Snare, K. Eranen and D. Y. Murzin, *Energy Fuels*, 2006, **21**, 30.
42. M. O. Miranda, A. Pietrangelo, M. A. Hillmyer and W. B. Tolman, *Green Chem.*, 2012, **14**, 490.
43. M. Snare, I. Kubickova, P. Maki-Arvela, D. Chichova, K. Eranen and D. Y. Murzin, *Fuel*, 2008, **87**, 933.
44. L. J. Goossen and N. Rodriguez, *Chem. Commun.*, 2004, 724.
45. H. Adkins, *J. Am. Chem. Soc.*, 1922, **44**, 2175.
46. R. T. Arnold, O. C. Elmer and R. M. Dodson, *J. Am. Chem. Soc.*, 1950, **72**, 4359.
47. R. T. Arnold and M. J. Danzig, *J. Am. Chem. Soc.*, 1957, **79**, 892.
48. Y. Takemura, A. Nakamura, H. Taguchi and K. Ouchi, *Ind. Eng. Chem. Prod. Res. Dev.*, 1985, **24**, 213.
49. J. J. Scheibel, and R. T. Reilman, *US Pat. Appl.*, 2007/0281875, 2007.
50. A. Berenblyum, R. Shamsiev, T. Podoplelova and V. Danyushevsky, *Russ. J. Phys. Chem. A*, 2012, **86**, 1199.
51. A. S. Berenblyum, T. A. Podoplelova, R. S. Shamsiev, E. A. Katsman and V. Y. Danyushevsky, *Pet. Chem.*, 2011, **51**, 336.
52. J. Dupont, P. A. Z. Suarez, M. R. Meneghetti and S. M. P. Meneghetti, *Energy Environ. Sci.*, 2009, **2**, 1265.
53. S. Bezergianni, A. Kalogianni and I. A. Vasalos, *Bioresour. Technol.*, 2009, **100**, 3036.
54. W. Charusiri and T. Vitidsant, *Energy Fuels*, 2005, **19**, 1783.
55. S. Bezergianni and A. Kalogianni, *Bioresour. Technol.*, 2009, **100**, 3927.
56. M. Galia, L. M. de Espinosa, J. C. Ronda, G. Lligadas and V. Cadiz, *Eur. J. Lipid Sci. Technol.*, 2010, **112**, 87.
57. L. I. Placek, *J. Am. Oil Chem. Soc.*, 1963, **41**, 319.
58. J. C. Cowan, *Ind. Eng. Chem.*, 1949, **41**, 294.
59. N. Komesvarakul, M. D. Sanders, E. Szekeres, E. J. Acosta, J. F. Faller, T. Mentlik, L. B. Fisher, G. Nicoll, D. Sabatini and J. F. Scamehorn, *J. Cosmet. Sci.*, 2006, **57**, 309.
60. L. Alonso, E. P. Cuesta and S. E. Gilliland, *J. Dairy Sci.*, 2003, **86**, 1941.
61. R. Quirino and R. Larock, *J. Am. Oil Chem. Soc.*, 2012, **89**, 1113.
62. D. Grotjahn, *US Pat. Appl.*, 2009/0143585, 2009.
63. D. Grotjahn, *US Pat. Appl.*, 2010/0228031, 2010.
64. D. B. Grotjahn, C. R. Larsen, J. L. Gustafson, R. Nair and A. Sharma, *J. Am. Chem. Soc.*, 2007, **129**, 9592.
65. R. C. Larock, *Lipid Technol.*, 2003, **15**, 58.
66. R. J. Angelici, and K.-C. Shih, *US Pat.*, 5859268, 1999.
67. G. R. Crooks, R. R. G. Johnson, J. Lewis, I. G. Williams and G. Gamlen, *J. Chem. Soc. A*, 1969, 2761.

68. A. Salvini, P. Frediani and F. Piacenti, *J. Mol. Catal. A: Chem.*, 2000, **159**, 185.
69. A. Sivaramakrishna, P. Mushonga, I. L. Rogers, F. Zheng, R. J. Haines, E. Nordlander and J. R. Moss, *Polyhedron*, 2008, **27**, 1911.
70. Y. Gao, J. Kuncheria, G. P. A. Yap and R. J. Puddephatt, *Chem. Commun.*, 1998, 2365.
71. R. E. Murray, E. L. Walter and K. M. Doll, *ACS Catal.*, 2014, **4**, 3517.
72. R. E. Murray, K. M. Doll, and Z. Liu, *US Pat.*, 9868679, 2018.
73. R. E. Murray, K. M. Doll, and Z. Liu, *US Pat.*, 10457615, 2018.
74. B. R. Moser, G. Knothe, E. L. Walter, R. E. Murray, R. O. Dunn and K. M. Doll, *Energy Fuels*, 2016, **30**, 7443.
75. G. Knothe, K. R. Steidley, B. R. Moser and K. M. Doll, *ACS Omega*, 2017, **2**, 6473.
76. W. C. Wang and L. Tao, *Renewable Sustainable Energy Rev.*, 2016, **53**, 801.
77. R. Mawhood, E. Gazis, S. de Jong, R. Hoefnagels and R. Slade, *Biofuels, Bioprod. Biorefin.*, 2016, **10**, 462.
78. C. Gutierrez-Antonio, F. I. Gomez-Castro, J. A. de Lira-Flores and S. A. Hernandez, *Renewable Sustainable Energy Rev.*, 2017, **79**, 709.
79. J. Holladay, A. Abdullah, and J. Heyne, *U.S. Department of Energy Sustainable Aviation Fuels Technical Review*, 2020.
80. ASTM International, ASTM D1655-19, Standard specification for aviation turbine fuels, 2019.
81. S. Maetani, R. Fukuyama, N. Suzuki, D. Ishihara and I. Ryu, *Organometallics*, 2011, **30**, 1389.
82. K. G. Caulton, M. G. Thomas, B. A. Sosinsky and E. L. Muetterties, *Proc. Natl. Acad. Sci.*, 1976, **73**, 4374.
83. A. Ray, K. Zhu, Y. V. Kissin, A. E. Cherian, G. W. Coates and A. S. Goldman, *Chem. Commun.*, 2005, 3388.
84. K. M. Doll, B. R. Moser, and G. H. Knothe, *US Pat.*, 11186802, 2012.
85. K. M. Doll, B. R. Moser and G. Knothe, *Int. J. Sustainable Eng.*, 2021, **14**, 2018.
86. K. Sudesh, H. Abe and Y. Doi, *Prog. Polym. Sci.*, 2000, **25**, 1503.
87. P. H. Nguyen, S. Spoljaric and J. Seppala, *Polymer*, 2018, **153**, 183.
88. D. M. Alonso, S. G. Wettstein and J. A. Dumesic, *Chem. Soc. Rev.*, 2012, **41**, 8075.
89. B. Thomas, M. C. Raj, B. K. Athira, H. M. Rubiyah, J. Joy, A. Moores, G. L. Drisko and C. Sanchez, *Chem. Rev.*, 2018, **118**, 11575.
90. K. J. Edgar, C. M. Buchanan, J. S. Debenham, P. A. Rundquist, B. D. Seiler, M. C. Shelton and D. Tindall, *Prog. Polym. Sci.*, 2001, **26**, 1605.
91. A. Dondoni and A. Marra, *Chem. Soc. Rev.*, 2012, **41**, 573.
92. G. Lligadas, J. C. Ronda, M. Galia and V. Cadiz, *Biomacromolecules*, 2010, **11**, 2825.
93. B. R. Moser, J. C. Lansing, and R. E. Murray, *US Pat. Appl.*, 2018/0105481, 2018.
94. J. C. Lansing, R. E. Murray and B. R. Moser, *ACS Sustainable Chem. Eng.*, 2017, **5**, 3132.

CHAPTER 9

A Novel Application of Chemically Modified Eleostearic Acid Methyl Ester (EAME) for Improving Diesel Lubricity†

ZENGSHE LIU,*[a] JING LI,[a,b] GERHARD KNOTHE,[a] BRAJENDRA K. SHARMA[c] AND JIANGCHUNG JIANG[b]

[a] Bio-Oils Research Unit, National Center for Agricultural Utilization Research, Agricultural Research Service, U.S. Department of Agriculture, 1815 N. University Street, Peoria, Illinois 61604, USA; [b] Institute of Chemical Industry of Forestry Products, CAF, Nanjing, Jiangsu 210042, China; [c] Illinois Sustainable Technology Center, University of Illinois at Urbana-Champaign, 1 Hazelwood Drive, Champaign, Illinois 61820, USA
*Email: kevin.liu@usda.gov

9.1 Introduction

The use of vegetable oils for lubrication purposes has been in practice for many years. Due to the discovery of petroleum and the availability of low-cost oils, this idea was scrapped. However, there is new interest in the use of

† Mention of trade names or commercial products in this publication is solely for the purpose of providing specific information and does not imply recommendation or endorsement by the U.S. Department of Agriculture. USDA is an equal opportunity provider and employer.

Green Chemistry Series No. 83
Green Chemistry and Green Materials from Plant Oils and Natural Acids
Edited by Zengshe Liu and George Kraus

Published by the Royal Society of Chemistry, www.rsc.org

lubricants from vegetable oils due to growing concern over the environmental impact of fossil fuels. Bio-based lubricants are promising alternatives to mineral oils since they retain the technical specifications of conventional lubricants. Bio-based lubricants are biodegradable lubricants, derived from edible and non-edible vegetable oils, and they have high lubricity, viscosity index and flash point. Recently, the use of ultra-low-sulfur diesel (ULSD) fuels, as required by regulations in the United States, Europe and elsewhere, has led to the failure of diesel engine parts such as fuel injectors and pumps because they are lubricated by the fuel itself (1–10). The lubricity issue is significant, because to regain its lubricity, ULSD requires additives or blending with another material of sufficient lubricity.[1–11] Blending with another fuel increases the cost as, for example, the recent price of biodiesel is >$4 per gal compared to <$3 per gal for petrodiesel (in most locations in the United States). For example, improvement of lubricity by blending with biodiesel typically requires at least 1% (10 000 ppm), or even better 2% (20 000 ppm), of such fuel.[1] Therefore, economically it is better to enhance the lubricity of ULSD by adding additives at low additive levels. There is currently a need for better additives for ULSD fuels.

Methods that have been approved as standards for testing diesel fuel lubricity include the scuffing load ball-on-cylinder lubricity evaluator (SL-BOCLE) (ASTM D-6078) and the high-frequency reciprocating rig (HFRR) (ASTM D-6079, ISO 12156) lubricity tester. The high-frequency reciprocating rig (HFRR) (ASTM D-6079, ISO 12156) lubricity tester has been used for lubricity tests because this method is more user-friendly than other tests and can be pressurized to study the lubricity of fuels.[12] The HFRR test is a ball-on-disc method and the maximum acceptable wear scar on the ball is 460 μm (60 °C) as described in the European petrodiesel standard EN 590 and 520 μm (60 °C) as described in the American petrodiesel standard ASTM D-975. These standards are used to indicate fuels with sufficient lubricity for practical use in a diesel engine, whereas fuels generating wear scars above those limits may or may not be acceptable.

Vernicia fordii, usually known as the tung tree, is a species of *Vernicia* in the spurge family native to southern China, Burma and northern Vietnam. Tung tree is also grown in the southern and western parts of the United States. It is valued for tung oil, which is derived from the seeds of the tree. Tung oil, also called China wood oil, has traditionally been used as an ingredient in paint, varnish and caulk. It is also used as a wood finish for furniture and other wooden objects. Because it is rich in unsaturated fatty acids, tung oil is classified as a "drying oil", and when exposed to air in thin films, it is converted to a firm and nonsticky film. Eleostearic acid as a major fatty acid in tung oil shows increased functionality, which can be chemically modified. Here, we reported the synthesis of new compounds based on tung oil. These new compounds can serve as lubricity-improving additives for ULSD. Thus, as shown in Scheme 9.1, eleostearic acid methyl ester (EAME) from tung oil gave EAME/MA *via* a Diels–Alder reaction with maleic anhydride (MA) and subsequent esterification of the MA moiety with a short-chain

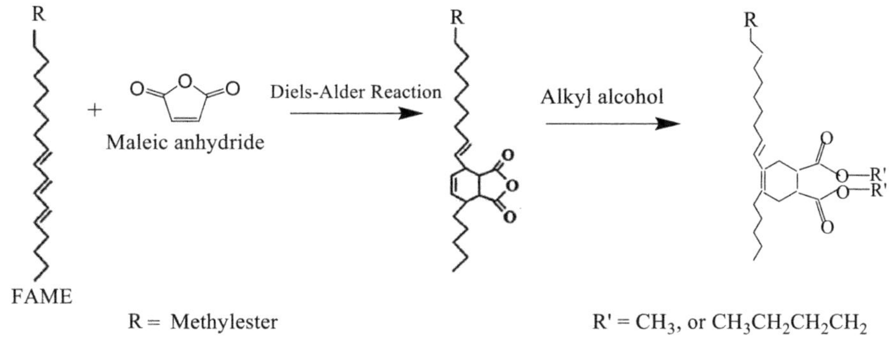

Scheme 9.1 The synthesis route of EAME/MA and its esters.

alcohol gave the compounds termed EAME/MA/ME and EAME/MA/BU. The lubricity of neat ULSD and petrodiesel fuels additized with EAME/MA/ME and EAME/MA/BU was evaluated using the HFRR lubricity test.

9.2 Characterization

9.2.1 FT-IR

FT-IR spectra were recorded on a Thermo Nicolet Nexus 470 FT-IR system (Madison, WI, USA) coupled with a Smart ARK accessory for liquid samples in the scanning range of 650–4000 cm^{-1} for 32 scans at a spectral resolution of 4 cm^{-1}.

9.2.2 NMR

^{1}H-NMR and ^{13}C-NMR spectra for extracted soluble substances from RPEO samples were recorded using a Bruker (Rheinstetten, Germany) ARX-500 NMR spectrometer operating at a frequency of 500 and 125 MHz, respectively.

9.3 Methods

9.3.1 Preparation of EAME from Tung Oil

Tung oil was purchased from Alnor Oil Company, Inc. (Valley Stream, NY, USA). Tung oil fatty acid esters (EAME) were prepared and purified in accordance with ref. 13. Typically, to a three-neck round-bottom flask fitted with a reflux condenser, thermometer and addition port, 50 g of tung oil and 12 mL of methanol solution were added and stirred at 60 °C and then 2.78 mL of sodium methoxide (25 wt% in methanol solution) was added and allowed to react for 1 h. After cooling the solution to room temperature, 50 mL of hexane was added. Using a separatory funnel, the lower layer methanol and glycerol were removed. The methyl esters were washed with deionized water (DI) three times or until the pH was near neutral. The solution was dried over magnesium sulfate and filtered. A rotary evaporator was used to remove hexane and any remaining methanol.

9.3.2 Preparation of EAME/MA and Esterification with Methanol and Butanol

To a 250 mL round-bottom flask equipped with a magnetic stirrer and reflux condenser, EAME (9.0 g, 30.8 mmol) and MA (3.17 g, 32.4 mmol) were added. The mixture was bubbled with N_2 for 5 min at room temperature and then stirred at 150 °C for 2 h under an N_2 atmosphere to give a clear, dark yellow and viscous liquid (EAME/MA). PTS (180 mg, 2 wt% to EAME/MA) and methanol (10.0 mL, 247.4 mmol, *i.e.*, molar ratio of methanol to EAME/MA 8 : 1) were added to the flask. The reaction mixture was stirred and heated to reflux for 3 h, and then excess methanol and produced water were removed through a vacuum-rotary evaporation procedure. 10.0 mL of fresh methanol was added, and the same operation was repeated twice. Eventually, the final product, methylated EAME/MA, was obtained through neutralization, water washing, separation and vacuum drying. The product was termed EAME/MA/ME. EAME/MA/BU was prepared in the same way using butanol instead of methanol at the same molar ratio.

9.3.3 Lubricity Determination

The scuffing load ball-on-cylinder lubricity evaluator (SL-BOCLE) (ASTMD-6078) and the high-frequency reciprocating rig (HFRR) (ASTM D-6079, ISO 12156) lubricity tester have been approved as standards for testing diesel fuel lubricity. The HFRR method was selected in a round-robin evaluation of several lubricity-testing methods for an ISO standard, because it has been included in the European petrodiesel standard EN 590, which utilizes the ISO 12156 method, and the American petrodiesel standard ASTM D-975. The HFRR method is more user-friendly and is also suitable for pressurization to study the lubricity of volatiles or fuels that are gases under ambient conditions. Lubricity determinations were performed at 60 °C (controlled to 1 °C), according to the standard method ASTM D-6079,[14] with an HFRR lubricity tester obtained from PCS Instruments (London, England) *via* Lazar Scientific (Granger, IN). Controlling the humidity to 30–50% is necessary for the HFRR test to give reproducible results[15] which was accomplished here, according to the standard[14] with a potassium carbonate bath (50% humidity). In addition to the usual wear scar data of the HFRR ball, we report the friction data which involve the coefficient of friction (16) and film data which involve the electrical resistance[16] recorded by the software during the experiments.

9.4 Results and Discussion

9.4.1 Characterization of EAMA/MA/ME and EAMA/MA/BU

To gain further knowledge of the chemical groups involved in the synthesis of ULSD additives, the chemically modified tung oil fatty acid methyl ester (EAME), the maleation product, EAMA/MA, and the esterification products, EAME/MAME and EAMA/MA/BU, were characterized using their FT-IR, ^{1}H-NMR and ^{13}C-NMR spectra as shown in Figures 9.1, 9.2 and 9.3, respectively.

The FT-IR spectra of tung oil, EAME, EAME/MA and EAME/MA/BU are illustrated in Figure 9.1. For EAME/MA, the strong peak observed at 1055 cm^{-1} was assigned to the double bond C–H bending of MA. The strong band at 993 cm^{-1} should be attributed to the conjugated double bonds of tung oil and EAME, but this band indicated the weak absorption of EAME/MA and EAME/MA/esters (named the bio-lubricant) due to the maleation reaction. The typical anhydride C=O stretching of MA was found at 1849 and 1776 cm^{-1} for EAME/MA. However, these two characteristic peaks disappeared after esterification, with the characteristic ester C=O stretching peak at 1720 cm^{-1} becoming strong as seen in the EAME/MA/BU ester spectra.

The compounds EAME, EAME/MA, EAME/MA/ME and EAME/MA/BU were characterized by ^{1}H-NMR and ^{13}C-NMR spectroscopy (shown in Figures 9.2 and 9.3). There is no peak at 7.1 ppm in the ^{1}H-NMR spectrum of EAMA/MA, indicating that no unreacted MA was left after purification. It is observed that the new peak at a chemical shift of 3.5 ppm represents the protons at the formed structures by the Diels–Alder reaction between MA and EAME. The chemical shifts at 5.3–6.4 ppm, which correspond to the protons on the

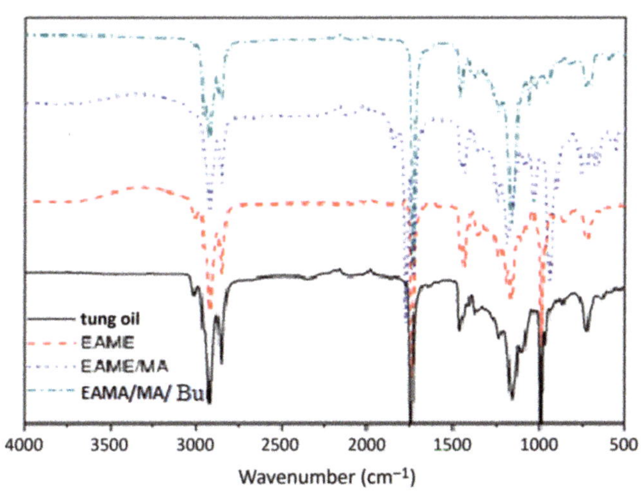

Figure 9.1 FT-IR spectra of tung oil, EAME, EAME/MA and its ester.

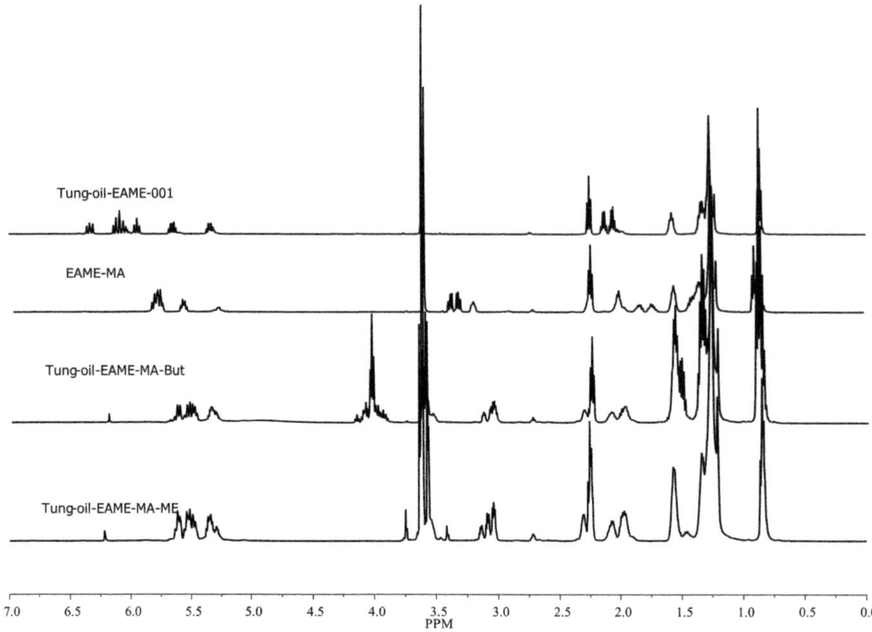

Figure 9.2 ^{1}H-NMR spectra of EAME, EAME/MA, EAME/MA/ME and EAME/MA/BU.

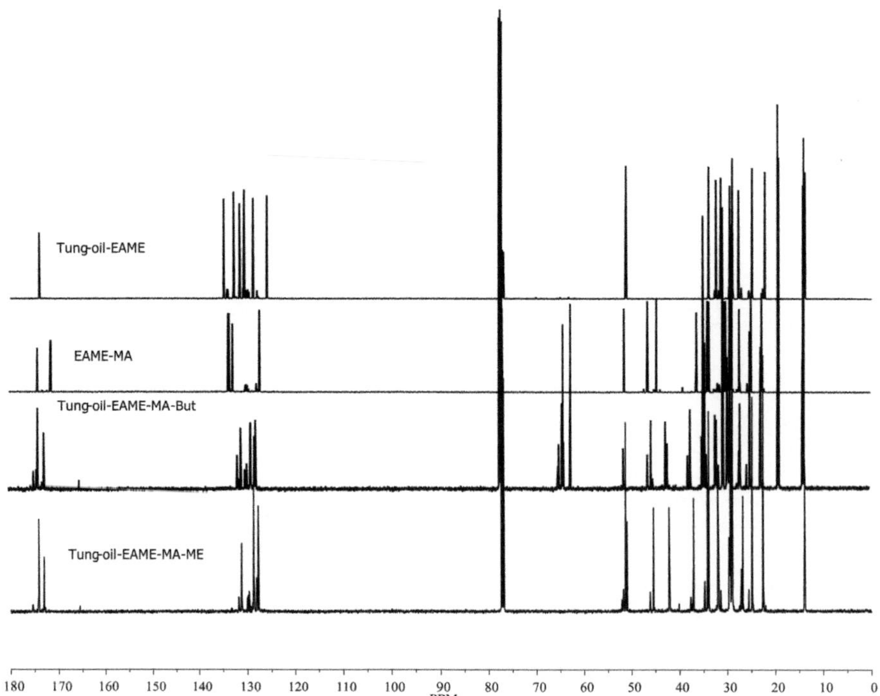

Figure 9.3 ^{13}C-NMR spectra of EAME, EAME/MA, EAME/MA/ME and EAME/MA/BU.

conjugated triene structures of EAME, decreased significantly because of the maleation reaction between MA and the C=C bonds on the EAME chain. In the ^{13}C-NMR spectrum (Figure 9.3), new band signals appeared at a chemical shift of 44–48 ppm representing the connection between MA and the EAME chain. The peaks around 126–136 ppm denoting the C=C bonds on EAME chains decreased significantly due to the Diels–Alder reaction. Finally, two peaks at chemical shifts around 172–174 ppm, which denote the carbonyl carbons on the attached anhydride groups, appeared for EAME/MA and disappeared for EAMA/MA esters, *i.e.*, EAME/MA/ME and EAME/MA/BU, because of the esterification with methanol and butanol. All the FT-IR and NMR results indicate that MA grafted onto EAME effectively and the esterification products were confirmed.

9.4.2 Lubricity

Lubricity was assessed using the ASTM D-6079 (HFRR) method at 60 °C. A variety of compounds, ULSD and petrodiesel base oils were studied including neat as well as with additives. Table 9.1 shows the wear scar values of the ULSD neat and ULSD with additives. Lubricity, as determined by the

Table 9.1 High-frequency reciprocating rig (HFRR) data of petrodiesel with additives.

HFRR (60 °C)	Wear scars (μm)					Results	
Sample ID	Ball *X*	Ball *Y*	Avg.	Disc *X*	Disc *Y*	Film (%)	Friction
ULSD neat	557	542	550	540	1318	9	0.448
	559	500	530	585	1272	10	0.408
EAME-MA-ME 100 ppm	515	458	487	535	1264	14	0.373
	555	485	520	564	1295	16	0.353
EAME-MA-ME 500 ppm	279	277	278	321	1117	83	0.343
	390	330	360[a]	407	1167	68	0.255
EAME-MA-ME 1000 ppm	371	278	325	400	1110	74	0.233
	377	281	329	389	1125	77	0.232[b]
EAME-MA-BU 100 ppm	513	451	482	514	1231	12	0.382
	559	484	522	546	1255	12	0.405
EAME-MA-BU 500 ppm	348	283	316	372	1130	71	0.242
	321	257	289[a]	318	1112	87	0.219
EAME-MA-BU 1000 ppm	304	227	266	312	1053	89	0.231
	337	242	290	372	1081	86	0.220[c]
EAME-MA-ME neat	98	83	91	80	1082	97	0.065
	116	99	108	85	968	93	0.066
EAME-MA-BU neat	252	157	205	296	1022	79	0.135
	243	142	193	270	1010	87	0.125

[a] Reduction by 40%.
[b] Reduction by 46%.
[c] Reduction by 47%.

HFRR test, would be evaluated using the lubricity specification in the petrodiesel standard ASTM D975. The low-lubricity ULSD used here exhibits very poor lubricity in the neat form (Table 9.1) with a wear scar value of 550 μm. The prescribed maximum wear scars are 460 μm in the EN 590 standard and

Table 9.2 High-frequency reciprocating rig (HFRR) data of polyalphaolefin (Durasyn 166 PAO) and Kendex 0150H GP1 with additives.

HFRR (60 °C)	Wear scars (μm)					Results	
Sample ID	Ball *X*	Ball *Y*	Avg.	Disc *X*	Disc *Y*	Film (%)	Friction
0150H GP1 Base Oil – neat	304	293	299	275	103	64	0.215
	301	284	293	288	1108	63	0.213
0150H GP1 Base Oil + 100 ppm EAME-MA-ME	313	269	291	296	1096	72	0.211
	309	272	291	302	1089	77	0.203
0150H GP1 Base Oil + 500 ppm EAME-MA-ME	308	259	284	331	1099	75	0.205
	298	217	258	321	1053	82	0.197
0150H GP1 Base Oil + 1000 ppm EAME-MA-ME	304	255	280 (redu.)	284	1089	81	0.204
	264	230	247 (11%)	266	1095	89	0.194
0150H GP1 Base Oil + 100 ppm EAME-MA-BU	301	253	277	276	1094	72	0.212
	330	286	308	303	1124	61	0.221
0150H GP1 Base Oil + 500 ppm EAME-MA-BU	316	259	288	319	1135	76	0.205
	274	256	265	270	1073	80	0.211
0150H GP1 Base Oil + 1000 ppm EAME-MA-BU	267	125	196 (redu.)	299	1025	92	0.195
	282	220	251 (25%)	301	1044	86	0.196
Durasyn 166 PAO – neat	320	290	305	281	1099	42	0.256
	312	281	297	294	1092	42	0.261
Durasyn 166 PAO + 100 ppm EAME-MA-ME	313	280	297	262	1108	64	0.217
	314	288	301	295	1110	47	0.246
Durasyn 166 PAO + 500 ppm EAME-MA-ME	319	277	298	303	1095	59	0.224
	317	297	307	303	1098	58	0.226
Durasyn 166 PAO + 1000 ppm EAME-MA-ME	259	131	195 (redu.)	262	1039	91	0.198
	257	206	232 (29%)	299	1036	90	0.189
Durasyn 166 PAO + 100 ppm EAME-MA-BU	305	271	288	269	1093	52	0.231
	317	263	290	270	1079	64	0.215
Durasyn 166 PAO + 500 ppm EAME-MA-BU	339	287	313	328	1130	69	0.216
	327	263	295	322	1102	68	0.216
Durasyn 166 PAO + 1000 ppm EAME-MA-BU	261	176	219 (redu.)	254	1025	94	0.193
	286	172	229 (26%)	285	1027	93	0.194

520 μm in the ASTM D-975 standard. A wear scar value of 550 μm is not acceptable. Neat samples of EAME-MA-ME and EAME-MA-BU (Table 9.1) showed excellent lubricity, as demonstrated by their low wear scar values, about 100 μm and 200 μm, respectively. These wear scar values are much smaller than 550 μm. These results prompted us to prepare samples in which, initially, 100 ppm, 500 ppm and 1000 ppm of EAME/MA/ME and EAME/MA/BU were added to the low-lubricity ULSD fuel. The effect of the lubricity additives on the ULSD is clearly visible. Both compounds EAME/MA/ME and EAME/MA/BU clearly perform extremely well. For example, adding the two compounds at 500 ppm appears to induce sufficient lubricity to the low-lubricity ULSD fuel. Addition of the two compounds reduced the wear scar value of ULSD by 40%. There appears to be little to no advantage in applying at a 1000 ppm level. On the other hand, for friction data, the 1000 ppm level performs a little better than the 500 ppm level, as it was reduced by 46% and 47%, respectively.

Lubricity for the other two types of diesel, Durasyn 166 PAO and Kendex 0150H GP1, with EAME/MA/ME and EAME/MA/BU as additives was also tested. Polyalphaolefin (Durasyn 166 PAO) with a specific gravity of 0.828 $g\,mL^{-1}$ (ASTM D 4052), kinematic viscosities at 40 and 100 °C of 31.13 and 5.91 cSt, respectively (ASTM D 445) and a pour point of −66 °C (ASTM D 97) was obtained from Ineos Oligomers (League City, TX, USA). Hydrotreated heavy paraffinic mineral oil (Kendex 0150H), a Group I base oil with a specific gravity of 0.864 $g\,mL^{-1}$ (ASTM D 4052), kinematic viscosities at 40 and 100 °C of 27 and 5.2 cSt, respectively (ASTM D 445), a pour point of <−9 °C (ASTM D 97) and a sulfur content of <300 ppm, ASTM D 5183, was received from American Refining Group (Bradford, PA, USA). As the HFRR data show in Table 9.2, with 1000 ppm of EAME-MA-BU, the wear scar values are reduced by 25% and 26% for 0150H GP1 Base Oil and Durasyn 166 PAO, respectively. However, EAME-MA-ME reduced it by 11.0% and 29.0% for 0150H GP1 Base Oil and Durasyn 166 PAO, respectively. EAME-MA-ME responded better for 166 PAO than for 150H GP1 Base Oil.

The present results not only show that EAME/MA/ME and EAME/MA/BU have excellent lubricity, as demonstrated by very low wear scar values, but also show that, as additives, both compounds effectively reduced the wear scar values of ULSD and other petrodiesels.

9.5 Conclusions

Tung oil-based methyl ester EAME and its maleation compound, *i.e.*, EAME-MA were synthesized by the transesterification of tung oil with methanol and by reacting with maleic anhydride *via* the Diels–Alder reaction. The formed cyclic anhydrides have been successfully reacted with short chain alcohols to form the esters, *i.e.*, EAME/MA/ME and EAME/MA/BU. EAME/MA/ME and EAME/MA/BU were characterized by FT-IR and NMR to confirm the structures of these compounds. The present results obtained with a high-frequency reciprocating rig (HFRR) lubricity tester give clear data that EAME/MA/ME and EAME/MA/BU at low additive levels (500–1000 ppm) significantly improved the lubricity of ULSD. Adding 500 ppm to low-lubricity ULSD decreased the wear scar and friction in the HFRR test by 40–47%, respectively. The additive concentrations are 20 and 40 times lower than blending biodiesel at 1–2%. EAME/MA/BU at a low additive level of 1000 ppm to 0150H GP1 and Durasyn 166 PAO base oils would reduce wear scar by 25% and 26%, respectively.

References

1. G. Knothe and K. R. Steidley, Lubricity of components of biodiesel and petrodiesel. The Origin of Biodiesel Lubricity, *Energy Fuels*, 2005, **19**, 1192–1200.
2. P. I. Lacey and S. R. Westbrook, Lubricity requirement of low sulfur diesel fuels, SAE Technical Paper Series, 1995, 950248.
3. D. Wei and H. A. Spikes, The lubricity of diesel fuels, *Wear*, 1986, **111**, 217–235.
4. P. I. Lacey and S. J. Lestz, Effect of low-lubricity fuels on diesel injection pumps – Part I: Field performance, SAE Technical Paper Series, 1992, 920823.
5. P. I. Lacey and S. J. Lestz, Effect of low-lubricity fuels on diesel injection pumps – Part II: Laboratory evaluation, SAE Technical Paper Series, 1992, 920824.
6. M. Nikanjam and P. T. Henderson, Lubricity of low sulfur diesel fuels, SAE Technical Paper Series, 1993, 932740.
7. J. C. Wang and D. J. Reynolds, The lubricity requirement of low sulfur diesel fuels, SAE Technical Paper Series, 1994, 942015.
8. R. F. Tucker, R. J. Stradling, P. E. Wolveridge, K. J. Rivers and A. Ubbens, The lubricity of deeply hydrogenated diesel fuels – The Swedish experience, SAE Technical Paper Series, 1994, 942016.
9. S. W. Wall, R. A. Grill and W. D. Byfleet, The no-harm performance of lubricity additives for low-sulfur diesel fuels, *Pet. Coal*, 1999, **41**, 38–42.
10. W. J. Dimitrakis, The importance of lubricity, *Hydrocarbon Eng.*, 2003, **8**, 37–39.
11. J. H. Van Gerpen, S. Soylu and M. E. Tat, Evaluation of the Lubricity of Soybean Oil-Based Additives in Diesel Fuel, in Proceedings of the 1999 ASAE/CSAE-SCGR Annual International Meeting, 1999, Paper No. 996134.
12. P. I. Lacey, D. W. Naegeli, J. L. De La Cruz and M. V. Whalen, Lubricity of volatile fuels for compression ignition engines, SAE Technical Paper Series, 2000, 2000-01-1804.
13. G. Knothe and R. O. Dunn, Biofuels derived from vegetable oils and fats, ed. F. D. Gunstone and R. J. Hamilton, in *Oleochemical Manufacture and Applications*, Sheffield Academic Press, Sheffield UK, 2001, pp. 106–163.

14. Standard Test Method for Evaluating Lubricity of Diesel Fuels by the High-Frequency Reciprocating Rig (HFRR), ASTM D-6079-99, *ASTM Annual Book of Standards*, American Society for Testing and Materials, West Conshohocken, PA, 1999.
15. M. Nikanjam and E. Burk, Diesel fuel lubricity additive study, SAE Technical Paper Series, 1994, 942014.
16. R. M. Crockett, M. P. Derendinger, P. L. Hug and S. Roos, Wear and electrical resistance on diesel lubricated surfaces undergoing reciprocating sliding, *Tribol. Lett.*, 2004, **16**, 187–194.

CHAPTER 10

Sustainable Lubricant Formulations from Natural Oils: A Short Review†

BRAJENDRA K. SHARMA,*[a] GOBINDA KARMAKAR,[b] RAJ SHAH,[c] PRANAB GHOSH,[b] MAJHER I. SARKER[a] AND SEVIM Z. ERHAN[a]

[a] United States Department of Agriculture, Agricultural Research Service Eastern Regional Research Center, Wyndmoor, PA 19038-8598, USA;
[b] Natural Product and Polymer Chemistry Laboratory, Department of Chemistry, University of North Bengal, Darjeeling 734013, India;
[c] Koehler Instrument Company, Holtsville, NY, USA
*Email: Brajendra.Sharma@usda.gov

10.1 Introduction

With increasing environmental consciousness, the use of ecofriendly lubricants prepared from natural oils has gained special attention due to the harmful impacts of petro-based lubricants. Vegetable oil (VO) based lubricants have several advantages compared to petroleum-based ones. VOs have higher biocompatibility, antifriction/antiwear properties, flash/fire points, and viscosity indexes. However, due to certain limitations, such as poor thermo-oxidative stability, poor hydrolytic stability, poor fluidity at low temperatures, high heterogeneity, and low abundance, their broad

† Mention of trade names or commercial products in this article is solely for the purpose of providing specific information and does not imply recommendation or endorsement by the U.S. Department of Agriculture. USDA is an equal opportunity provider and employer.

Green Chemistry Series No. 83
Green Chemistry and Green Materials from Plant Oils and Natural Acids
Edited by Zengshe Liu and George Kraus

Published by the Royal Society of Chemistry, www.rsc.org

application in the lubricant industry is restricted. The alternative option, *i.e.* petroleum-based lubricants, is readily available, leading to its dominance in the market. But, due to the rapid depletion of non-renewable resources and potential pollution problems from spills and total loss applications, VOs and natural oils, after overcoming their limitations by structural modifications, have come into focus recently in the formulation of biolubricants. VOs can be used for both boundary and hydrodynamic lubrication due to their amphiphilic character (polar ester group and non-polar hydrocarbon chain). The global biolubricant market size was 1.9 billion USD in 2021 and is set to surpass 2.3 billion USD by 2028, exhibiting a CAGR of 3.9% during the forecast period 2022–2028.[1] The reason for the fast growth of the biolubricant market is the biocompatibility of plant-based oils and the implementation of several environmental related regulations, like the Vessel General Permit (VGP) in the US, which is recently being replaced by the Vessel Incidental Discharge Act (VIDA), National Fire Protection Association (NFPA), Environment Protection Agency (EPA), and EcoLabel in Europe, which has made the use of biolubricants mandatory in shipping vessels.

VOs are classified as either food or non-food. Since food VOs are used in the food-chain, the use of non-food/waste cooking oils (WCO)/bio-oils from other natural resources became a more popular option for the majority of the lubricant industry, except for the food processing industry, for which food VOs are generally preferred. The use of waste-derived oils also addresses the problems associated with the waste management system. VOs, WCO, microalgae oils, and liquids from lignocellulosic biomass (by pyrolysis) can also be used as base fluids for lubricant formulations. However, the choice of bio-oil obviously depends on the specific area of lubricant application, as shown in Table 10.1. Additionally, Figure 10.1 shows the total lubricant consumption in different industrial sectors.

Although vegetable oil-based lubricants have great potential in the biolubricant industry, they are not widely commercialized due to certain limitations of vegetable oils, as mentioned before. Different chemical ways that improve the potentiality of VOs by mitigating their limitations typically involve reactions at the olefinic bonds, reactions with the polar ester ends, and blending of different chemical additives with the base oils. The most commonly applied chemical reactions to modify VOs as base stocks for biolubricant formulation are listed below:

(1) Transesterification
(2) Hydrolysis
(3) Epoxidation and opening of epoxy rings with different nucleophiles
(4) Cyclic carbonation and ring opening
(5) Partial hydrogenation
(6) Estolide formation/oligomerization
(7) Blending different transesterified or epoxidized vegetable oils/free fatty acids

Table 10.1 Application areas of different types of Vegetable oils.

VOs	Specific application areas
Palm oil	Greases, metalworking fluids (MWFs)
Soybean oil	Hydraulic oils, biodiesel fuel, engine oils, food processing lubricating fluids, transmission fluids, printing inks, paints, detergents, coatings, pesticides, shampoos
Sunflower oil	Diesel fuels, greases
Castor oil	Greases, gear lubricants
Jatropha oil	Biodiesel
Coconut oil	Engine oil
Olive oil	Automotive lubricants
High oleic sunflower oil/soybean oil	Marine lubricant
Canola oil	Food processing lubricants, metalworking fluids, transmission fluids, hydraulic fluids, penetrating oils
Rapeseed oil	Power transformer applications, hydraulic fluids, greases, chainsaw oils
Liquids from pyrolysis of natural resources (wood biomass/sugarcane)	Biofuel for energy applications, engine oil, heating, steel and cement industries

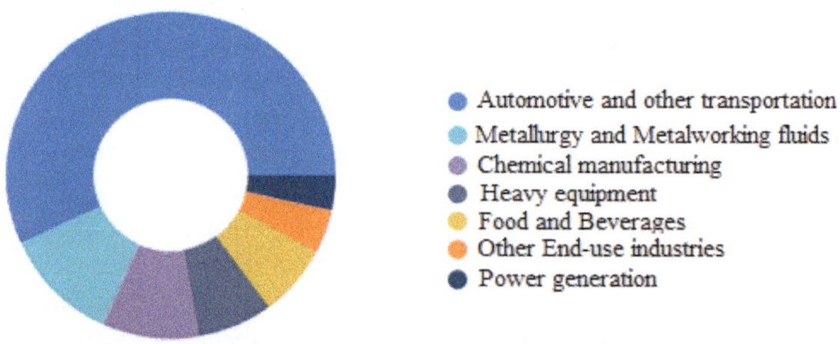

Figure 10.1 Global lubricant market, volume share (%), 2021. Reproduced from https://www.mordorintelligence.com/industry-reports/global-lubricants-market-industry with permission from Mordor Intelligence.

Blending of additives is a crucial way to improve the overall performance of lubricating oils, which helps to increase the lifespan and longevity of engines. Additives generally improve the thermo-oxidative stability, pour point, anti-wear performance, and the viscosity index of the lubricating oils. Nowadays, with the advent of advanced additive technology, the blending of additives in the formulation of high-performance lubricants for modern engines has become essential. Additives may be synthetic, semi-synthetic, or biobased. Commercially used additives are generally petro-based synthetics (*e.g.*, acrylate-based polymers, polycaprolactones, polyesteramides, and aliphatic copolyesters). Although their performance is satisfactory, they are nonbiodegradable and harmful to the biosphere.[2–4] Organic additives such as zinc dialkyldithiophosphate (ZDDP) and molybdenum dithiocarbamate (Mo-DTC), which are widely used as efficient antifriction additives in turbine oil lubricants, may result in harmful emissions and can pollute soil and groundwater. The application of bioadditives is an emerging area of research in the lubricant sector that minimizes several environmental hazards associated with commercial petroleum-based additives and improves the antiwear performance of lubricants and therefore lifetime of engines by providing satisfactory lubrication of machinery parts. The suitability of vegetable oils, waste cooking oils, and other bio-resources such as microalgae oil, plant-based materials, such as cellulose, lignin, chitosan, and biobased ionic liquids is being investigated to prepare efficient bio-additives in the field of tribology. There are a substantial number of reports, where chemically modified VOs or other biobased materials have been used as additives in the formulation of environmentally acceptable lubricants. In the future, the application of biobased additives in the formulation of biodegradable lubricants will be more significant, as legislatures have become more proactive towards ecosystem protection.

The use of bio-additives based on VOs such as soybean oil, sunflower oil, castor oil, rice bran oil, palm oil, olive oil, jojoba oil, *etc.* to prepare sustainable lubricant compositions has already been explored.[5–11] Recently the use of palmitoylated lignin as an antioxidant additive in castor oil has been reported.[12] It showed better performance compared to butylated hydroxytoluene, a commercially used antioxidant. The advantages of vegetable oil-based antioxidant additives, such as tocopherols, propyl gallate, and ascorbyl palmitate, over synthetic additives [4,4′-methylenebis(2,6-di-*tert*-butylphenol] have been disclosed.[13] It has been observed that the antioxidants, even at low concentrations, defer the onset of oxidation or slow the oxidation rate, thus improving the thermos-oxidative stability of the studied vegetable oils. Oils obtained from microalgae have recently been reported to be used in the formulation of biolubricants.[14] Waste cooking oils have attracted attention towards the manufacture of high value biolubricants along with biofuels recently. Except for these natural oils, liquids obtained from the pyrolysis of wood biomass have also come into focus towards the formulation of biolubricants. In this short review, we have discussed a few chemical ways for the preparation of additives/base stocks from bio-oils and the formulation of different sustainable lubricant compositions with them.

10.2 Characteristics of Vegetable Oils

The chemical constituents of vegetable oils include triacylglycerols (92–98%), diacylglycerols/monoacylglycerols (0.5%), sterols (0.3%), free fatty acids (0.1%), and tocopherols/tocotrienols (0.1%). The triglyceride structure is basically the ester of glycerol with different long chain saturated and unsaturated fatty acids. The physical and chemical properties of VOs mainly vary with the type of fatty acids (FAs) in the triglyceride. FAs with low percentages of unsaturation are more suitable for formulating lubricating oils.[15] There are a lot of VOs that contain a higher percentage of unsaturation, such as soybean oil, sunflower oil, safflower oil, cottonseed oil, and linseed oil. These VOs require functionalization/chemical modification to improve oxidative stability for preparing base stocks/additives for lubricants. Non-edible vegetable oils such as pennycress, cress, cuphea, lesquerella, meadowfoam, castor, neem, karanja, mahua, jatropha, linseed, *etc.*, being comparatively less expensive, have an advantage over edible oils as a bioresource for producing additives for biofuels/biolubricants.[16] Except for the percentage of unsaturation, the performance of a biolubricant also depends on the physical characteristics of the VOs from which it was derived. A VO with a higher viscosity index, flash point, thermo-oxidation stability, shear stability, lubricity, and lower pour point and cloud point typically produces biolubricants with enhanced performance. Standard ASTM methods are normally used to evaluate these properties. Table 10.2 showcases the major physical properties of some common vegetable oils reported earlier[15] and the associated ASTM methods for their evaluation.

10.3 Chemical Modification of VOs

Recently, a large portion of green additives have been synthesized from chemical modification of VOs. Soybean oil, sunflower oil, palm oil, and rice bran oil, due to their abundance, are often chosen to produce biolubricants/additives. However, non-edible vegetable oils, such as castor, jatropha, neem, mahua, karanja, and linseed oils, have a distinct advantage in the production of additives for biofuels/biolubricants, as they do not impact the food supply chain. Recently, additives/bio-oils have been synthesized from waste cooking oils and other natural resources, like microalgae oils and pyrolysis liquids from lignocellulosic biomass. These also have the advantage of being cheaper. However, the choice of oil depends on the specific application of the synthesized additives and the physiochemical characteristics of the oils. The derivatives of natural oils, which are used as multifunctional additives for biolubricants, should have a higher viscosity index, flash point, thermo-oxidation stability, shear stability, and lubricity and a lower pour point and cloud point. Standard ASTM methods are applied to evaluate these properties. The chemical methods that are widely applied to synthesize base fluids are also utilized to derive additives from VOs. These processes include transesterification, hydrolysis, epoxidation, formation of cyclic carbonates, incorporation of different nanoparticles in the backbone of VO, polymerization (homopolymerization and copolymerization with suitable comonomers), preparation of polymer composites by reinforcing different organic and inorganic nanofillers in the vegetable oil polymer matrix, formation of ionic liquids, and derivatization of vegetable fatty esters into fatty amines, fatty amides, or fatty alcohols. The possible chemical routes for modification of VOs are shown in Scheme 10.1.

Table 10.2 Physical properties of vegetable oils.

Vegetable oils	Iodine value	Pour point (°C)	Cloud point (°C)	Kinematic viscosity at 40 (°C) $mm^2 s^{-1}$	Viscosity index	Flash point (°C)	Density at 15 (°C)
	ASTM D 5554-95	ASTM D 97	ASTM D 2500	ASTM D 445	ASTM D 2270	ASTM D 56/93	ASTM D 1657
Soybean oil	128–143	−12	−4	29	170	254	0.914
Sunflower oil	125–140	−15	−9.5	36	163	274	0.916
Rapeseed oil	98–105	−15	−2	35	143	246	0.912
Palm oil	48–58	23.6	25.2	39.4	195	252	0.919
Rice bran oil	103	13	16	38.2	169	184	0.906
Olive oil	75–94	−14	−11	39	160	177	0.918
Safflower oil	145	−7	−2	28.3	—	260	0.914
Coconut oil	8–11	12.7	13.1	27	165	266	0.918
Jatropha oil	82–98	−6	11	34	220	225	0.94
Rubber seed oil	104	18	25	33.89	—	228	0.928
Castor oil	83–86	−21	−18	251	91	229	0.960
Karanja oil	81–90	−4	2	38.8	136	212	0.9358
Cotton seed oil	90–119	−4.5	−0.5	34	211	234	0.918
Linseed oil	168–204	−15	5	26–29	248	241	0.938
Neem oil	81	7	13	35.8	175	200	0.918
Sesame oil	104–116	−11	−8	36	210	260	0.918
Mahua oil	58–70	11	20	37.18	—	238	0.945

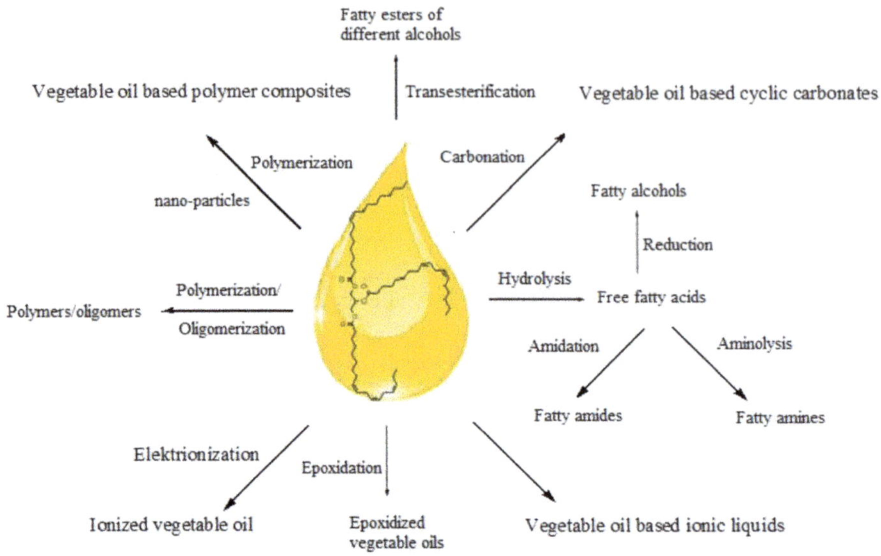

Scheme 10.1 Different chemical routes for the modification of VOs to produce additives/base stocks.

10.3.1 Transesterification

In transesterification, triglyceride esters of vegetable oils are converted into fatty acid alkyl esters by reacting with smaller alcohols such as methanol, ethanol, or propanol in the presence of a homogeneous or heterogeneous (acid or base) or enzyme catalyst.[17] Although the selection of catalyst depends a lot on the quality of the feedstock, the use of homogeneous alkali catalysts (KOH/NaOH) is very common. For base-catalyzed transesterification, the VO should be of high quality and should contain a low percentage of free fatty acids; otherwise saponification reactions with free fatty acids will produce soap-like materials causing poor separation and recovery. The acid-catalyzed transesterification is generally applied for the low quality feedstocks (*e.g.*, waste cooking oils, non-edible oils, grease, and animal fats) that have comparatively large amounts of free fatty acid content. In this case, high temperature and acid-resistant reactors, which are costly, will have to be used. Due to several disadvantages of both homogeneous acid or base catalysts, the use of heterogeneous solid catalysts, photocatalysts, and enzymes and non-catalytic transesterification under subcritical and supercritical conditions have been focused on instead.[18]

Fatty acid alkyl esters (FAAE) produced in the transesterification step can be easily converted into fatty amines, amides, alcohols, polyols, epoxides, or other essential compounds (by various reactions such as oxidation, hydrogenation, metathesis, isomerization, *etc.*), key components for the production of biolubricants/bio-additives. The FAAE can easily be further transesterified with different long chain alcohols and polyols like trimethylolpropane (TMP), trimethylolethane (TME), trimethylolhexane (TMH), neopentylglycol (NPG), and pentaerythritol (PE) for enhanced lubricant/additive performances.[19] The fatty acid methyl ester (FAME) of canola oil was used as an excellent lubricity additive for biodiesel.[20] A transesterification product of olax scandens oil was used as a biobased additive in petroleum oil.[21] The blends showed better engine performance, higher flash points, and higher lubricity. Desert date oil, modified by a two-step transesterification using methanol and trimethylolpropane respectively, was reported to exhibit efficient tribological performance as a biolubricant.[22] It was further observed that the addition of 0.9% Cu nanoparticles increases the performance of the biolubricant significantly. In another work, a mass-transfer-limited transesterification reaction of a promising industrial crop, camelina oil, with trimethylolpropane was investigated by microwave irradiation at 800 W to prepare a biolubricant (TMP ester).[23]

Epoxidation followed by transesterification of *Calophyllum inophyllum* oil, a non-edible tree seed oil, produced a potential bio-base stock having improved oxidation stability, viscosity index, and lower wear scar diameter.[24] Wafti *et al.* conducted a kinetic study of transesterification of high oleic palm oil methyl esters and trimethylolpropane (TMP) by an immobilized lipase enzyme, Novozyme 435, to prepare biolubricant base stocks having improved physicochemical properties required for an efficient biolubricant.[25] Recently, Naushad *et al.*[26] have found a way for the transesterification of palm oil to produce biodiesel with 86.4% yield using fabricated novel magnetic ionic liquid-based nanocomposite catalysts prepared from 1-butylimidazole, (3-bromopropyl)-trimethoxysilane, and $NiFe_2O_4$ nanoparticles. It was noticed that the tribological properties of FAME of castor oil additized with ethylene glycol and 0.5% Fe_3O_4 nanoparticles showed a reduction in the coefficient of friction and wear to almost 50% and 40% respectively compared to raw castor seed oil.[27] The trimethylolpropane oleic acid triester synthesized by transesterification of methyl oleate (MO) with TMP using K_2CO_3 as the catalyst has superior thermal stability, excellent rheological and antifrictional properties, and low volatility compared to the free fatty acid itself.[28]

10.3.2 Hydrolysis

Hydrolysis of triglyceride esters using different acid or base catalysts produces free fatty acids. These free FAs then can be used as an oleochemical feedstock to prepare esters of different mono-ols/polyols (which are used as biodiesels/base stock for biolubricants), fatty amides, amines, alcohols, and ionic liquids that are used as multifunctional additives for lubricants. Recently, a hydro-esterification process (*i.e.* hydrolysis followed by the esterification process), a modern alternative technique, has been employed for the production of biodiesel. It allows the use of any fatty esters with high acidity and moisture content and mitigates several problems associated with the two separate chemical reactions, *i.e.* hydrolysis and esterification. The use of hydro-esterification reaction on residual frying oil using vermiculite as a heterogeneous catalyst to prepare biodiesel was reported by Lacerda *et al.*[29] Fatty acids are used for the production of ionic liquids also. Tetrabutylammonium ionic liquids having fatty acid anions of different chain lengths and unsaturation are synthesized by Gusain *et al.* These are used as excellent antiwear additives for mineral base stock.[30]

Recently, new catalysts, such as enzymes, subcritical water, ultrasound-assisted enzymes, heterogeneous catalysts, and SO_3H functional Bronsted acids,[31–34] have been developed for the hydrolysis of triglyceride esters. Baena *et al.* explored the complete hydrolysis of waste fats, oils, and grease into free fatty acids using lipase as a catalyst.[35] Machado *et al.* reported the satisfactory performance of castor bean lipase (CBL) as a catalyst in the hydrolysis of non-edible high-acidity oil and microbial oils.[36] It was further noticed that CBL is effective in catalyzing the hydrolysis of macaw pulp oil with 55% conversion in only 3 h of reaction.

10.3.3 Catalytic Partial Hydrogenation

The low oxidative stability of vegetable oils is mainly due to the presence of bis-allylic protons in the unsaturated FA chains. Saturation increases thermo-oxidative stability as evidenced by higher oxidative stability of saturated FA compared to unsaturated ones. Catalyzed partial hydrogenation using suitable catalysts is a technique by which we can transform highly unsaturated FAs into saturated or partially saturated ones which have a wide range of applicability for use in biocompatible products including biolubricants (base stock/additive).[37–40] Saturated branched chain fatty acids (sbc-FA), commonly known as 'isostearic acids' or their ester derivatives, prepared by isomerization of linear chain unsaturated FAs (which make it branched chain unsaturated FAs) followed by catalytic hydrogenation, can be used as biodegradable lubricants due to their higher oxidative stability, lower pour point, and higher viscosity index.[41] Ester derivatives of sbc-FA showed better results compared to sbc-FA. Shomchoam *et al.* carried out partial hydrogenation of palm oil using $Pd/\gamma\text{-}Al_2O_3$ as a catalyst. The hydrogenated oil showed improved oxidative stability and was used for formulating biolubricants.[42] Selective partial hydrogenation of canola, rapeseed oil, and sunflower oil under different reaction conditions using a Lindlar catalyst was carried out by Lavendura *et al.* to produce monounsaturated oils with improved oxidative stability that are used as precursors for biolubricant production.[43] Aelst *et al.* carried out partial hydrogenation of soybean oil using a shape selective Pt/ZSM-5 catalyst to improve the oxidative stability and pour point of the biolubricant formulated.[44]

10.3.4 Amidation/Amination

Amines and amide derivatives of VO/fatty acids are widely used as friction modifiers or anti-wear additives in engine oils. Fatty acids/esters are converted into amides by catalytic aminolysis with ammonia or other primary/secondary amines. For the preparation of amines, aminolysis followed by dehydration and hydrogenation with suitable catalysts is carried out. These additives showed better tribological performance compared to commercial ZDDP-type anti-wear additives.

Partially neutralized fatty amine salts are used as antifriction additives for lubricants used in internal combustion engines. Fatty amides obtained from soybean oil are used as a rheology modifier in coatings. The chemistry of tribology of fatty amines on steel surfaces was discussed by Nalam *et al.*[45] Amide derivatives synthesized from non-edible oils such as waste cooking oil, karanja oil, and jatropha oil using a Li doped $CaO–Ca(OH)_2$ heterogeneous nanocatalyst were used as diesel fuel additives. The cetane number and lubricity of the diesel fuel improved significantly due to the blending of these additives.[46] Reyes-Dorantes *et al.* demonstrated the use of fatty amides synthesized from rice bran oil as corrosion inhibitors with an inhibition efficiency of at least 95% at the optimum inhibitor concentration.[47] Jamil *et al.* used different heterogeneous catalysts to transform the triglycerides of VO into fatty amides, amines, and nitriles.[48]

10.3.5 Epoxidation and Carbonation

Epoxidation is a very crucial step through which the oxidation stability of the VOs is improved. Epoxidized derivatives are used to prepare base stocks and anti-wear/antifriction additives having considerably better performance than conventional petroleum-based lubricating oils. Epoxidation is generally carried out by reacting the triglycerides/fatty esters/acids with a peroxy acid like performic, peracetic, or *m*-chloroperbenzoic acid in the presence of a catalyst. The extent of the reaction can be controlled by tuning the experimental parameters such as the reaction time, temperature, type of peracid and catalyst, reactant molar ratio, solvent, and contacting patterns (batch/semi-batch mode/azeotropic distillation).

Abdel-Hameed *et al.* discussed the synthesis of bio-lubricants from soybean and jojoba oil by epoxidation of the carbon–carbon double bonds using Amberlite IR 120 and peroxide as a combined catalyst followed by transesterification, epoxy ring-opening, and finally acetylation of the hydroxylated obtained products. The percentage of epoxidized products in this method was about 80%.[49] Epoxidation of *Euphorbia lathyris* oil improved the lubricity of unmodified oil as studied by Singh *et al.*[50] They also observed significant enhancement in tribological performance, viscosity index, and pour point values of the lubricating oils due to the addition of 0.5% ZnO nanoparticles. Epoxidation of palm stearin oil methyl ester (PSME) using acetic acid and hydrogen peroxide (H_2O_2) in the presence of a lipase biocatalyst produces a biolubricant. The epoxidized derivative decreased the coefficient of friction (CoF) by 44.5% and the wear scar diameter (WSD) by 16.6% compared to those of PSME itself.[51]

The epoxidation reaction of methyl oleate has been carried out by Suarez *et al.* using heterogeneous alumina doped with Lewis acidic metals as catalysts in the presence of an aqueous hydrogen peroxide solution.[52] Lubricant properties such as antiwear performance, pour point, and oxidative stability of epoxy derivatives of different oleochemicals like pure VO and methyl ester of different fatty acids have been discussed by Doll *et al.*[53] The performances shown are very promising. Epoxidation of methyl esters of different fatty acids such as oleic, linoleic, and linolenic acids has been performed to produce epoxidized derivatives of the fatty acid methyl esters.[54] These epoxidized derivatives showed improved oxidative stability compared to the corresponding fatty esters and potential performances as additive/bio-base fluids in some cases. Tribological properties and oxidation stability of a TMP-based tri-ester and epoxidized biolubricants prepared from karanja oil (KO) and rice bran oil (RBO) were analyzed by Edla *et al.*[55] The modified oils showed improved oxidative stability, VI, PP, and AW properties compared to unmodified oils. The epoxidized derivatives are more effective than the transesterified oils. The opening of oxirane rings of the epoxy triglycerides/fatty esters/acids with different nucleophiles [alcohols, thiols, amines (aliphatic/aromatic), acids, anhydrides, azides, water, cyclic borates, dialkyl phosphates] in the presence of catalysts also produces versatile products that are used as biolubricants (base stocks/additives), bio-plasticizers, and other industrially useful chemicals.[56–60]

Epoxidized methyl oleate glyceride and palm stearin methoxy ester ethoxylate were used as surfactants for preparing emulsions of vegetable oil-based metalworking fluids that improve the overall performance of the emulsion.[61] Opening of epoxy rings of epoxidized soybean oil by different cyclic borate compounds using a titanium isopropoxide catalyst produced triglyceride borate adducts. These compounds are used as antiwear additives having performances comparable with that of the ZDDP type of additives when blended with different base stocks such as vegetable oils, PAO, Gr. III base oil, and hexadecane.[62,63] Opening of epoxy rings of an oleic acid methyl ester with dialkyl phosphates using heterogeneous titanium/zirconium catalysts produced sulfur and metal-free extreme pressure additives which may be an excellent green alternative to ZDDP, commercially used as an antiwear additive.[64] Biswas *et al.* have disclosed the synthesis of various nitrogen containing derivatives of soybean oil/fatty esters such as fatty amides, fatty amines, fatty imidazolines, and soybean oil-based urethane. They also discussed the opening of the epoxy rings of epoxidized soybean oil with different nucleophiles such as water, amines, HCN, $RCONH_2$, and azide to produce a large number of industrially important functional compounds.[65] The epoxidation of methyl linoleate, methyl oleate, and methyl soyate followed by epoxy ring opening with sodium azide to form azide derivative fatty esters using ionic liquids as catalysts was also studied by the same group.[66]

The opening of the epoxy rings of VOs using organic thiols produced thioether hydroxyl derivatives of vegetable oils in which the unsaturation in the FA chains is removed due to the addition of polar functional groups of thiol.[58,67,68] These derivatives are used as an effective and excellent green alternative to commercially available antifriction additives in the automotive industry. Epoxidation of SBO followed by opening of the epoxy rings with different acid anhydrides using suitable catalysts and solvents produced acyl derivatives of SBO that showed significant improvement in cold flow properties, thermal stability, oxidative stability, and most importantly antiwear performance. The epoxy ring opening was found to be maximum in the presence of hexanoic anhydride catalyzed by boron trifluoride etherate.[69,70] The opening of epoxy rings of methyl oleate or other alkyl oleates using different organic acids such as propionic, levulinic, octanoic, hexanoic, or 2-ethylhexanoic acids produced different alkyl α-hydroxy stearate esters.[71–74] The synthesized hydroxyl esters were used as additives in VOs and polyalphaolefin. They showed better antifriction performances, improved thermal stability, and low temperature fluidity compared to the corresponding epoxy methyl or other alkyl oleates. The reaction of epoxy derivatives of soyabean oil, soybean oil methyl esters, and methyl oleate with acetone in the presence of ferric chloride produced acetonide derivatives (a cyclic ketal) of the corresponding compounds which can be used as a starting material to produce various compounds including lubricants.[75]

Free fatty acids can also be converted into potential biolubricants through this method. Xu *et al.* prepared a linoleic acid-based hydroxytetrahydrofuran ester through regioselective enzymatic hydration of linoleic acid followed by selective chemo-enzymatic epoxidation with an intramolecular epoxide ring-opening reaction and esterification of the product with methyl laurate or methyl palmitate. It was found that the final product has excellent lubricity properties, a lower pour point, a higher viscosity index, and thermal stability compared to mineral lubricants used conventionally.[76]

Epoxy derivatives of oils may be converted into cyclic carbonated derivatives by reacting with CO_2 in the presence of catalysts.[77] These carbonated derivatives are also used to prepare lubricant additives and other essential value-added materials. These derivatives have an advantage as they consume and utilize atmospheric CO_2 towards producing biolubricants. The synthesized compounds showed potential application as industrial lubricants or fuel additives. The opening of the epoxy or carbonate rings by different nucleophilic reagents such as amines, acids, ethers, alcohols, thiols, *etc.* produces a wide variety of products that have different industrial applications along with lubricant additives. Figure 10.2 shows the epoxidation and carbonation reactions of VO and the ring opening reactions.

10.3.6 Elektrionization™

It is a unique technique to improve the viscosity index, extreme pressure, and anti-wear properties. Ionized vegetable oils are used as bio-additives for

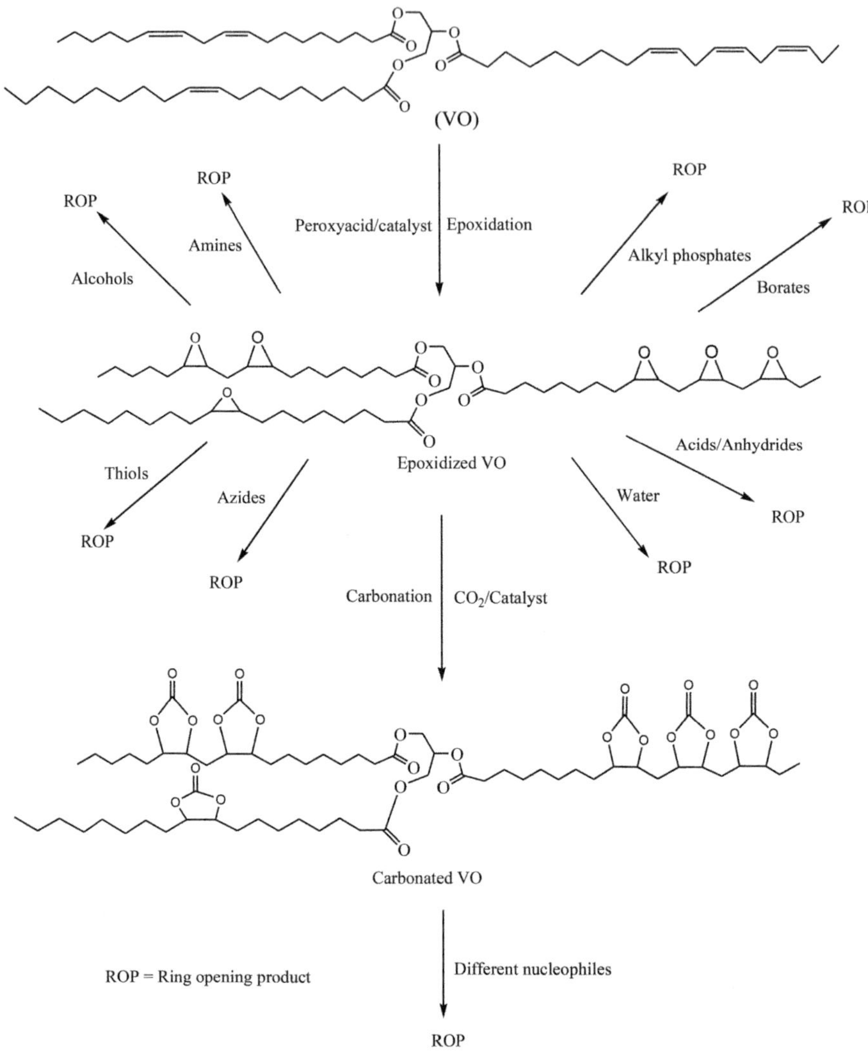

Figure 10.2 Epoxidation, ring opening reactions, and carbonation of VO.

lubricants. Ionized vegetable oils are more polar. They form sufficiently thick and resilient protective layers on the rubbing surface areas and reduce friction. This method of modification of vegetable oils does not require any catalyst.[78]

10.3.7 Estolide Formation/Oligomerization

Estolides are oligomers of fatty acids/esters containing fatty acid repeat units, with secondary ester linkages on the alkyl backbone. There are three different classes: estolide of free fatty acids, esters of fatty acids, and triglyceride of fatty acids (VO). The structures of different types of estolides are discussed in detail by Cermak *et al.*[79] This modification improves the lubricity, oxidative stability, load-carrying characteristics, and viscosity index and also lowers the pour point. Thermal stability and functional properties of the estolides depend on their structure and branch chains. The chemistry of the functioning of different classes of estolides as lubricants has been explored by the USDA. Zapateiro-García *et al.* have evaluated the tribological performance of high-oleic sunflower and castor oils by blending various estolides synthesized from both oleic and ricinoleic acids through acid catalytic synthetic processes. They have found that the reduction of wear is more for estolides prepared from ricinoleic acid using sulfuric and perchloric acid-catalyzed methods.[80] Triglyceride estolides of castor and lesquerella oils are potentially used in a wide range of lubricant applications.[81] Estolides synthesized from the fatty acid of castor oil (ricinoleic acid) can also be further converted into their epoxy and cyclic carbonated derivatives (by epoxidation and carbonation) with improved viscosities, oxidation onset

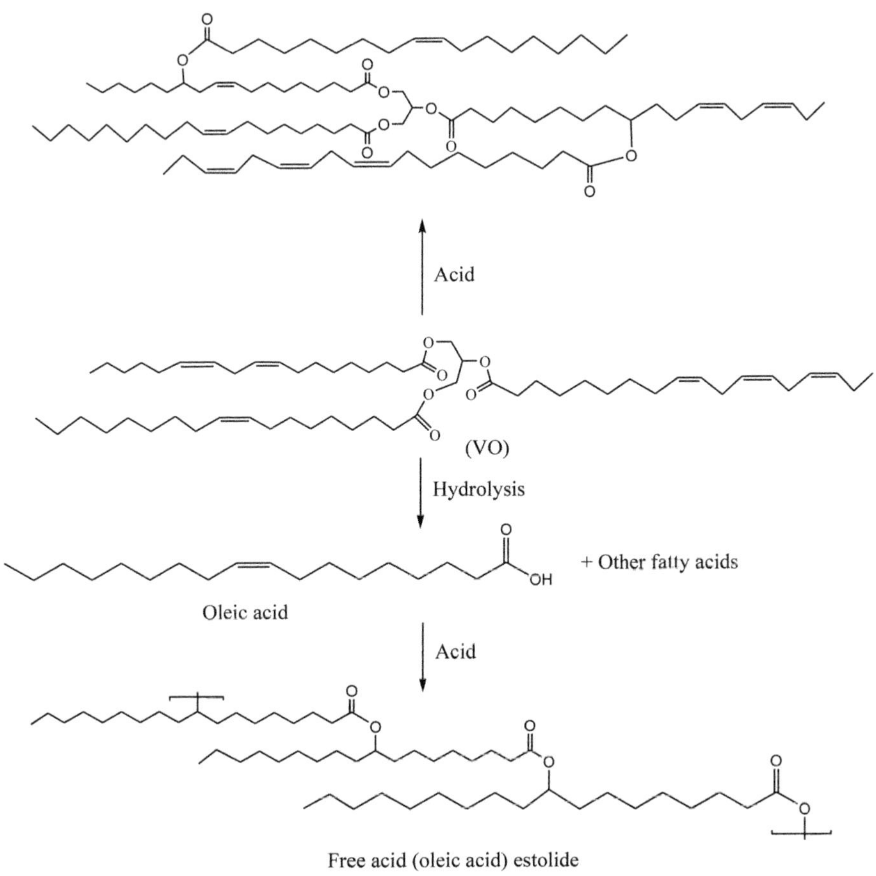

Figure 10.3 Formation of estolides from VO.

temperatures, and pour points compared to their corresponding estolides.[82] Figure 10.3 shows the structures of estolides.

10.3.8 Polymerization

Polymerization is a potential way to improve the oxidation stability of VOs. Polymerized plant oils have extensive applications as additives in the formulation of biolubricants. Depending upon the end application, polymerization of vegetable oil fatty esters/acids can be performed in different ways, such as cationic, free radical, ring-opening, metathesis, and condensation polymerization. Many lubricant additives are manufactured by free radical polymerization.[5–10] The macrostructure of the polymers, upon which their potentiality as biolubricants (additive/base stock) depends, is determined also by the type of monomers and conditions of the polymerization reaction, like duration of polymerization, heating procedure (thermal, microwave), temperature, and control on the reaction (living polymerization/atom transfer radical polymerization). It was found that the antifriction performance increases due to the incorporation of polar functionalities in the VO skeleton. The homopolymers of VOs and their copolymers with different functional monomers, such as methyl methacrylate, stearyl methacrylate, alkyl acrylates (methyl/decyl/dodecyl acrylates), and α-olefins (α-pinene, 1-dodecene, 1-hexadecene, and 1-octadecene), were reported as multifunctional additives in biolubricants. Singh *et al.* synthesized copolymers of palm oil with stearyl methacrylate and evaluated their performance as a multifunctional additive in mineral base oil (N-150) in three different concentrations, *viz.* 0.5%, 1.0% and 1.5% w/v. The additives enhanced the performance of formulated lubricants.[83] Homopolymers of rice bran oil and its copolymers with dodecyl acrylate, styrene, and stearic acid have potential multifunctional performance, especially as an antiwear additive.[84] The homopolymers of soybean oil, sunflower oil, castor oil, and their copolymers with different acrylates, such as methyl/decyl/dodecyl acrylates and methyl methacrylates, showed excellent multifunctional additive performances as well. These polymerization reactions are shown in Figure 10.4. Thermal and microwave irradiation methods have been applied to synthesize all these polymers by a free radical polymerization method using benzoyl peroxide or AIBN initiator. A few polymers were synthesized by the ATRP (atom transfer radical polymerization) method also to produce additives having desired performances. Polymers of soybean oils having lower molecular weights were prepared by cationic polymerization using supercritical carbon dioxide in the presence of a boron trifluoride diethyl etherate initiator.[85,86] The lower molecular weight polymers were used as lubricants or hydraulic fluids after blending suitable additives with them. The oxidative stability, thermal stability, and pour point values of the polymerized soy oil were studied after blending antioxidant additives (ZDDP and BHT). It was observed that the onset temperature and thermal stability of the lubricant compositions were better compared to SBO. These

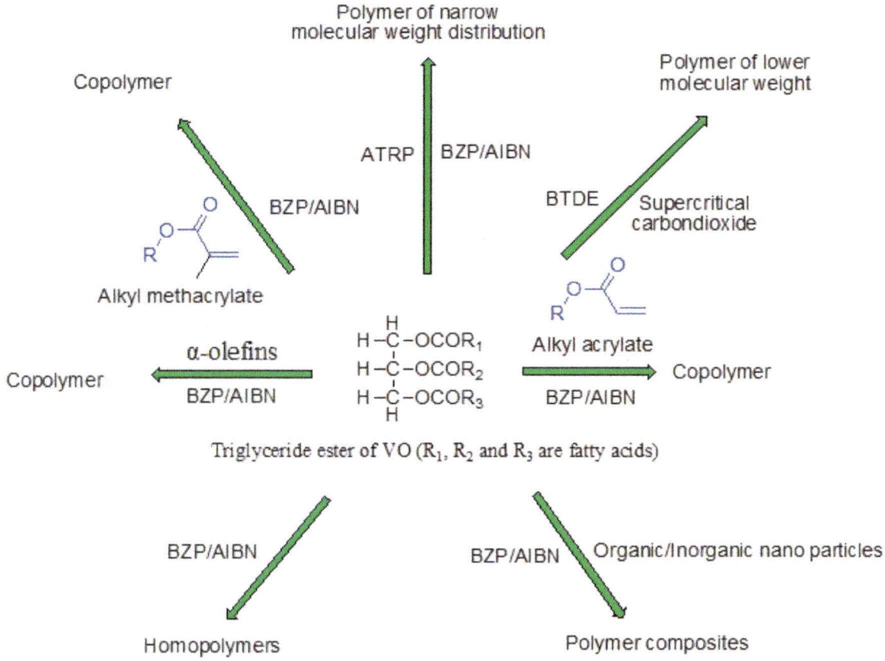

Figure 10.4 Polymerization reactions using VO.

sustainable formulations after adding diluents can be used as biobased gear oil.

Addition of different organic or inorganic nanofillers during polymerization has a noticeable impact on increases in additive performance and thermal stability.[87,88] It was reported that organic nanoparticles, such as nanocellulose, nanographene, and nanoparticles of different elements, like zinc, boron, iron, *etc.*, significantly enhanced the antifriction properties when introduced into the macro-structure of vegetable oil-based polymers. The nanoparticles present in the macrostructure of the polymers help the lubricant to bind the metal surface more strongly and therefore reduce friction.

10.4 Lubricants (Base Oil/Additives) from Other Natural Resources

Different natural substances other than vegetable oils, such as microalgae oil, lignin, curcumin, chitosan, and cellulose, are used as natural additives for lubricants. The red microalga *Dixoniella grisea* has been shown to reduce friction in a tribosystem.[89] Azomethine-functionalized chitosan showed excellent tribological performance when added to paraffin base oil.[90] Palmitoylated lignin and curcumin were used as antioxidant additives[12] in VO/waste cooking oils. The lignin fractions obtained from straw and birch wood after palmitoylation exhibited excellent antioxidant performance compared to butylated hydroxytoluene (BHT), a commonly used petroleum-based antioxidant, when blended with castor oil.

Lignocellulose-derived furan compounds were used to produce high-performance biolubricant base oils.[91] In this work, a highly symmetrical and branched T-shaped biolubricant was derived by hydroalkylation involving C–C coupling followed by a hydrodeoxygenation reaction between cellulose-derived 5-hydroxymethylfurfural and hemicellulose-derived alkyl furan. This process is also cost effective for the preparation of biolubricants. Lipoate esters prepared by catalytic esterification of lipoic acid with different alcohols displayed good properties as a viscosity index improver, an antioxidant, and an extreme pressure additive in high oleic sunflower oil.[92]

The crude extract of Lanzhou lily (LLCE) exhibited a higher load-carrying capacity above 600 N and notable lubricity at temperatures above 50 °C. The coefficient of friction at the stable stage was about 0.12 and the wear rate of the steel discs was on the order of 10^{-7}–10^{-8} $mm^3 Nm^{-1}$. Phenol esters or glycosides in LLCE were used as functional base fluids to provide boundary lubrication. It was observed that the extreme pressure and lubricity performances were improved further due to blending of components such as phospholipids, alkaloids, and organic sulfides with the LLCE bio-base stock.[93]

10.5 Conclusion

This review mainly focused on various ways of chemical modification of vegetable oils through which we could mitigate their shortcomings and use them for the formulation of potential bio-based stocks and or additives. The preparation of biolubricants from bio-resources other than vegetable oils has also been discussed. The application of biolubricants in industry is very significant given the increasing environmental hazards from petro-based synthetic lubricants and strict environmental protection regulations. The performance of the formulated biolubricants (base stock or additive), with respect to viscosity index, antiwear, pour point, flash point, and cloud point, is very promising and comparable to or in some cases better than that of commercially available lubricants. The global biolubricant market is gaining momentum very rapidly. The use of biolubricants in certain applications has already become mandatory in developed countries. But due to economic reasons and shortage of raw materials, the use of biolubricants is not widely commercialized still. Research in this field in the near future will develop new green and economical methodologies for manufacturing biolubricants for their vast applicability.

References

1. N. Salih and J. Salimon, A Review on Eco-Friendly Green Biolubricants from Renewable and Sustainable Plant Oil Sources, *Biointerface Res. Appl. Chem.*, 2021, **11**(5), 13303.

2. P. Ghosh, M. Das, M. Upadhyay, T. Das and A. Mandal, Synthesis and Evaluation of Acrylate Polymers in Lubricating Oil, *J. Chem. Eng. Data*, 2011, **56**(10), 3752.
3. D. W. Johnson and J. E. Hils, Phosphate Esters, Thiophosphate Esters and Metal Thiophosphates as Lubricant Additives, *Lubricants*, 2013, **1**, 132.
4. S. Boyde, Green lubricants. Environmental benefits and impacts of lubrication, *Green Chem.*, 2002, **4**, 293.
5. G. Karmakar and P. Ghosh, Green Additives for Lubricating Oil, *ACS Sustainable Chem. Eng.*, 2013, **1**, 1364.
6. P. R. Campos Flexa Ribeiro Filho, M. Rocha do Nascimento, S. S. Otaviano da Silva, F. M. Tavares de Luna, E. Rodríguez-Castellón and C. Loureiro Cavalcante, Jr., Synthesis and Frictional Characteristics of Bio-Based Lubricants Obtained from Fatty Acids of Castor Oil, *Lubricants*, 2023, **11**, 57.
7. N. Sapawe, M. F. Hanafi and S. Samion, The Use of Palm Oil as New Alternative Biolubricant for Improving Anti-Friction and Anti-Wear Properties, *Mater. Today: Proc.*, 2019, **19**(4), 1126.
8. S. Rani, M. L. Joy and K. P. Nair, Evaluation of physiochemical and tribological properties of rice bran oil – biodegradable and potential base stoke for industrial lubricants, *Mater. Today: Proc.*, 2015, **65**, 328.
9. P. Ghosh and M. Upadhyay, Isodecyl acrylate – olive oil copolymers as potential biodegradable additive for lubricating oil, *J. Polym. Res.*, 2016, **23**, 100.
10. F. J. Owuna, Stability of vegetable based oils used in the formulation of ecofriendly lubricants – a review, *Egypt. J. Pet.*, 2020, **29**, 251.
11. S. K. Tulashie and F. Kotoka, The potential of castor, palm kernel, and coconut oils as biolubricant base oil *via* chemical modification and formulation, *Therm. Sci. Eng. Prog.*, 2020, **16**, 100480.
12. M. A. Jedrzejczyk, S. V. Bosch, J. V. Aelst, K. V. Aelst, P. D. Kouris, M. Moalin, G. R. M. M. Haenen, M. D. Boot, E. J. M. Hensen, B. Lagrain, B. F. Sels and K. V. Bernaerts, Lignin-Based Additives for Improved Thermo-Oxidative Stability of Biolubricants, *ACS Sustainable Chem. Eng*, 2021, **9**(37), 12548.
13. L. A. Quinchia, M. A. Delgado, C. Valencia, J. M. Franco and C. Gallegos, Natural and Synthetic Antioxidant Additives for Improving the Performance of New Biolubricant Formulations, *J. Agric. Food Chem.*, 2011, **59**(24), 12917.
14. A. P. T. Da Silva, E. H. Bredda, H. F. de Castro and P. C. M. Da Rós, Enzymatic catalysis: An environmentally friendly method to enhance the transesterification of microalgal oil with fusel oil for production of fatty acid esters with potential application as biolubricants, *Fuel*, 2020, **273**, 117786.
15. G. Karmakar, P. Ghosh and B. K. Sharma, Chemically modifying vegetable oils to prepare green lubricants, *Lubricants*, 2017, **5**, 44.
16. G. Karmakar, K. Dey, P. Ghosh, B. K. Sharma and S. Z. Erhan, A Short Review on Polymeric Biomaterials as Additives for Lubricants, *Polymers*, 2021, **13**, 1333.

17. R. Jothiramalingam and M. K. Wang, Review of Recent Developments in Solid Acid, Base, and Enzyme Catalysts (Heterogeneous) for Biodiesel Production *via* Transesterification, *Ind. Eng. Chem. Res.*, 2009, **48**(13), 6162.
18. A. S. Belousov, A. L. Esipovich, E. A. Kanakov and K. V. Otopkova, Recent advances in sustainable production and catalytic transformations of fatty acid methyl esters, *Sustainable Energy Fuels*, 2021, **5**, 4512.
19. F. J. Owuna, M. U. Dabai, M. A. Sokoto, S. M. Dangoggo, B. U. Bagudo, U. A. Birnin-Yauri, L. G. Hassan, I. Sada, A. L. Abubakar and M. S. Jibrin, Chemical modification of vegetable oils for the production of bio-lubricants using trimethylolpropane: A review, *Egypt. J. Pet.*, 2020, **29**, 75.
20. S. L. Dmytryshyn, A. K. Dalai, S. T. Chaudhari, H. K. Mishra and M. J. Reaney, Synthesis and characterization of vegetable oil derived esters: evaluation for their diesel additive properties, *Bioresour Technol.*, 2004, **92**(1), 55.
21. V. K. Garlapati, S. B. Mohapatra, R. C. Mohanty and P. Das, Trans-esterified Olax scandens oil as a bio-additive: Production and engine performance studies, *Tribol. Int.*, 2021, **153**, 106653.
22. Y. Singh, A. Sharma, N. Singh and W. H. Chen, Development of bio-based lubricant from modified desert date oil (balanites aegyptiaca) with copper nanoparticles addition and their tribological analysis, *Fuel*, 2020, **259**, 116259.
23. K. Rokni, M. Mostafaei, M. Dehghani-Soufi and D. Kahziri, Microwave-assisted synthesis of trimethylolpropane triester (bio-lubricant) from camelina oil, *Sci. Rep.*, 2022, **12**, 11941.
24. P. Prasannakumar, S. Edla, A. D. Thampi, M. Arif and R. Santhakumari, A comparative study on the lubricant properties of chemically modified Calophyllum inophyllum oils for bio-lubricant applications, *J. Cleaner Prod.*, 2022, **339**, 130733.
25. N. S. Abd Wafti, T. S. Y. Choong, H. L. N. Lau, R. Yunus, S. Abd-Aziz and N. A. Raof, Kinetic study on the production of biodegradable lubricant by enzymatic transesterification of high oleic palm oil, *Process Biochem.*, 2023, **131**, 91.
26. M. Naushad, T. Ahamad and M. R. Khan, Fabrication of magnetic nano-particles supported ionic liquid catalyst for transesterification of vegetable oil to produce biodiesel, *J. Mol. Liq.*, 2021, **330**, 115648.
27. U. Ahmad, S. R. Naqvi, I. Ali, F. Saleem, M. T. Mehran, U. Sikandar and D. Juchelková, Biolubricant production from castor oil using iron oxide nanoparticles as an additive: Experimental, modelling and tribological assessment, *Fuel*, 2022, **324**(Part A), 124565.
28. Q. Xie, H. Zhu, P. Xu, K. Xing, S. Yu, X. Liang, W. Ji, Y. Nie and J. Ji, Transesterification of methyl oleate for sustainable production of bio-lubricant: Process optimization and kinetic study, *Mater. Today: Proc.*, 2022, **182**, 114879.
29. J. G. P. de Lacerda, R. A. Candeia, L. L. de Morais Sales, A. dos Santos Araújo, A. F. P. da Cunha, A. F. Wanderley and A. F. Campos,

Characterization of biodiesel from frying oil obtained by hydro-esterification using vermiculite as heterogeneous catalyst, *J. Therm. Anal. Calorim.*, 2019, **137**, 2045.

30. R. Gusain, A. Khan and O. P. Khatri, Fatty acid-derived ionic liquids as renewable lubricant additives: Effect of chain length and unsaturation, *J. Mol. Liq.*, 2020, **301**, 112322.
31. W. Zhang, H. Ji, Y. Song, S. Ma, W. Xiong, C. Chen, B. Chen and X. Zhang, Green preparation of branched biolubricant by chemically modifying waste cooking oil with lipase and ionic liquid, *J. Cleaner Prod.*, 2020, **274**, 122918.
32. J. Huang, Q. Zhao, W. Bu, C. Zhang, Z. Yang, X. Zhang and K. Zhang, Ultrasound-assisted hydrolysis of lard for free fatty acids catalyzed by combined two lipases in aqueous medium, *Bioengineered*, 2020, **11**(1), 241.
33. K. Teodora, J. Jozsef and C. Ján, Hydrolysis of fatty acid esters in subcritical water, *Eur. J. Lipid Sci. Technol.*, 2006, **108**, 652.
34. L. Hui, X. Kai, F. Weiyu, L. Chuan, N. Guozhi and L. Zhaomin, Hydrolysis of Vegetable Oils to Fatty Acids Using Brønsted Acidic Ionic Liquids as Catalysts, *Ind. Eng. Chem. Res.*, 2014, **53**, 11653.
35. A. Baena, A. Orjuela, S. K. Rakshit and J. H. Clark, Enzymatic hydrolysis of waste fats, oils and greases (FOGs): Status, prospective, and process intensification alternatives, *Chem. Eng. Process*, 2022, **175**, 108930.
36. S. A. Machado, P. C. M. Da Rós, H. F. de Castro and D. S. Giordani, Hydrolysis of vegetable and microbial oils catalyzed by a solid preparation of castor bean lipase, *Biocatal. Agric. Biotechnol.*, 2021, **37**, 102188.
37. N. Ravasio, F. Zaccheria, M. Gargano, S. Recchia, A. Fusi, N. Poli and R. Psaro, Environmental friendly lubricants through selective hydrogenation of rapeseed oil over supported copper catalysts, *Appl. Catal., A*, 2002, **233**, 1.
38. A. F. Trasarti, D. J. Segobia, C. R. Apesteguía, F. Santoro, F. Zaccheria and N. Ravasio, Selective Hydrogenation of Soybean Oil on Copper Catalysts as a Tool Towards Improved Bioproducts, *J. Am. Oil Chem. Soc.*, 2012, **89**, 2245.
39. B. R. Moser, M. J. Haas, J. K. Winkler, M. A. Jackson, S. Z. Erhan and G. R. List, Evaluation of partially hydrogenated methyl esters of soybean oil as biodiesel, *Eur. J. Lipid Sci. Technol.*, 2007, **109**, 17.
40. A. Bouriazos, K. Mouratidis, N. Psaroudakis and G. Papadogianakis, Catalytic Conversions in Aqueous Media. Part 2. A Novel and Highly Efficient Biphasic Hydrogenation of Renewable Methyl Esters of Linseed and Sunflower Oils to High Quality Biodiesel Employing Rh/TPPTS Complexes, *Catal. Lett.*, 2008, **121**, 158.
41. H. L. Ngo, R. O. Dunn, B. Sharma and T. A. Foglia, Synthesis and physical properties of isostearic acids and their esters, *Eur. J. Lipid Sci. Technol.*, 2011, **113**, 180.
42. B. Shomchoam and B. Yoosuk, Eco-friendly lubricant by partial hydrogenation of palm oil over Pd/γ-Al_2O_3 catalyst, *Ind. Crops Prod.*, 2014, **62**, 395.

43. U. P. Laverdura, L. Rossi, F. Ferella, C. Courson, A. Zarli, R. Alhajyoussef and K. Gallucci, Selective Catalytic Hydrogenation of Vegetable Oils on Lindlar Catalyst, *ACS Omega*, 2020, **5**, 22901.
44. J. V. Aelst, A. Philippaerts, E. Bartholomeeusen, E. Fayad, F. Thibault-Starzyk, J. Lu, D. Schryvers, R. Ooms, D. Verboekend, P. Jacobs and B. Sels, Towards biolubricant compatible vegetable oils by pore mouth hydrogenation with shape-selective Pt/ZSM-5 catalysts, *Catal. Sci. Technol.*, 2016, **6**, 2820.
45. P. C. Nalam, A. Pham, R. V. Castillo and R. M. Espinosa-Marzal, Adsorption Behavior and Nanotribology of Amine-Based Friction Modifiers on Steel Surfaces, *J. Phys. Chem. C*, 2019, **123**(22), 13672.
46. D. Kumar, C. H. Park and C. S. Kim, Sustainable heterogeneously catalyzed single-step and two-step amide derivatives of non-edible natural triglycerides as dual-functional diesel fuel additives, *Ind. Crops Prod.*, 2020, **158**, 113001.
47. E. Reyes-Dorantes, J. Zuniga-Diaz, A. Quinto-Hernandez, J. Porcayo-Calderon, J. G. Gonzalez-Rodriguez and L. Martinez-Gomez, Fatty amides from Crude Rice Bran Oil as Green Corrosion Inhibitors, *J. Chem.*, 2017, DOI: 10.1155/2017/2871034.
48. M. A. R. Jamil, S. M. A. H. Siddiki, A. S. Touchy, M. N. Rashed, S. S. Poly, Y. Jing, K. W. Ting, T. Toyao, Z. Maeno and K. Shimizu, Selective Transformations of Triglycerides into Fatty Amines, Amides, and Nitriles by using Heterogeneous Catalysis, *ChemSusChem*, 2019, **12**(13), 3115.
49. H. S. Abdel-Hameed, S. M. El-Saeed, N. S. Ahmed, A. M. Nassar, A. F. El-Kafrawy and A. I. Hashem, Chemical transformation of Jojoba oil and Soybean oil and study of their uses as bio-lubricants, *Ind. Crops Prod.*, 2022, **187**(A), 115256.
50. Y. Singh, N. K. Singh, A. Sharma, A. Singla, D. Singh and E. A. Rahim, Effect of ZnO nanoparticles concentration as additives to the epoxidized Euphorbia lathyris oil and their tribological characterization, *Fuel*, 2021, **285**, 119148.
51. A. N. Afifah, S. Syahrullail, N. I. W. Azlee and A. M. Rohah, Synthesis and tribological studies of epoxidized palm stearin methyl ester as a green lubricant, *J. Cleaner Prod.*, 2021, **280**, 124320.
52. P. A. Z. Suarez, M. S. C. Pereira, K. M. Doll, B. K. Sharma and S. Z. Erhan, Epoxidation of Methyl Oleate Using Heterogeneous Catalyst, *Ind. Eng. Chem. Res.*, 2009, **48**, 3268.
53. K. M. Doll, B. K. Sharma and S. Z. Erhan, Friction Reducing Properties and Stability of Epoxidized Oleochemicals, *Clean*, 2008, **36**, 700.
54. B. K. Sharma, K. M. Doll and S. Z. Erhan, Oxidation, friction reducing, and low temperature properties of epoxy fatty acid methyl esters, *Green Chem.*, 2007, **9**, 469.
55. S. Edla, A. D. Thampi, P. Prasannakumar and S. Rani, Evaluation of physicochemical, tribological and oxidative stability properties of chemically modified rice bran and karanja oils as viable lubricant base stocks for industrial applications, *Tribol. Int.*, 2022, **173**, 107631.

56. Z. R. Lozada, G. J. Suppes, Y. C. Tu and F. H. Hsieh, Soy-Based Polyols from Oxirane Ring Opening by Alcoholysis Reaction, *J. Appl. Polym. Sci.*, 2009, **113**(4), 2552.
57. H. Varshney, A. Ahmad and A. Rauf, Ring Opening of Epoxy Fatty Esters by Nucleophile to Form the Derivatives of Substituted β-Amino Alcohol, *Food Nutr. Sci.*, 2013, **4**, 21.
58. B. K. Sharma, A. Adhvaryu and S. Z. Erhan, Synthesis of Hydroxy Thioether Derivatives of Vegetable Oil, *J. Agric. Food Chem.*, 2006, **54**, 9866.
59. C. H. Chan, S. W. Tang, N. K. Mohd, W. H. Lim, S. K. Yeong and Z. Idris, Tribological behavior of biolubricant base stocks and additives, *Renew. Sustain. Energy Rev.*, 2018, **93**, 145.
60. A. Biswas, B. K. Sharma, K. M. Doll, S. Z. Erhan, J. L. Willett and H. N. Cheng, Synthesis of an Amine–Oleate Derivative Using an Ionic Liquid Catalyst, *J. Agric. Food Chem.*, 2009, **57**, 8136.
61. K. M. Doll and B. K. Sharma, Emulsification of Chemically Modified Vegetable Oils for Lubricant Use, *J. Surfact. Deterg.*, 2011, **14**, 131.
62. B. K. Sharma, K. M. Doll, G. L. Heise, M. Myslinska and S. Z. Erhan, Antiwear Additive Derived from Soybean Oil and Boron Utilized in a Gear Oil Formulation, *Ind. Eng. Chem. Res.*, 2012, **51**, 11941.
63. G. L. Heise, B. K. Sharma and S. Z. Erhan, *US Pat.*, US2012/0083433A1, 2012.
64. B. K. Sharma, K. M. Doll, M. S. C. Pereira, G. F. N. Santos, P. A. Z. Suarez and S. Z. Erhan, Production of phosphorus-containing oleochemicals through an epoxide route, *Int. J. Sustain. Eng.*, 2012, **5**, 280.
65. A. Biswas, B. K. Sharma, J. L. Willett, S. Z. Erhan and H. N. Cheng, Soybean oil as a renewable feedstock for nitrogen-containing derivatives, *Energy Environ. Sci.*, 2008, **1**, 639.
66. A. Biswas, B. K. Sharma, J. L. Willett, A. Advaryu, S. Z. Erhan and H. N. Cheng, Azide Derivatives of Soybean Oil and Fatty Esters, *J. Agric. Food Chem.*, 2008, **56**, 5611.
67. B. K. Sharma, A. Advaryu and S. Z. Erhan, Friction and wear behaviour of thioether hydroxyl vegetable oil, *Tribol. Int.*, 2009, **42**, 353.
68. S. Z. Erhan, A. Adhvaryu and B. K. Sharma, *US Pat.*, US2006/0009365A1, 2006.
69. S. Z. Erhan, B. K. Sharma, Z. Liu and A. Advaryu, Lubricant Base Stock Potential of Chemically Modified Vegetable Oils, *J. Agric. Food Chem.*, 2008, **56**, 8919.
70. B. K. Sharma, Z. Liu, A. Advaryu and S. Z. Erhan, One-Pot Synthesis of Chemically Modified Vegetable Oils, *J. Agric. Food Chem.*, 2008, **56**, 3049.
71. B. K. Sharma, K. M. Doll and S. Z. Erhan, Ester hydroxy derivatives of methyl oleate: Tribological, oxidation and low temperature properties, *Bioresour. Technol.*, 2008, **99**, 7333.
72. B. R. Moser, B. K. Sharma, K. M. Doll and S. Z. Erhan, Diesters from Oleic Acid: Synthesis, Low Temperature Properties, and Oxidation Stability, *J. Am. Oil Chem. Soc.*, 2007, **84**, 675.

73. K. M. Doll, B. K. Sharma and S. Z. Erhan, Synthesis of Branched Methyl Hydroxy Stearates Including an Ester from Bio-Based Levulinic Acid, *Ind. Eng. Chem. Res.*, 2007, **46**, 3513.
74. S. Z. Erhan, K. M. Doll and B. K. Sharma, *US Pat.*, US8173825B2, 2012.
75. A. Biswas, B. K. Sharma, K. Vermillion, J. L. Willett and H. N. Cheng, Preparation of Acetonides from Soybean Oil, Methyl Soyate, and Fatty Esters, *J. Agric. Food Chem.*, 2011, **59**, 3066.
76. J. Xu, L. Kong, L. Deng, G. Mazza, F. Wang, J. Baeyens and K. Nie, The conversion of linoleic acid into hydroxytetrahydrofuran-structured bio-lubricant, *J. Environ. Manage.*, 2021, **291**, 112692.
77. W. Y. Perez-Sena, K. Eränen, N. Kumar, L. Estel, S. Leveneur and T. Salmi, New insights into the cocatalyst-free carbonation of vegetable oil derivatives using heterogeneous catalysts, *J. CO_2 Util.*, 2022, **57**, 101879.
78. M. Roegiers, H. Zhang and Z. Boris, Elektrionized Vegetable Oils as Lubricity Components in Metalworking Lubricants, *FME Trans.*, 2008, **36**(3), 133.
79. S. C. Cermak, T. A. Isbell, J. W. Bredsguard and T. D. Thompson, Estolides: Synthesis and Applications, in *Fatty Acids: Chemistry, Synthesis, and Applications*, AOCS Press, 2017, ch. 14, p. 431, DOI: 10.1016/B978-0-12-809521-8.00015-5.
80. L. A. García-Zapateiro, J. M. Franco, C. Valencia, M. A. Delgado and C. Gallegos, Viscous, thermal and tribological characterization of oleic and ricinoleic acids-derived estolides and their blends with vegetable oils, *J. Ind. Eng. Chem.*, 2013, **19**(4), 1289.
81. T. A. Isbell, B. A. Lowery, S. S. DeKeyser, M. L. Winchell and S. C. Cermak, Physical properties of triglyceride estolides from lesquerella and castor oils, *Ind. Crops Prod.*, 2006, **23**(3), 256.
82. K. M. Doll, S. C. Cermak, J. A. Kenar, E. L. Walter and T. A. Isbell, Derivatization of castor oil based estolide esters: Preparation of epoxides and cyclic carbonates, *Mater. Today: Proc.*, 2017, **104**, 269.
83. N. Singh, P. Agarwal, J. Porwal and S. K. Porwal, Evaluation of block copolymer and homopolymer of stearyl methacrylate as multifunctional additives for lubricating oil, *Polym. Bull.*, 2023, DOI: 10.1007/s00289-023-04937-9.
84. S. Talukdar and P. Ghosh, Biodegradable vegetable oil polymer as a multifunctional lubricating oil additive, *J. Macromol. Sci., Part A: Pure Appl.Chem.*, 2019, **57**, 1.
85. Z. Liu, B. K. Sharma and S. Z. Erhan, From Oligomers to Molecular Giants of Soybean Oil in Supercritical Carbon Dioxide Medium: 1. Preparation of Polymers with Lower Molecular Weight from Soybean Oil, *Biomacromolecules*, 2007, **8**, 233.
86. Z. Liu, B. K. Sharma, S. Z. Erhan, A. Biswas, R. Wang and T. P. Schuman, Oxidation and low temperature stability of polymerized soybean oil-based lubricants, *Thermochim. Acta*, 2015, **601**, 9.
87. Y. Ren, L. Zhang and G. Xie, A review on tribology of polymer composite coatings, *Friction*, 2021, **9**, 429.

88. S. Bhuyan, S. Sundararajan, Y. Lu and R. C. Larock, A study of the physical and tribological properties of biobased polymer–clay nanocomposites at different clay concentrations, *Wear*, 2010, **268**, 797.
89. A. Gavalás-Olea, A. Siol, Y. Sakka, J. Köser, N. Nentwig, T. Hauser, J. Filser, J. Thöming and I. Lang, Potential of the Red Alga Dixoniella grisea for the Production of Additives for Lubricants, *Plants*, 2021, **10**, 1836.
90. M. Murmu, S. Sengupta, R. Pal, S. Mandal, N. C. Murmu and P. Banerjee, Efficient tribological properties of azomethine-functionalized chitosan as a bio-lubricant additive in paraffin oil: experimental and theoretical analysis, *RSC Adv.*, 2020, **10**(55), 33401.
91. S. Chen and C. Zhao, Production of Highly Symmetrical and Branched Biolubricants from Lignocellulose-Derived Furan Compounds, *ACS Sustainable Chem. Eng.*, 2021, **9**, 10818.
92. G. Biresaw, D. Compton, K. Evans and G. B. Bantchev, Lipoate Ester Multifunctional Lubricant Additives, *Ind. Eng. Chem. Res.*, 2016, **55**(1), 373.
93. M. Xie, H. Tan and G. Zhao, A clean and sustainable strategy to produce bio-lubricant with high-bearing and good anti-oxidation ability from Lanzhou lily, *J. Cleaner Prod.*, 2022, **371**, 133333.

CHAPTER 11

Plant Oil-based Plasticizers

YUN HU AND PUYOU JIA*

Institute of Chemical Industry of Forest Products, Chinese Academy of Forestry (CAF), Co-Innovation Center of Efficient Processing and Utilization of Forest Resources, Nanjing Forestry University, Key Lab of Biomass Energy and Materials, Jiangsu Province, 16 Suojin North Road, Nanjing 210042, P. R. China
*Email: jiapuyou@icifp.cn

11.1 Introduction

Plasticizers are usually one kind of additive with a low molecular weight, added to polymer materials especially polyvinyl chloride (PVC) to improve the flexibility, plasticity, extensibility and expansibility of the materials. Plasticizers are mainly ester compounds, which form uniform physical units with PVC materials through physical interaction *via* swelling and dissolution.[1] Furthermore, the addition of plasticizers can reduce the hardness, melt viscosity and elastic modulus of polymer materials, thus improving the fracture strain and impact resistance, without changing the chemical properties of PVC. At present, traditional phthalate plasticizers account for the largest consumption in industry due to the advantages of a high plasticizing efficiency, wide sources of raw materials, a simple synthesis process and low cost. Therefore, phthalate plasticizers have become the most widely used for total plasticizer production and sales.[2] However, the use of phthalate plasticizers has been limited due to the potential carcinogenic and reproductive toxicity, which may pose a threat to human health and the environment. Therefore, phthalate plasticizers are gradually prohibited in fields that have close contact with the human body, *e.g.* medical treatment, food packaging, children's toys and others.[3,4] In order to improve the safety

Green Chemistry Series No. 83
Green Chemistry and Green Materials from Plant Oils and Natural Acids
Edited by Zengshe Liu and George Kraus

Published by the Royal Society of Chemistry, www.rsc.org

and broaden the application fields of PVC, it is necessary to develop environmentally friendly and renewable plasticizers to substitute phthalate plasticizers.[5] Vegetable oils are composed of unsaturated fatty acids and glycerol and are widely distributed in nature with the advantages of low price, being non-toxic and harmless, environmental protection and sustainability. Glycerol, saturated and unsaturated fatty acids derived from vegetable oils have a large space for modification, and hence, vegetable oils are an ideal raw material for preparing environmentally friendly plasticizers.[6]

11.2 Plant Oil-based Plasticizers

11.2.1 Plant Oil-based Epoxidized Plasticizer

A vegetable oil-based plasticizer is a kind of compound based on vegetable oils and their derivatives modified with other chemical reagents, which can be used in food packaging, children's toys, medical devices, *etc.*[7] At present, epoxidized plasticizers are obtained *via* a relatively mature process, *i.e.* by selective epoxidation of vegetable oils in industrial production, such as in the plastic industry, in the rubber industry, in coatings and in new polymer materials.[8,9] Epoxidized plasticizers not only have an outstanding plasticizing effect, but also can absorb and neutralize the chlorine atom decomposed from the PVC chain, thus hindering the continuous decomposition of PVC, making the product more stable and extending its service life, including epoxy linseed oil, epoxy castor oil, epoxy rice bran oil, epoxy cottonseed oil, epoxy sunflower oil, epoxy soybean oil (ESO), *etc.* ESO is a typical representative of epoxidized plasticizers and is widely used in the market.[10]

ESO has excellent plasticizing effect and good thermal stability, which play a good synergistic role when used together with other main plasticizers. Ferrer *et al.* evaluated ESBO as the only plasticizer and stabilizer for PVC with the content from 30 to 50 wt% and used it in food processing at high temperatures without the risk of degradation.[11] Park *et al.*[12] reported that ESO toughens tetrafunctional epoxy resins. The results indicated that the glass transition temperature (T_g) of the resultant resins decreased, while the stress intensity factor increased with the addition of ESO. Zhao *et al.*[13] used ESO as a novel plasticizer with polybutylene succinate (PBS) in a twin-screw extruder. It is found that the elongation at break could reach the maximum value (63%) along with 5 wt% ESO, 15 times that of pure PBS. Xiong *et al.*[14] added ESO to toughen the entirely bio-sourced polymers poly(lactic acid) (PLA)/starch composites using a lab-scale co-extruder. The elongation at break of the blend material was 140%, which increased from 5% compared with pure PLA.

Our group fabricated a series of epoxidized vegetable oil-based plasticizers, *e.g.* epoxidized castor oil polyol esters, epoxidized soybean oil polyol esters and epoxidized tung oil methyl ester *via* alcoholysis and epoxidation, and their plasticizing effects on PVC films were investigated systematically.[15–17] As shown in Figure 11.1, the epoxidized castor oil polyol ester was prepared

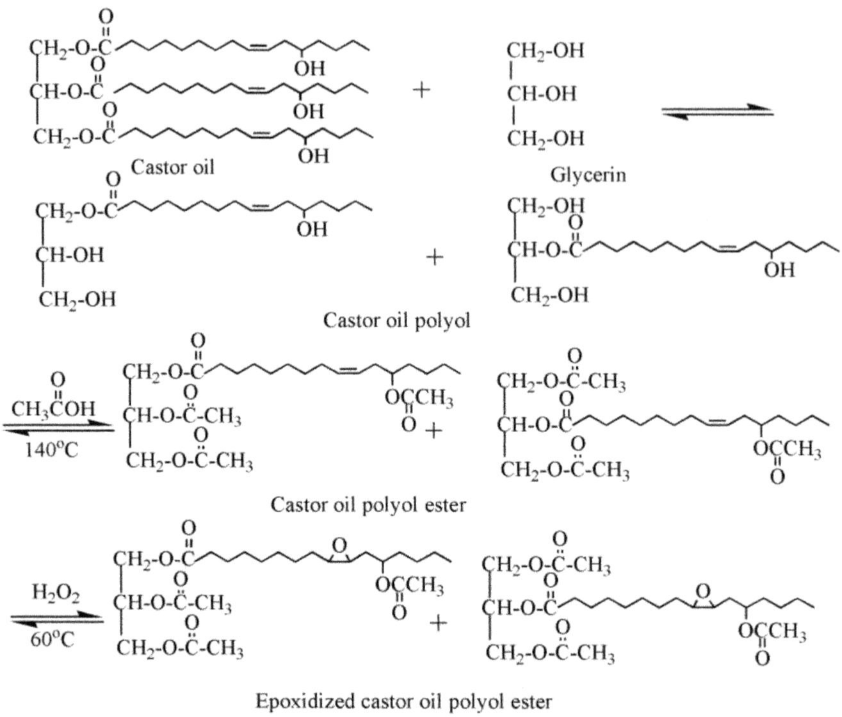

Figure 11.1 Synthesis of an epoxidized castor oil polyol ester. Reproduced from ref. 16 with permission from American Chemical Society, Copyright 2015.

and used as a plasticizer for PVC. The properties of the obtained PVC materials were investigated and compared to those of commercial plasticizers of dioctyl phthalate (DOP) and ESO. The results showed that castor oil-based polyol esters could significantly improve the thermal stability of PVC blends and can be used as main plasticizers to completely replace DOP in flexible PVC films.

Vieira *et al.*[18] reported a novel plasticizer obtained through the esterification reaction of rice fatty acids with polyols. The experiment indicated that the elongation at break of the resultant PVC materials could reach 371.2%. In addition, Faria-Machado *et al.*[19] also evaluated two epoxidant agents through a modified natural plasticizer obtained from a rice oil distillate. Even though both of them enhanced the performance of PVC materials, natural epoxidized plasticizers derived from peracetic acid (NP-Ac) possessed better ability than peroctanoicacid (NP-Oc), due to a higher degree of epoxidation in NP-Ac. Besides, a mixture of fatty acids obtained from rapeseed oil has been reported as a cheap and readily available plasticizer for PVC.[20] Silverajah *et al.*[21] prepared a fully biodegradable matrix PLA blended with three kinds of epoxidized palm oil (EPO) and found that 1 wt% of EPO was enough to improve the flexibility and strength of PLA. To obtain the best performance, EPO(3) (a mixture of EPO and soybean oil) was chosen and the tensile modulus reached 1214 MPa, while the value for pure PLA was only 1054 MPa.

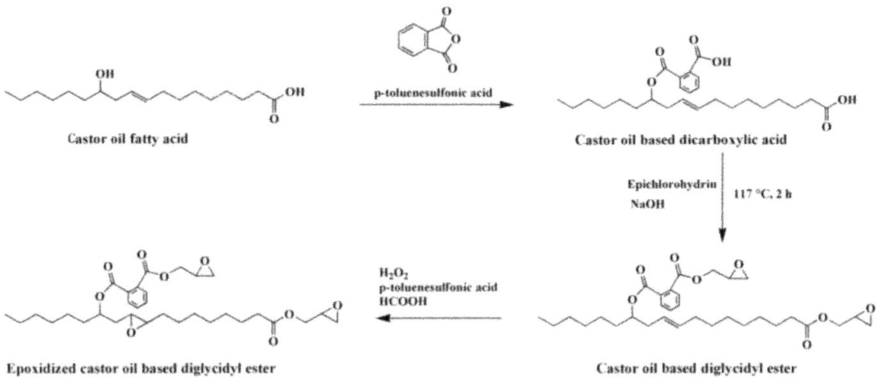

Figure 11.2 Synthesis of ECODP. Reproduced from ref. 22 with permission from John Wiley & Sons, Copyright © 2018 Wiley Periodicals, Inc.

Epoxidized plasticizers with benzene groups were also developed. Chen *et al.*[22] synthesized epoxidized castor oil-based diglycidyl-phthalate (ECODP) to replace the commercial plasticizer DOP as presented in Figure 11.2. It was found that the thermal stability, compatibility and flexibility of the PVC material were improved when incorporated with ECODP, *e.g.* the elongation at break was up to 332.9%.

Furthermore, a tung-oil-derived renewable plasticizer, tung-maleic triglycidyl ester (TMTE), was also developed by this group.[23] As shown in Figure 11.3, the chemical structure of TMTE is similar to that of ECODP and it exhibited excellent plasticizing effect and thermal stabilization for the PVC material.

Epoxidized sunflower oil (ESO) could be well applied to PLA and PVC. Benaniba *et al.*[24] investigated the stabilizing effect of ESO on the degradation of PVC and confirmed that the stabilizing performances of ESO were better than those of ESBO. Atek *et al.*[25] determined the migration of ESO from rigid PVC into food simulants by chromatography-mass spectrometry (GC–MS) with linoleic acid as an external standard, which provided a new method to detect the migration of plasticizers.

11.2.2 Plant Oil-based Polyester Plasticizers and Macromolecular Plasticizers

Polyester plasticizers are a kind of polymer with an ester group on the skeleton, which are prepared by polycondensation of diol and diacid, and they are an important series of plasticizers. On the one hand, the plasticizing efficiency of polyester plasticizers is lower than that of general plasticizers with high viscosity and less processability. On the other hand, polyester plasticizers exhibit excellent performance, *e.g.* strong migration resistance, low toxicity and safety and high solvent resistance, due to their large molecular weight.[5] Plastic prepared with polyester plasticizers possesses outstanding low-temperature impact resistance and high tear resistance and is applied in the

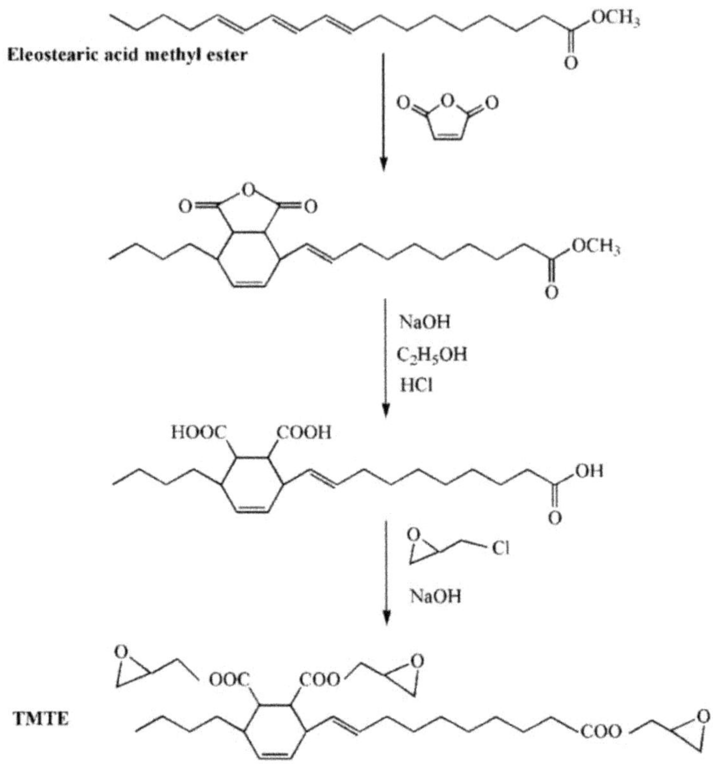

Figure 11.3 Synthesis of TMTE. Reproduced from ref. 23 with permission from American Chemical Society, Copyright 2018.

field of wires and cables, automotive interior products, tires and other indoor and outdoor products that need long-term use. Polyester plasticizers can solve the contradiction between durability, processability and low temperature. However, traditional polyester plasticizers are derived from fossil resources and are against requirements of sustainable development, *e.g.* adipic acid esters, sebacic acid esters and phthalic acid esters. Therefore, using environmentally friendly and low-cost raw materials to develop green polyester plasticizers has become a hot spot and an important direction in this field.

Bio-based polyester plasticizers have attracted a lot of attention in recent years.[26] Sander *et al.*[27] investigated the plasticizer action of oleic acid polyesters (OAP) in blends with polyethylene (PE) and polypropylene (PP). The results suggested that the OAP had a plasticizing effect on PE and PP. However, the processing torque, crystallinity and mechanical resistance of the obtained material were slightly reduced. Lindström *et al.*[28] assessed linear and branched poly(butylene adipate)s (PBA) with molecular weights ranging from 2000 to 10 000 g mol^{-1} plasticized for PVC along with the content of branching agent between 0 and 1.8%. They concluded that the degree of branching was most decisive for plasticizing efficiency.

Glycidol can be derived from glycerol, which is a kind of polyhydroxy compound with three hydroxyl groups, as a by-product of biodiesel synthesis.[29] Polyhydroxy compounds are recognized as a kind of plasticizer due to intramolecular hydrogen bonds and increase the activity space of polymer molecular chains. Lee *et al.*[30,31] developed a series of green hyperbranched plasticizers (alkyl terminal hyperbranched polyglycerol (alkyl-HPG) and butyl-esterified highly branched polycaprolactone (hbPCL)) from glycidol *via* a simple, low-cost method to produce a non-toxic, phthalate-free PVC material. As presented in Figure 11.4, alkyl-HPG was fabricated *via* anionic ring-opening polymerization followed by esterification. Alkyl-HPGs exhibited similar flexibility and stretchability to DEHP films with the corresponding properties of PVC/di(2-ethylhexyl) phthalate (DEHP), but presented considerably lower volatility than DEHP films. Additionally, the migration degree for PVC/HPG-C4 was 80% lower than that of PVC/DEHP according to leaching tests.[31] As illustrated in Figure 11.5, hbPCLs were developed *via* one-pot, solvent-free copolymerization of ε-caprolactone and glycidol, which showed a similar plasticizing effect and solvent extraction resistance to alkyl-HPG. Mixtures of hbPCLs and PVC exhibited outstanding flexibility comparable to that of DEHP. Furthermore, PVC/hbPCL-C4 showed outstanding migration stability with a weight loss after extraction being >85% lower than that of PVC/DEHP, and the elongation at break of hbPCLs was 397%, which was 21% greater than that of PVC/DEHP.

Polymers such as long-chain linear polyesters, poly(butylene 2-methylsuccinate) (PBM) and aromatic hyperbranched polyesters (HBPEs) were also considered as polymeric plasticizers to substitute for DOP.[32,33] Sunny *et al.* compared the performances of PVC plasticized with DEHP partially replaced with nitrile rubber (NBR), carboxylated nitrile rubber (XNBR) and epoxidized natural rubber (ENR). It was found that NBR enhanced the resistance and stability of PVC.[34] Moreover, the migration of plasticizers had raised many concerns, especially for food-grade PVC materials, which were investigated by experimental evaluation, legislation and theoretical prediction.[35,36] Starch is a low-price polysaccharide derived from agricultural plants and has been used as a filler to prepare biodegradable films. The effects of different sources of starch (such as maize, potato and wheat) and the type of plasticizers, *e.g.* glycerol, urea and formamide, on the physicochemical

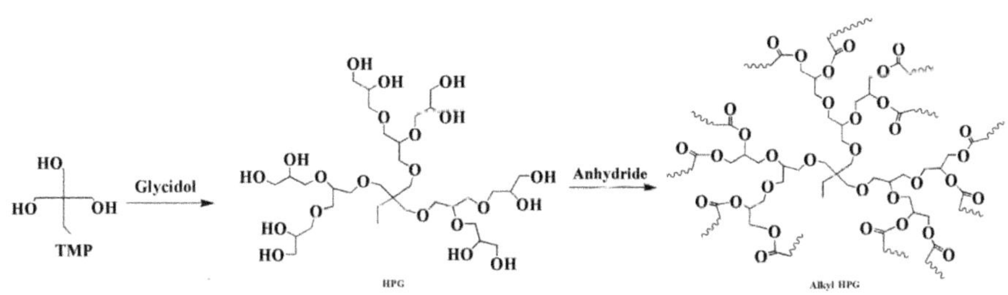

Figure 11.4 Synthesis of alkyl-HPG. Reproduced from ref. 30 with permission from the Royal Society of Chemistry.

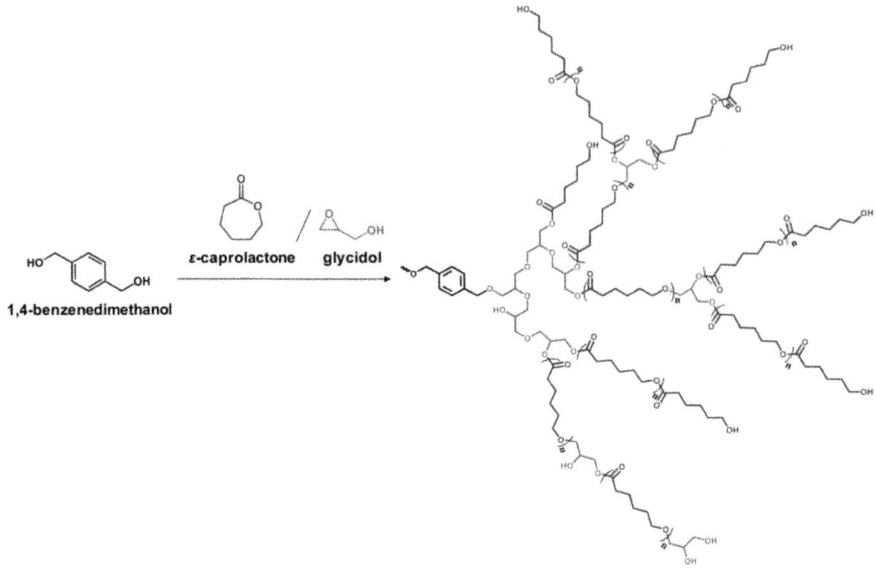

Figure 11.5 Synthesis of hbPCL. Reproduced from ref. 31 with permission from American Chemical Society, Copyright 2018.

properties were investigated by researchers. It was found that the film combination of high-amylose (>51%) starch and urea/formamide exhibited good elongational viscosity, high deformability of the melt and strain-hardening behaviour.[37] Pushpadass *et al.*[38] fabricated starch films with water, glycerol and stearic acid and assessed the effects of extrusion and plasticizers on gelatinization and molecular and structural properties. The results revealed that starch underwent fragmentation during extrusion under highly plasticized conditions. However, the degradation was not severe as compared to previous findings.

Diprotic acid, derived from the alcoholysis of vegetable oil, was utilized to synthesize polyester plasticizers. Our group also prepared a novel palm oil monoglyceride-based polyester plasticizer (POMP) and provided a general method for producing bio-based polyester plasticizers.[39] As the content of POMP increased, the PVC blends demonstrated better compatibility, the degradation temperature increased from 252.6 to 257.0 °C, the T_g dropped from 55 to 49.5 °C, the tensile strength decreased from 12.6 to 6.1 MPa, and the elongation at break decreased from 210.81% to 60.96%.

Fakhoury *et al.*[40] indicated that the development of edible and biodegradable films has attracted much attention for the protection of the environment and sustainable development, such as glycerol and sorbitol. The experiment indicated that the tensile strength and elongation at break of the obtained plasticizer with sorbitol were better than those of glycerol. In addition, our group[41,42] also developed several high-performance polyester plasticizers based on glycerol for PVC.

11.2.3 Plant Oil-based Flame Retardant Plasticizers

Phosphate esters have been used as plasticizers for a long time, which can not only endow the material with plasticizing properties, but also greatly improve its flame retardancy, including alkyl diphenyl phosphate and triphenyl phosphate.[43] Phosphate ester plasticizers are widely used because of their variety and excellent flame retardancy, in rubber, plastics, military products, textiles, electrical appliances, conveyor belts and various building materials. Due to the disadvantages of high volatility, high toxicity and low thermal stability, the development of green and high-performance phosphate ester plasticizers will be essential in the future. The chlorine content of pure PVC was 56.8% and it passed most flammability tests without flame retardants. However, plasticized PVC usually cannot present self-extinguishing behaviour.[44]

Feng *et al.*[45] synthesized a phosphorus-containing plasticizer based on castor oil (FRC) and applied it for the flame retardancy of PVC (Table 11.1(a)). With the addition of FRC, the torque reduced by 33.6%, and the flame retardancy of LOI enhanced by 31.9%. Meanwhile, the tensile strength and elongation at break decreased by 61.94% and approximately 60%, respectively, indicating that FRC

Table 11.1 Flame retardant elements (phosphate, chlorine and nitrogen) containing vegetable oil-based plasticizers.

Flame retardant elements (phosphate, chlorine and nitrogen) containing vegetable oil-based plasticizers	Ref.	No.
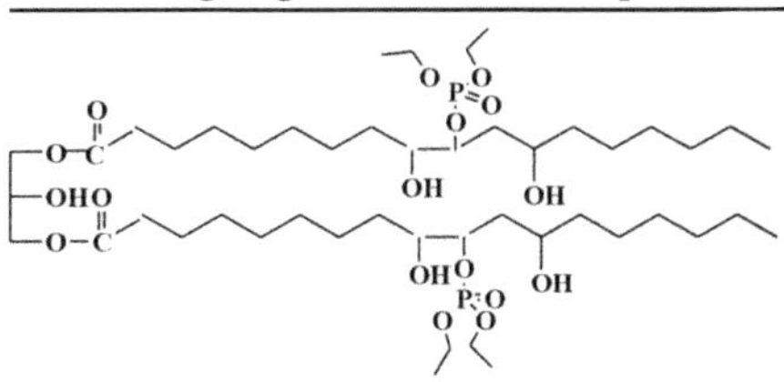	49	a
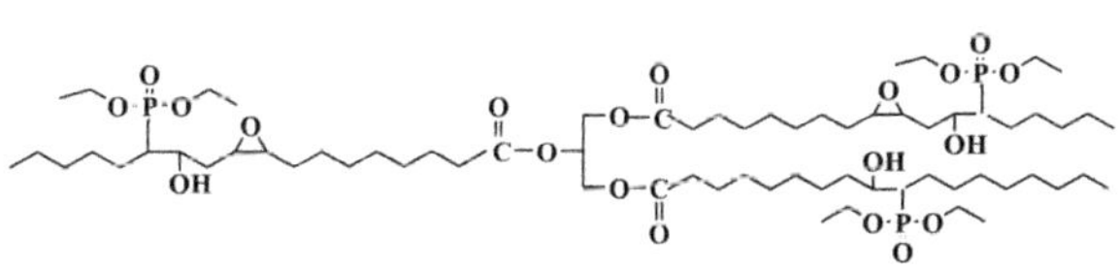	50	b
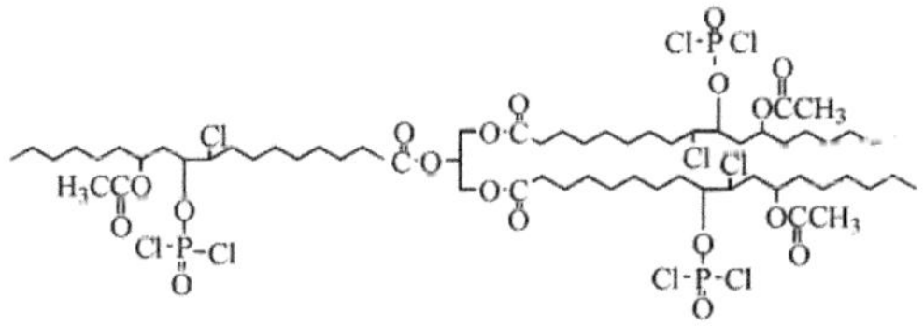	51	c
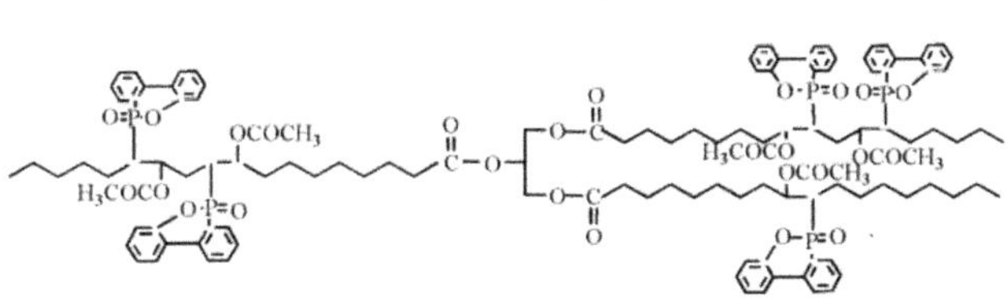	52	d

Table 11.1 (*Continued*)

Flame retardant elements (phosphate, chlorine and nitrogen) containing vegetable oil-based plasticizers	Ref.	No.
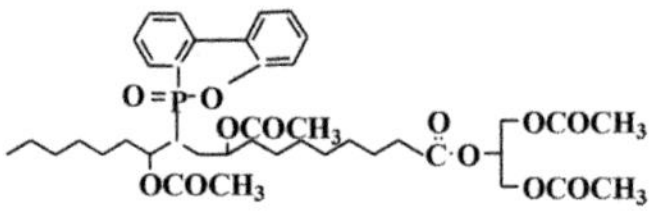	53	e
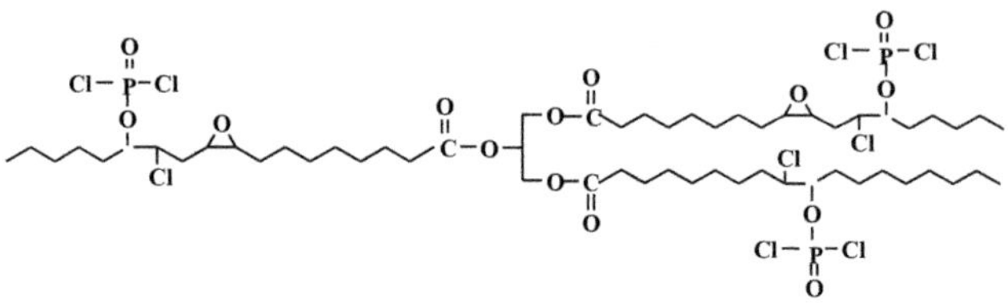	54	f
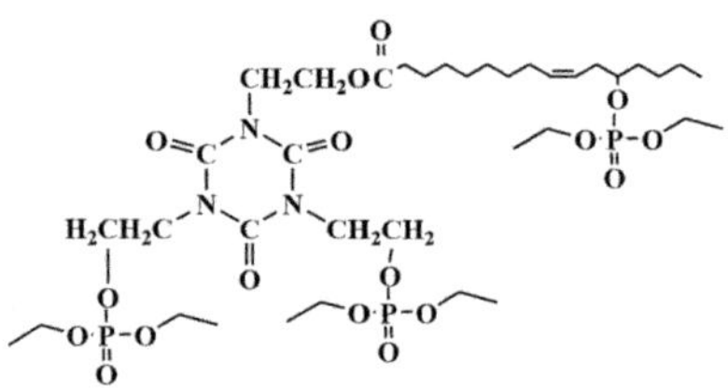	55	g
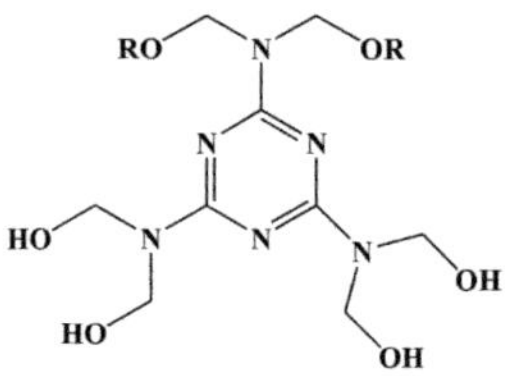	56	h
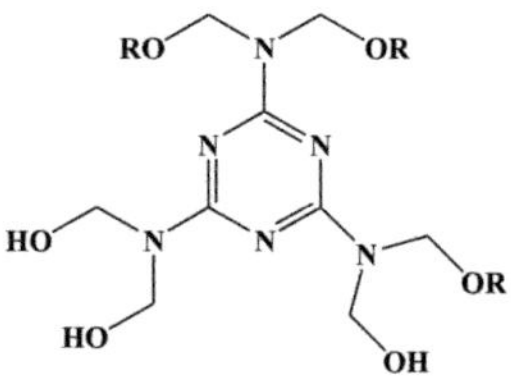	56	i
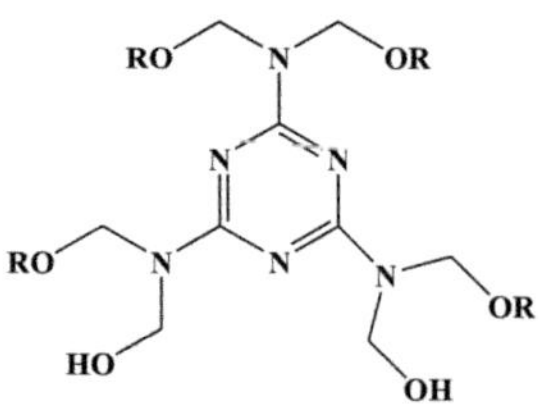	56	j
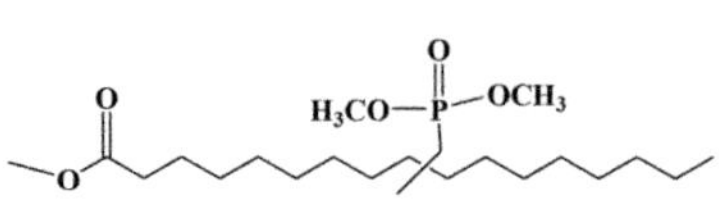	57	k

Table 11.1 (*Continued*)

Flame retardant elements (phosphate, chlorine and nitrogen) containing vegetable oil-based plasticizers	Ref.	No.
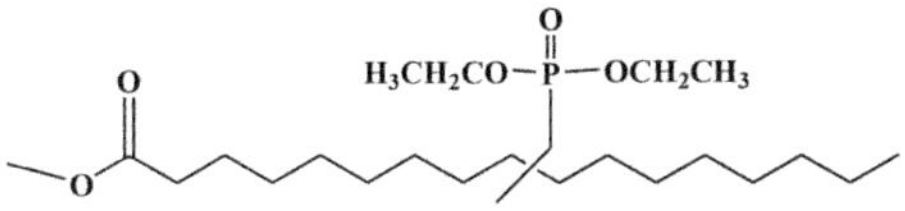	57	l
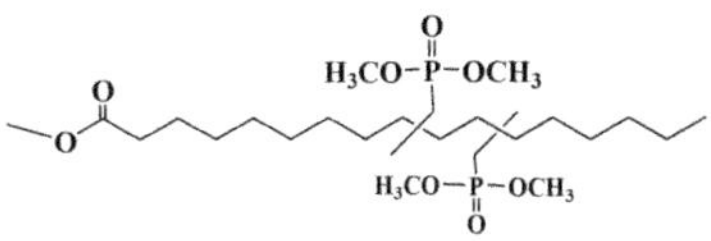	57	m
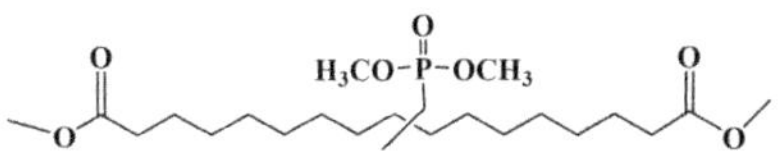	57	n

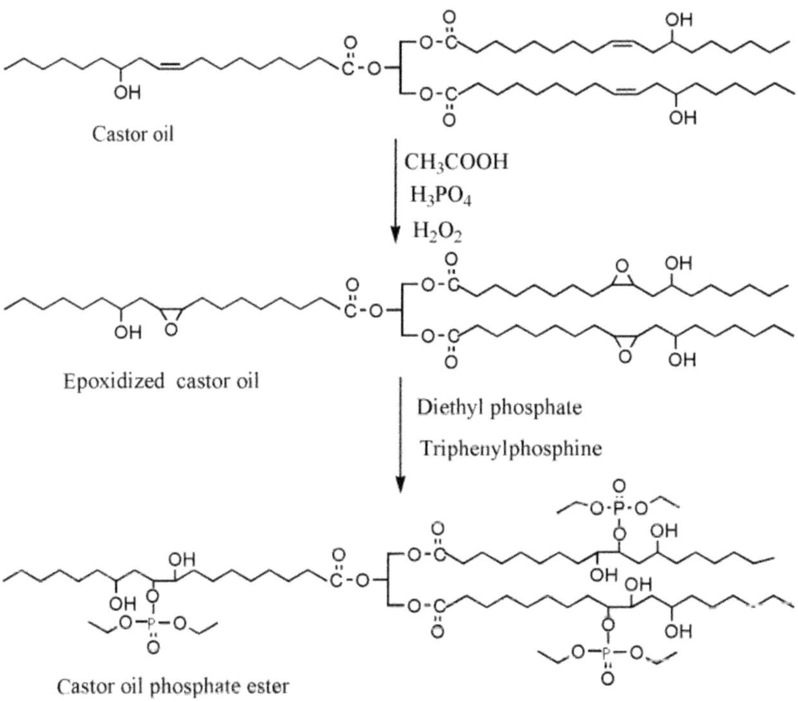

Figure 11.6 Synthesis of PPC. Reproduced from ref. 46 with permission from Springer Nature, Copyright 2015.

had a plasticizing effect on PVC. The addition of FRC caused a significant improvement in PVC materials. Our groups[46–49] developed a series of flame retardant vegetable oil-based plasticizers. As shown in Figure 11.6 and Table 11.1(b), a fame-retardant castor oil-based plasticizer (PPC) was synthesized from phosphate and diethyl phosphate.[46] As the content of PPC increased from 0 to 20 g, the results indicated that the T_g increased from 26.5 to 52.6 °C and the thermal stability was maintained by PPC until 350 °C. Furthermore, the results indicated that the solvent extraction resistance of PPC was better than that of DOP according to the leaching test.

As shown in Figure 11.7 and Table 11.1(c), a castor oil-based plasticizer containing chlorine and phosphorus (CPECO) was also prepared.[47] The LOI of PVC was 23.6% with 40 wt% of DOP, and the value of LOI reached 35.4% as half of the DOP was replaced with CPECO. Meanwhile, the peak heat release (pHRR) dropped from 379 to 289 kW m^{-2}, and the total heat release (THR) declined from 31.78 to 19.12 MJ m^{-2}.

A soybean oil-based flame retardant plasticizer (PSPE) was also prepared as presented in Figure 11.8 and Table 11.1(d).[50] The value of LOI reached 36.25%, indicating that PVC with PSPE possessed better flame retardancy. PSPE provided abundant carbon sources and acid sources, such as HPO_2 for

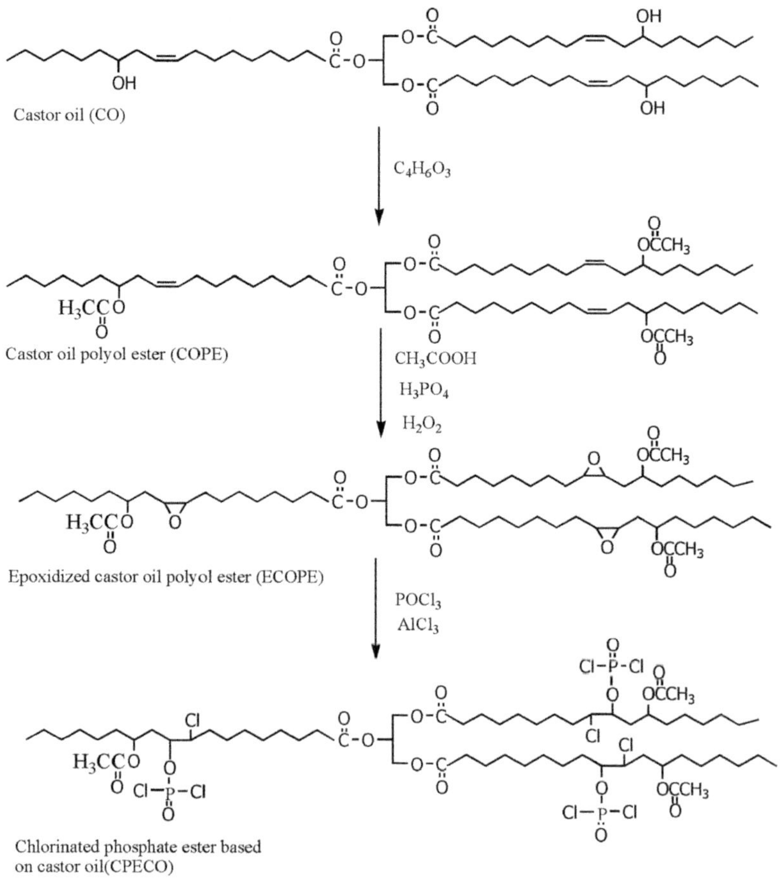

Figure 11.7 Synthesis of CPECO. Reproduced from ref. 47 with permission from the Royal Society of Chemistry.

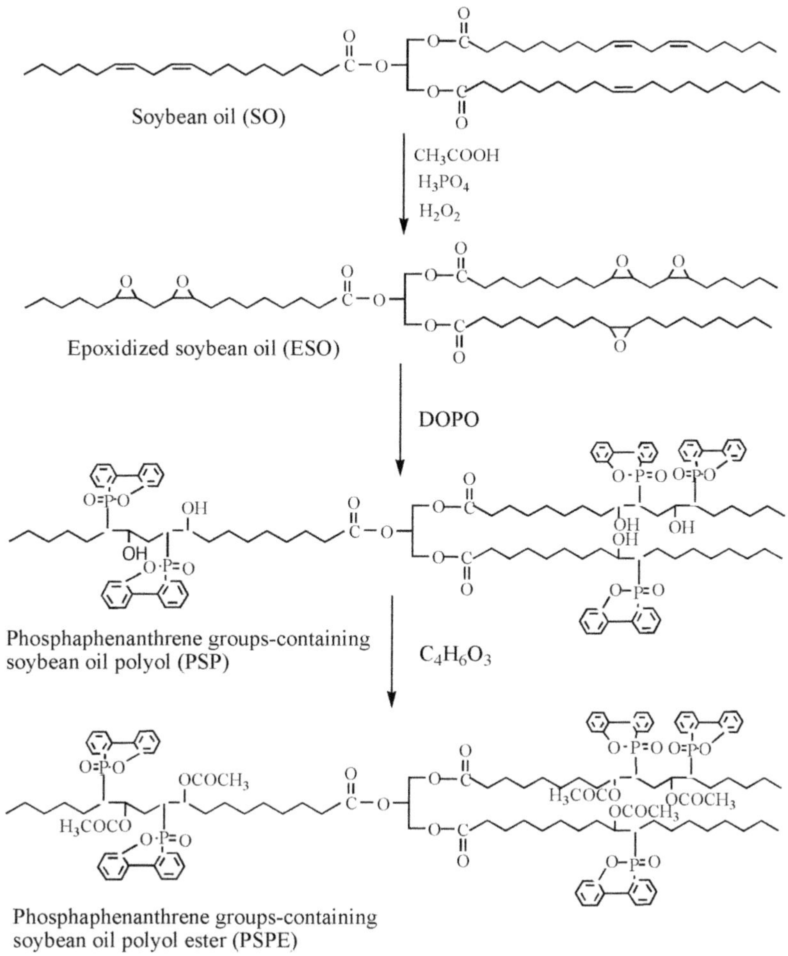

Figure 11.8 Synthesis of PSPE. Reproduced from ref. 50 with permission from Elsevier, Copyright 2015.

PVC blends, were mainly for the flame retardant function. In addition, a phosphaphenanthrene group-containing soybean-oil-based plasticizer (PSOPE) was prepared as presented in Table 11.1(e).[51] It was found that the value of LOI reached 36.2% plasticized with 40 wt% PSOPE, and the flame retardancy of PSOPE was similar to that of PSPE.

A series of synergistic flame retardant plasticizers were developed. Synergistic flame retardant plasticizers (THEIC-MA phosphate) were obtained by combining nitrogen, phosphorus and chlorine from raw materials of vegetable oil[48,49] (Figure 11.9 and Table 11.1(f), (g)). The flame retardancy of PVC was improved using THEIC-MR-phosphate *via* the formation of a char residue in the solid phase and dilution and isolation of oxygen in the gas phase, which was confirmed by cone tests of the decreased HRR.

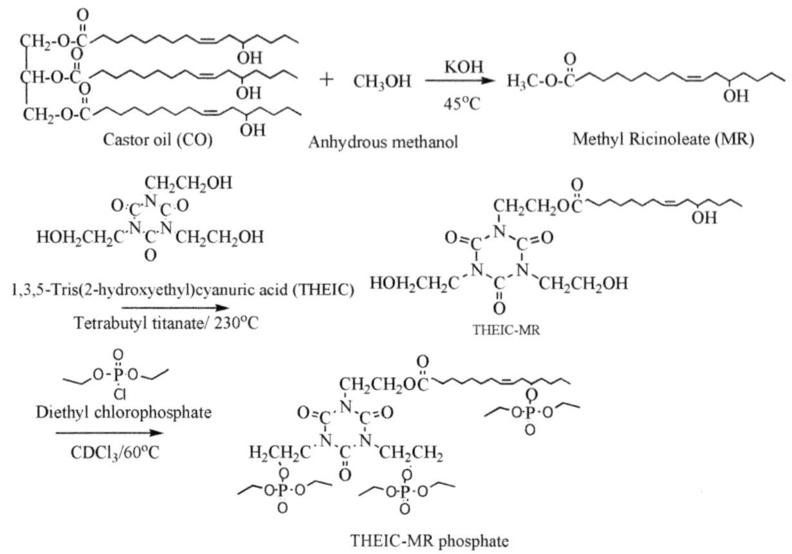

Figure 11.9 Synthesis of THEIC-MA. Reproduced from ref. 49 with permission from the Royal Society of Chemistry.

Four kinds of tung-oil-based ester plasticizers containing nitrogen (GEHTMA-1, GEHTMA-2, GEHTMA-3 and GEHTMA-4) were prepared to replace dioctyl terephthalate (DOTP) (Figure 11.10 and Table 11.1(h)–(j)).[52] GEHTMA-3 displayed better mechanical properties and well-balanced flexibility and strength compared with other tung-oil-based ester plasticizers, due to the hydroxyl, epoxy, benzene ring, ester and nitrogen groups in the structure of GEHTMA-3.

11.2.4 Glyceryl Ester Plasticizer

Glycerol, as a by-product of biodiesel production, was also applied in the field of plasticizers.[53] All the glycerol triesters were compatible with PVC and exhibited an effective plasticizing effect on PVC. Palacios *et al.*[54] prepared eight kinds of glycerol triesters from butanoic (butyric), propanoic, isobutanoic (isobutyric), isopentanoic (isovaleric) and benzoic acids (Figure 11.11). Glyceryl ester was also a perfect addition for starch plasticizers.[55,56] Chang *et al.*[57] developed a glycerol plasticized-wheat starch (GPS)/CN nanocomposite *via* a casting process. As the content of nanoparticles increased from 0 to 5 wt%, the tensile strength of the material increased from 3.15 to 10.98 MPa.

11.2.5 Plant Oil-based Reactive Plasticizers and Internally Plasticized PVC

Using internal plasticizers, attached to PVC chains by covalent bonds, is an effective strategy to avoid the migration of plasticizers *via* increasing the

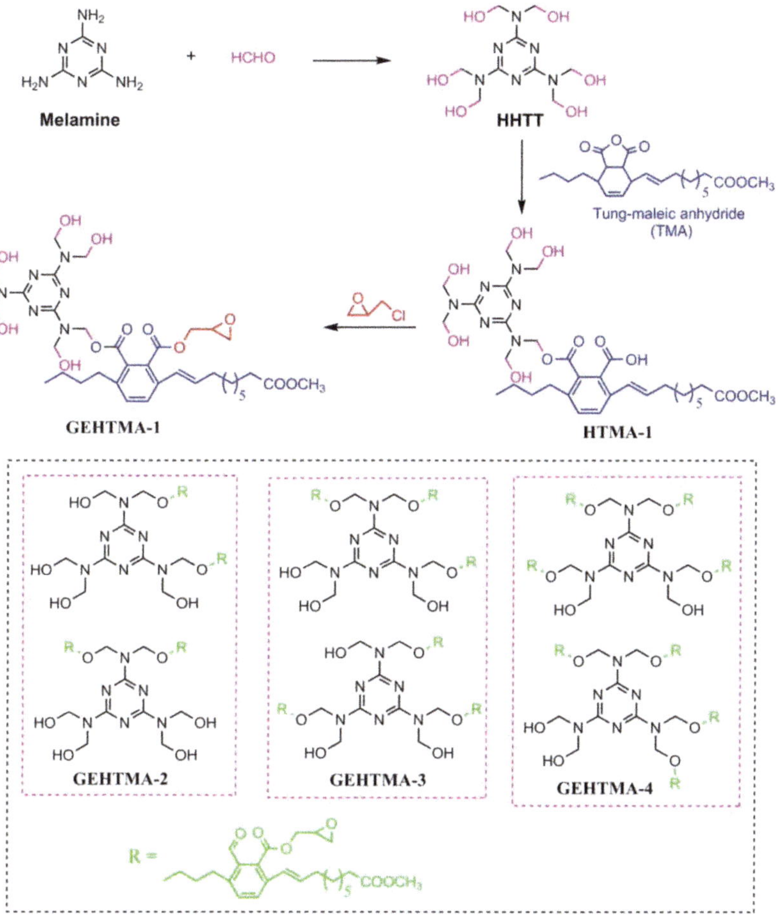

Figure 11.10 Synthesis of GEHTMA. Reproduced from ref. 52 with permission from the Royal Society of Chemistry.

mutual movement of the polymer chains.[58,59] Potential non-migratory bio-based plasticizers were prepared by a click reaction of azide functionalized PVC and an alkynyl group.[60,61] As presented in Figure 11.12. Alkynylation EAMR-DOPO was grafted onto a PVC chain. Leaching tests indicated that the plasticizer exhibited non-migration and its solvent extraction resistance and volatility resistance were excellent, which were suitable for potential applications with the high requirement of non-migration.

11.2.6 Other Plant Oil-based Plasticizers

Li *et al.*[62] synthesized a tung oil-derived plasticizer (epoxidized-C21-DAE) *via* transesterification, Diels–Alder reaction and epoxidation (Figure 11.13). The obtained epoxidized-C21-DAE was utilized as the main plasticizer and

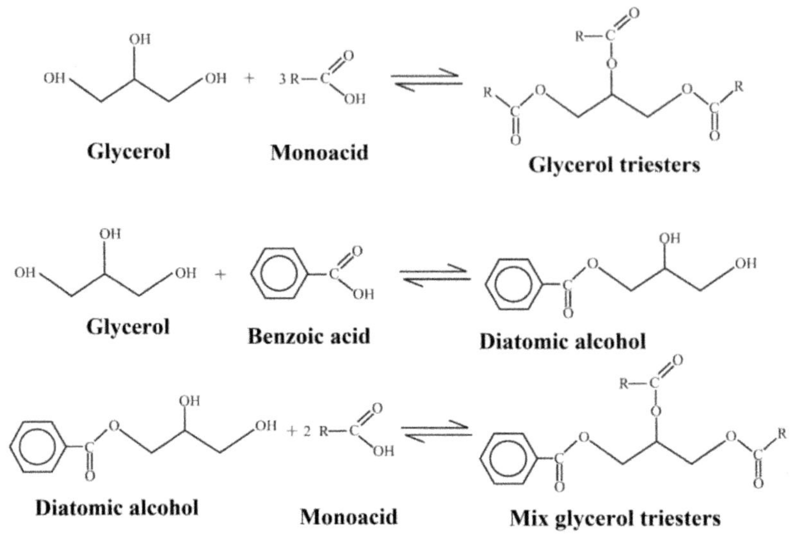

Figure 11.11 Synthesis of glycerol triesters and mix glycerol triesters. Reproduced from ref. 54 with permission from John Wiley & Sons, Copyright © 2014 Society of Plastics Engineers.

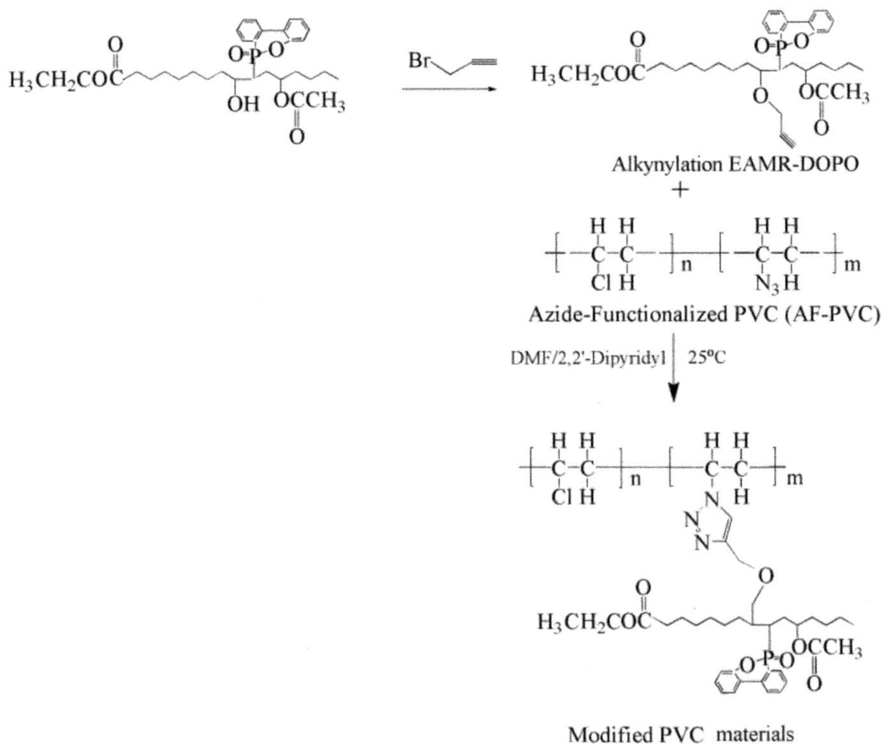

Figure 11.12 Synthesis of internally plasticized PVC. Reproduced from ref. 60 with permission from Elsevier, Copyright 2017.

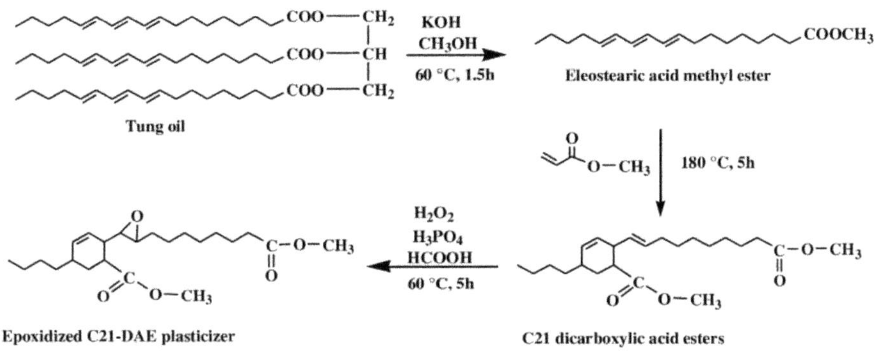

Figure 11.13 Synthesis of epoxidized-C21-DAE. Reproduced from ref. 62 with permission from Elsevier, Copyright 2017.

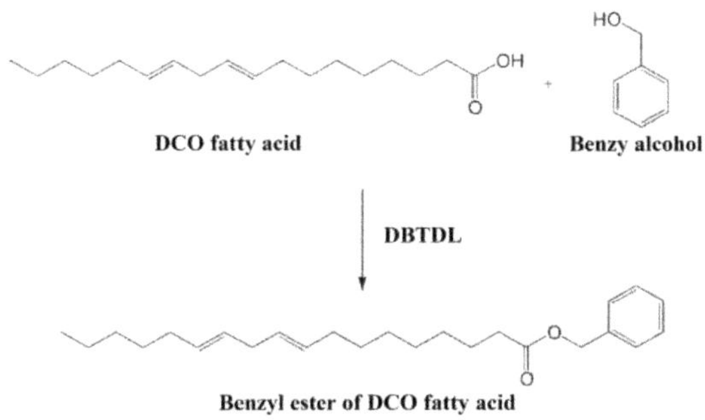

Figure 11.14 Synthesis of the benzyl ester of DCO fatty acid. Reproduced from ref. 63 with permission from John Wiley & Sons, Copyright © 2013 Society of Chemical Industry.

auxiliary stabilizer, and it exhibited excellent thermal stability and plasticizing ability with advanced migration and volatility stabilities than DOTP.

Mehta *et al.*[63] investigated the castor oil derivative, benzyl ester of DCO fatty acid (BE). It was used as a biodegradable and non-toxic plasticizer and an auxiliary plasticizer to prepare PVC materials (Figure 11.14). Its mechanical properties and thermal degradation were improved along with its anti-leaching properties simultaneously, and therefore it is an environmentally friendly plasticizer.

11.2.7 Conclusion

As traditional phthalate ester plasticizers are prone to migration and have potential toxicity, the development of green plasticizers is beneficial for environmental protection and sustainable development. The preparation of vegetable oil-based plasticizers with advantages of good biodegradability and low price has become a hot spot. For example, ESO has been industrialized and used as a primary plasticizer in food packing materials.

The flexible long fatty acid chains and unsaturated bonds of vegetable oils can be modified to produce general plasticizers, multifunctional plasticizers, such as flame retardant plasticizers by promoting the formation of char residues, internal plasticizers without plasticizer migration and so on. However, the types and total amount of vegetable oil-based plasticizers are far from meeting the level of replacing phthalate plasticizers. Therefore, considerable efforts should be devoted to designing and fabricating non-toxic and environmentally friendly bio-based plasticizers from renewable resources using the synthesis strategy.

References

1. A. Bodaghi, *Polym. Adv. Technol.*, 2019, **31**(3), 355–367.
2. Y. Feng, Z. Chu, L. Man, Y. Hu, C. Zhang, T. Yuan and Z. Yang, *ACS Sustainable Chem. Eng.*, 2019, **7**(23), 18976–18984.
3. W. Schutyser, S.-F. Koelewijn, M. Dusselier, S. Van de Vyver, J. Thomas, F. Yu, M. J. Carbone, M. Smet, P. Van Puyvelde, W. Dehaen and B. F. Sels, *Green Chem.*, 2014, **16**(4), 1999–2007.
4. Y. Yang, J. Huang, R. Zhang and J. Zhu, *Mater. Des.*, 2017, **126**, 29–36.
5. P. Jia, H. Xia, K. Tang and Y. Zhou, *Polymers*, 2018, **10**(12), 1303.
6. B. W. Chieng, N. A. Ibrahim, Y. Y. Then and Y. Y. Loo, *Polymers*, 2017, **9**(6), 204.
7. V. Volpe, G. De Feo, I. De Marco and R. Pantani, *Ind. Crops Prod.*, 2018, **122**, 545–552.
8. P. G. Nihul, S. T. Mhaske and V. V. Shertukde, *Iran. Polym. J.*, 2014, **23**(8), 599–608.
9. Y. Yong, X. Zhu, L. Zhang, Z. Tang, R. Zhang and Z. Jin, *Plast. Addit.*, 2016, **6**(3), 57–62.
10. S. G. Tan and W. S. Chow, *Polym.-Plast. Technol. Eng.*, 2010, **49**(15), 1581–1590.
11. C. Bueno-Ferrer, M. C. Garrigós and A. Jiménez, *Polym. Degrad. Stab.*, 2010, **95**(11), 2207–2212.
12. S.-J. Park, F.-L. Jin and J.-R. Lee, *Mater. Sci. Eng. A*, 2004, **374**(1–2), 109–114.
13. Y. Zhao, J. Qu, Y. Feng, Z. Wu, F. Chen and H. Tang, *Polym. Adv. Technol.*, 2012, **23**(3), 632–638.
14. Z. Xiong, Y. Yang, J. Feng, X. Zhang, C. Zhang, Z. Tang and J. Zhu, *Carbohydr. Polym.*, 2013, **92**(1), 810–816.
15. P. Jia, M. Zhang, L. Hu and Y. Zhou, *Korean J. Chem. Eng.*, 2016, **33**(3), 1080–1087.
16. P. Jia, M. Zhang, L. Hu, G. Feng, C. Bo and Y. Zhou, *ACS Sustainable Chem. Eng.*, 2015, **3**(9), 2187–2193.
17. P. Jia, Y. Ma, H. Xia, M. Zheng, G. Feng, L. Hu, M. Zhang and Y. Zhou, *ACS Sustainable Chem. Eng.*, 2018, **6**(11), 13983–13994.
18. M. G. A. Vieira, M. A. D. Silva, A. C. G. Maçumoto, L. O. D. Santos and M. M. Beppu, *Mater. Res.*, 2014, **17**(2), 386–391.

19. A. F. Faria-Machado, M. A. da Silva, M. G. A. Vieira and M. M. Beppu, *J. Appl. Polym. Sci.*, 2013, **127**(5), 3543–3549.
20. L. Montero de Espinosa, A. Gevers, B. Woldt, M. Graß and M. A. R. Meier, *Green Chem.*, 2014, **16**(4), 1883–1896.
21. V. S. Giita Silverajah, N. A. Ibrahim, W. M. Yunus, H. A. Hassan and C. B. Woei, *Int. J. Mol. Sci.*, 2012, **13**(5), 5878–5898.
22. J. Chen, Z. Liu, K. Wang, J. Huang, K. Li, X. Nie and J. Jiang, *J. Appl. Polym. Sci.*, 2019, **136**(9), 47142–47151.
23. J. Chen, Y. Wang, J. Huang, K. Li and X. Nie, *ACS Sustainable Chem. Eng.*, 2018, **6**(1), 642–651.
24. M. T. Benaniba, N. Belhaneche-Bensemra and G. Gelbard, *Polym. Degrad. Stab.*, 2001, **74**(3), 501–505.
25. D. Atek, N. Belhaneche-Bensemra and M. Turki, *Int. J. Polym. Mater.*, 2010, **59**(5), 342–352.
26. W. J. Lin, H. K. Lee and D. M. Wang, *J. Controlled Release*, 2004, **99**(3), 415–421.
27. M. M. Sander, A. Nicolau, R. Guzatto and D. Samios, *Polym. Test.*, 2012, **31**(8), 1077–1082.
28. A. Lindström and M. Hakkarainen, *J. Appl. Polym. Sci.*, 2006, **100**(3), 2180–2188.
29. M. Koller and G. Braunegg, *Polimery*, 2015, **60**(5), 298–308.
30. K. W. Lee, J. W. Chung and S.-Y. Kwak, *Green Chem.*, 2016, **18**(4), 999–1009.
31. K. W. Lee, J. W. Chung and S.-Y. Kwak, *ACS Sustainable Chem. Eng.*, 2018, **6**(7), 9006–9017.
32. M. G. McKee, S. Unal, G. L. Wilkes and T. E. Long, *Prog. Polym. Sci.*, 2005, **30**(5), 507–539.
33. T. Xie, C. Gao, C. Wang, S. E. Shen and Y. Wu, *Polym.-Plast. Technol. Eng.*, 2014, **53**(5), 465–471.
34. M. C. Sunny, P. Ramesh and K. E. George, *J. Elastomers Plast.*, 2016, **36**(1), 19–31.
35. A. Fankhauser-Noti and K. Grob, *Trends Food Sci. Technol.*, 2006, **17**(3), 105–112.
36. J. L. Audic, D. Reyx and J. C. Brosse, *J. Appl. Polym. Sci.*, 2003, **89**(5), 1291–1299.
37. R. Zullo and S. Iannace, *Carbohydr. Polym.*, 2009, **77**(2), 376–383.
38. H. A. Pushpadass, A. Kumar, D. S. Jackson, R. L. Wehling, J. J. Dumais and M. A. Hanna, *Starch - Stärke*, 2009, **61**(5), 256–266.
39. P. Y. Jia, M. Zhang, L. H. Hu and Y. H. Zhou, *Pol. J. Chem. Technol.*, 2016, **18**(1), 9–14.
40. F. M. Fakhouri, S. Maria Martelli, L. Canhadas Bertan, F. Yamashita, L. H. Innocentini Mei and F. P. Collares Queiroz, *LWT.*, 2012, **49**(1), 149–154.
41. P. Jia, C. Bo, L. Hu, M. Zhang and Y. Zhou, *J. Vinyl Addit. Technol.*, 2016, **22**(4), 514–519.
42. P. Jia, C. Bo, L. Hu and Y. Zhou, *Korean J. Chem. Eng.*, 2014, **32**(3), 547–551.

43. B. L. Wadey, *J. Vinyl Addit. Technol.*, 2003, **9**(4), 172–176.
44. S. V. Levchik and E. D. Weil, *Polym. Adv. Technol.*, 2005, **16**(10), 707–716.
45. G. Feng, P. Jia, L. Zhang, L. Hu, M. Zhang and Y.-H. Zhou, *Korean J. Chem. Eng.*, 2015, **32**(6), 1201–1206.
46. P. Jia, M. Zhang, L. Hu, C. Bo and Y.-H. Zhou, *J. Therm. Anal. Calorim.*, 2015, **120**(3), 1731–1740.
47. P. Jia, M. Zhang, C. Liu, L. Hu, G. Feng, C. Bo and Y. Zhou, *RSC Adv.*, 2015, **5**(51), 41169–41178.
48. P. Jia, M. Zhang, L. Hu, C. Bo and Y. Zhou, *Thermochim. Acta*, 2015, **613**, 113–120.
49. P. Jia, L. Hu, G. Feng, C. Bo, J. Zhou, M. Zhang and Y. Zhou, *RSC Adv.*, 2017, **7**(2), 897–903.
50. P. Jia, M. Zhang, L. Hu, J. Zhou, G. Feng and Y. Zhou, *Polym. Degrad. Stab.*, 2015, **121**, 292–302.
51. P. Jia, M. Zhang, L. Hu, G. Feng, C. Bo, C. Liu and Y. Zhou, *Ind. Crops Prod.*, 2015, **76**, 590–603.
52. M. Wang, X. Song, J. Jiang, J. Xia, H. Ding and M. Li, *New J. Chem.*, 2018, **42**(4), 2422–2431.
53. P. D. Vaidya and A. E. Rodrigues, *Chem. Eng. Technol.*, 2009, **32**(10), 1463–1469.
54. O. Y. S. Palacios, P. C. N. Rincon, J. P. Corriou, M. C. Pardo and C. Fonteix, *J. Vinyl Addit. Technol.*, 2014, **20**(2), 65–71.
55. K. S. Mikkonen, S. Heikkinen, A. Soovre, M. Peura, R. Serimaa, R. A. Talja, H. Helén, L. Hyvönen and M. Tenkanen, *J. Appl. Polym. Sci.*, 2009, **114**(1), 457–466.
56. S. Sun, Y. Song and Q. Zheng, *Food Hydrocolloids*, 2007, **21**(7), 1005–1013.
57. P. R. Chang, R. J. Jian, P. W. Zheng, J. G. Yu and X. F. Ma, *Carbohydr. Polym.*, 2010, **79**(2), 301–305.
58. P. Jia, L. Hu, G. Feng, C. Bo, M. Zhang and Y. Zhou, *Mater. Chem. Phys.*, 2017, **190**, 25–30.
59. P. Jia, L. Hu, X. Yang, M. Zhang, Q. Shang and Y. Zhou, *RSC Adv.*, 2017, **7**(48), 30101–30108.
60. P. Jia, L. Hu, M. Zhang, G. Feng and Y. Zhou, *Eur. Polym. J.*, 2017, **87**, 209–220.
61. K. W. Lee, J. W. Chung and S. Y. Kwak, *Macromol. Rapid Commun.*, 2016, **37**(24), 2045–2051.
62. M. Li, S. Li, J. Xia, C. Ding, M. Wang, L. Xu, X. Yang and K. Huang, *Mater. Des.*, 2017, **122**, 366–375.
63. B. Mehta, M. Kathalewar and A. Sabnis, *Polym. Int.*, 2014, **63**(8), 1456–1464.

CHAPTER 12

Genetic and Process Engineering for Select Glycolipid Biosynthesis from Plant/Algal Oils or Their Derivatives[†]

JOSEPH MSANNE AND RICHARD D. ASHBY*

United States Department of Agriculture, Agricultural Research Service, Eastern Regional Research Center, Sustainable Biofuels and Coproducts Research Unit, 600 East Mermaid Lane, Wyndmoor, PA 19038, USA
*Email: Rick.Ashby@usda.gov

12.1 Introduction

Commercial bioproducts and biomaterials are largely derived from plants, many of which are typically used as food or feed, resulting in direct competition for arable land and potable water, thus threatening food security and increasing demand worldwide. Microorganisms offer potential sources for sustainable production of bioproducts and specialty materials; they usually have fast growth rates with high biomass yields and do not affect the supply of food or other crop products. Microbial biosurfactants are of great interest due to their various industrial and medical applications. They are

[†]Mention of trade names or commercial products in this publication is solely for the purpose of providing specific information and does not imply recommendation or endorsement by the U.S. Department of Agriculture (USDA). USDA is an equal opportunity provider and employer.

Green Chemistry Series No. 83
Green Chemistry and Green Materials from Plant Oils and Natural Acids
Edited by Zengshe Liu and George Kraus

Published by the Royal Society of Chemistry, www.rsc.org

surface-active bioproducts with amphipathic structures composed of both hydrophobic and hydrophilic moieties.[1–3] Considerable amounts may be produced by a variety of microorganisms, including yeasts and bacteria,[4–6] and some biosurfactants have significant commercial prospects and can be supplied through various fermentation methods. Biosurfactants encompass low molecular weight compounds (*i.e.*, glycolipids and lipopeptides) and high molecular weight polymers (*i.e.*, lipoproteins).[4] A few examples include the sophorolipid glycolipids produced by the yeast *Starmerella bombicola*,[7] surfactin lipopeptides produced by the Gram-positive bacterium *Bacillus subtilis*,[8] and lipoproteins produced by the fungus *Streptomyces* sp.[9] They are either secreted into the growth medium as extracellular compounds or tethered to the microbial cell surface as structural components.[1,5,10]

Due to their amphipathic attributes, biosurfactants may function as emulsifiers that can reduce surface tensions.[11] They have been utilized in industries including food manufacturing, textile, cosmetics, oil recovery and bioremediation, and nanotechnology.[12–18] In addition, some biosurfactants have shown important medical and pharmaceutical properties including anticancer, antibiofilm, and antimicrobial activities.[19–24] Currently, most chemical surfactants are produced from non-renewable sources, particularly the petrochemical industry.[25,26] The application of chemical surfactants in synthetic detergents *per se* has detrimental effects on the ocean and wildlife due to their high toxicity and accumulation in the environment.[27] Therefore, large-scale production of biosurfactants as sustainable and effective alternatives is gaining more attention,[1] since they have a lower environmental impact and a higher biodegradability, and they retain their activity and specificity under non-optimal conditions (*e.g.*, extreme temperature or pH, *etc.*).[27–33] More importantly, microbial biosurfactants can be produced using inexpensive substrates including simple sugars, oils, glycerol, and various hydrocarbon molecules, as well as renewable low-value agricultural feedstocks.[1,4,5,8] This is essential for the economic viability and sustainability of commercial biosurfactant production. However, it is also important to understand the basic biology of the producing microbes, particularly the effects of environmental conditions on cellular metabolism and the regulation of biosynthetic pathways.

Despite the wealth of literature investigating the optimization of microbial production of biosurfactants *via* fermentation, the cellular and molecular attributes of several of the producing microbes remain poorly understood. More knowledge at the genetic level is a prerequisite to identify and characterize potential targets for metabolic engineering strategies. For instance, the availability of nuclear genome sequences is essential for the identification of *cis*-regulatory elements, such as specific promoters used in vector cloning and transgene regulation, as well as the development of dependable transformation and selection procedures for sustainable production of specialty bioproducts. So far, the biosynthetic pathways and genetic regulation at the molecular level have been described for only a few biosurfactants, particularly low molecular weight glycolipids.[34,35] Some of these pathways will be further discussed in this chapter, but in general the biosynthesis of glycolipid biosurfactants involves enzymes such as glycosyltransferases and acyltransferases, which catalyze the transfer of sugar and lipid moieties to the final glycolipid structure, while in some instances cytochromes and other *trans*-regulatory elements (*e.g.*, transcription factors) are also involved in the process. In addition to the overview of known metabolic regulations involved in glycolipid biosynthesis, the knowledge available on relevant microbial strains and yield and quality improvements upon introduction of novel traits through genetic engineering of biosurfactant-producing microorganisms are discussed.

12.2 Synthesis and Regulation of Select Glycolipid Biosurfactants

Several low molecular weight glycolipid biosurfactants commonly found in nature have been well described and documented.[12,18] They are composed of hydrocarbon chains of fatty acids or hydroxy fatty acids that are directly linked to a carbohydrate moiety (*e.g.*, rhamnose, glucose, mannose, sophorose, trehalose, *etc.*) *via O*-glycosidic or ester bonds.[4,36,37] As a result, these associations yield glycolipids such as rhamnolipids, sophorolipids, mannosylerythritol lipids, cellobiose lipids, and trehalose lipids among others synthesized by a variety of bacteria and yeasts.[12,18,34,38] Environmental conditions may significantly alter glycolipid titers, and particularly growth substrates and deficiency in some nutrients or trace elements (*e.g.*, nitrogen, phosphorus, iron, *etc.*) were found to enhance production.[18] Secreted glycolipids are not required for cell survival, and it is not clear why they are abundantly produced. However, they may function as repositories for carbon storage and/or defense mechanisms against invading microorganisms, since they can intercalate and destabilize biological membranes resulting in cell destruction.[12,39,40] In the following sections, the genetic regulation of select representative glycolipid biosynthetic pathways is further discussed.

12.2.1 Rhamnolipids (RLs)

The interest in RL biosurfactants is due to their excellent properties and high content in some microbes, which is of great importance for various industries. This class was among the early glycolipids described in the Gram-negative bacterium *Pseudomonas aeruginosa*,[41] an opportunistic human pathogen that was found to produce high titers of shorter chain RLs.[42] Under growth-limiting conditions, particularly nitrogen deprivation, *P. aeruginosa* could alter its lipid metabolic pathways towards the formation and accumulation of RLs.[4,43,44] Under optimal conditions for RL production, titers as high as 112 $g\,L^{-1}$ have been previously reported.[45] In addition to *Pseudomonas* sp., considerable amounts of RLs with longer chain fatty acids were identified from cultures of pathogenic *Burkholderia* sp.[46–48] Moreover, several RL forms were characterized in non-pathogenic microbes including the thermophilic bacteria *Thermus* and *Meiothermus* and the gliding bacteria *Myxococcus* that is capable of synthesizing an unusual form of RL termed myxotyrosides,[1] which carry a tyrosine residue in addition to rhamnose and fatty acids.

12.2.1.1 Rhamnolipid Biosynthetic Pathway

Rhamnolipid biosynthesis has been extensively studied in *P. aeruginosa*,[41,49,50] and several key enzymes have been identified.[4,51] However, more research is needed to understand signal perception and transduction under environmental stress conditions, as well as different molecular mechanisms leading to RL biosynthesis and accumulation. RLs are structurally diverse with over 60 molecular forms identified containing either one or two rhamnose sugar moieties that are linked to one, two, or seldom three β-hydroxy fatty acids (Figure 12.1).[17,52–56] Rhamnose may be produced *in vivo* from glucose,[57–59] whereas fatty acids, the building blocks of various types of lipids, are either synthesized by the fatty acid synthase II (FAS II) complex[60–62] or produced from lipid turnover through the β-oxidation pathway.[57,63] The enzyme PhaG in *Pseudomonas* could redirect β-hydroxyacyl-ACP from *de novo* fatty acid biosynthesis towards RL production.[61,64] To initiate RL biosynthesis, two intermediates of hydroxylated fatty acids are dimerized by esterification to form 3-(3-hydroxyalkanoyloxy) alcanoic acids (HAAs), commonly the hydrophobic

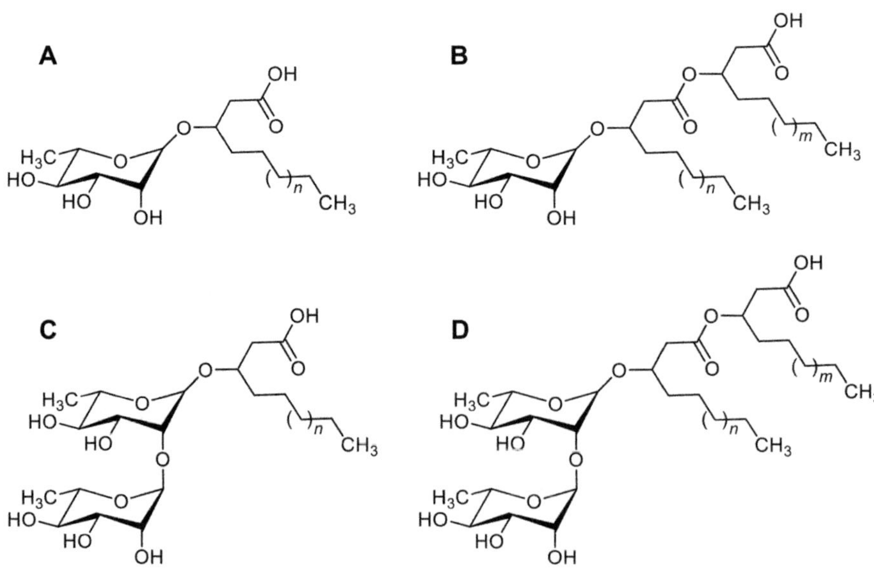

Figure 12.1 Chemical structures of rhamnolipid biosurfactants comprising mono-rhamno-mono-lipidic congeners (A), mono-rhamno-di-lipidic congeners (B), di-rhamno-mono-lipidic congeners (C), and di-rhamno-di-lipidic congeners (D) (*n* and *m* typically range between 1 and 7 carbon units resulting in overall lipid chain lengths of between 8 and 14 carbons but can be shorter or longer depending on the microbial source).[71,72]

β-hydroxydecanoyl-β-hydroxydecanoate backbone. This reaction is catalyzed by the acyltransferase enzyme RhlA.[60,61] A thymidine diphospho-rhamnose (or dTDP-L-rhamnose) sugar moiety is subsequently linked to β-hydroxydecanoyl-β-hydroxydecanoate to generate mono-rhamnolipids, a reaction carried out by the enzyme rhamnosyltransferase I (RhlB).[4,65] Finally, a sequential transfer of a dTDP-L-rhamnose by rhamnosyltransferase II (RhlC) generates di-rhamnolipids,[51] with rhamnose residues usually bound together by an α-1,2-glycosidic linkage.[6] The di-rhamnolipid L-rhamnosyl-L-rhamnosyl-β-hydroxydecanoyl-β-hydroxydecanoate (Rha_2-C_{10}-C_{10}) and mono-rhamnolipid L-rhamnosyl-β-hydroxydecanoyl-β-hydroxydecanoate (Rha-C_{10}-C_{10}) classes were identified as the main RL congeners from *P. aeruginosa* liquid cultures.[44] Moreover, mono-rhamno-mono-lipid (Rha-C_{10}) and di-rhamno-mono-lipid (Rha_2-C_{10}) classes were also identified from resting *Pseudomonas* cultures.[4,23,66,67] The formation of the latter may be catalyzed by hydrolase enzymes,[65] which can cleave a β-hydroxydecanoic acid (10 carbons) from the RL structure. Although β-hydroxydecanoic acid was reported as the most common fatty acid in RLs, other molecules having 6 to 24 saturated or unsaturated carbon units have also been identified in bacteria,[4,52,68–70] namely *Pseudomonas desmolyticum* can produce RLs with β-hydroxy fatty acids containing 6 and 8 carbon units,[71] while *Thermus* sp. can produce RLs with long-chain fatty acids of 24 carbons.[72]

12.2.1.2 Genetic Regulation and Heterologous Expression of RL Genes

Environmental and growth conditions seem to have major impacts on cellular metabolism and regulation of genes involved in microbial RL biosynthesis. Using a reverse genetics approach, the *rhl* operon was identified in *P. aeruginosa*,[4,44] with *rhlA* and *rhlB* as the key genes involved in mono-rhamnolipid metabolism.[36,38,44] However, the *rhlC* gene required for di-rhamnolipid biosynthesis was found on another operon that also encodes for proteins of the major facilitator superfamily (MFS), usually involved in substrate uptake and secretion.[51,65] An *rhlC* gene knock-out in *P. aeruginosa* resulted in about 23% increase in mono-rhamnolipid classes, whereas di-rhamnolipids were not detected in the mutants.[70] It is also worth noting that wild-type *P. chlororaphis*, exclusively producing mono-rhamnolipids, could synthesize di-rhamnolipids following the introduction of the *rhlC* gene.[73] In contrast to *P. aeruginosa*, other bacteria capable of RL biosynthesis including *Burkholderia pseudomallei* and *Burkholderia thailandensis* retain the *rhl* genes on the same operon.[74,75] Downstream of *rhlA/B*, the regulatory genes *rhlR* and *rhlI*, encoding for a putative transcriptional activator and an autoinducer synthesis protein respectively, are involved in signal transduction and quorum sensing regulation required for the induction of RL production.[44,76–78] *P. aeruginosa* mutants that did not express *rhlR* activity could not synthesize RLs, whereas the overexpression of this gene resulted in a slight increase in glycolipid synthesis.[4] Therefore, RhlR is essential for the expression of *rhlA/B* genes found to increase 20-fold upon the onset of the stationary phase of cell growth, when nitrogen levels severely dropped.[44] Subcellular localization analyses of RhlA, RhlB, and RhlC enzymes showed that these proteins possessing signal peptides and membrane-spanning domains are mainly targeted to the inner cell membrane where RLs are most likely synthesized prior to export outside of the cell.[1,44,51]

The annotation of *rhl* operon(s) involved in RL metabolic pathways allowed the identification of targets for metabolic engineering and heterologous gene expression that may have the potential for increasing microbial glycolipid production. Since the main producers (*i.e., Pseudomonas* and *Burkholderia*) are potential disease-causing agents,[17,79] finding competent host organisms with appropriate genetic and biochemical background is an essential first step for target heterologous RL production using genetic material from either of the microbial genera.[56,80] The first experiments were an attempt to engineer *Escherichia coli* DH5α with *rhlA/B* genes regulated by the inducible *tac* promoter. Although the enzyme activity was confirmed in the mutant cells, RL production was not detected.[44] Furthermore, *E. coli* was engineered using *rhlA* genes from *Pseudomonas fluorescens* and *Dickeya dadantii*.[81] These manipulations resulted in the biosynthesis of the 3-(3-hydroxyalkanoyloxy) alcanoic acid RL backbone molecules with variable fatty acid lengths of 10, 12, and 14 carbon units.[81] Different hosts were also engineered to heterologously produce RLs with various levels of success. Results from these studies were previously reviewed.[79] Briefly, a noticeable increase in RL production was determined in hosts including *Cellvibrio japonicus*,[82] *Pseudomonas putida*,[83–85] and *Pseudomonas stutzeri*[86] following the expression of a different combination of *rhl* genes with native, synthetic, or inducible promoters.[79] Overall, the substrate availability (*e.g.,* fatty acids, rhamnose) was essential and a requirement for RL biosynthesis in heterologous hosts.[44] In addition to *rhl* genes, researchers co-expressed the *rmlBDAC* operon to improve the conversion of glucose to rhamnose. This resulted in increased RL biosynthesis in *E. coli*[87] and in *Saccharomyces cerevisiae*.[88]

12.2.1.3 Functional Properties and Characteristics of RL Biosurfactants

As determined in *P. aeruginosa*, rhamnolipids exhibit important *in vivo* biological activities, and they are essential for cell-to-cell interactions, quorum sensing, and biofilm formation.[89–91] As emulsifiers, they can facilitate the uptake of various hydrocarbon molecules thus allowing the cells to use them as growth substrates.[4,92–94] RLs also have wide applications with functional properties affected by the molecule structures, particularly chain length and the number of rhamnose residues.[70,95–97] As biosurfactants, they can replace chemical surfactants in washing detergents and cleaning supplies[98] and can significantly lower the surface tension in water as well as the critical micelle concentration values.[43] Due to their potency and physico-chemical properties, RLs have also found applications in soil bioremediation and the simultaneous removal of heavy metals and hydrophobic pollutants.[99–103] It was determined that mono-rhamnolipids were more effective than di-rhamnolipids in the biodegradation of crude oil at higher ambient temperatures.[104,105] Furthermore, mono-rhamnolipids have found applications in post-harvesting crop handling and storage as antimicrobial agents since they can inhibit the growth of several bacterial and fungal species.[6,70,106] On the other hand, di-rhamnolipids showed potent therapeutic properties; they were able to inhibit *in vitro* proliferation of breast cancer cells thus showing anti-cancer effects.[6]

12.2.2 Sophorolipids (SLs)

Compared to rhamnolipids, high titers of sophorolipid glycolipids can be produced from non-pathogenic aerobic yeasts including *Candida apicola*,[107] *Pseudohyphozyma* (*Rhodotorula*) *bogoriensis*,[108] and *Starmerella* (*Candida*) *bombicola*.[109] Among those species, *S. bombicola* has been widely studied and tested for a variety of industrial applications (*e.g.*, uptake of a renewable substrate, utilization of waste streams, *etc.*) due to its high sophorolipid yield (reportedly over 400 $g L^{-1}$ under certain defined process conditions) and biomass productivity.[7,27,110] These glycolipids are potent biosurfactants that are also marked by low toxicity and high biodegradability as opposed to chemical surfactants.[27] They naturally occur as open acidic structures or closed lactones (Figure 12.2),[7,111] and this influences their functional properties and characteristics.[27] Similar to other glycolipid-producing microorganisms, excess carbon availability along with nitrogen or phosphorous deprivation at late exponential or early stationary growth phases seems to trigger biochemical changes in the metabolic pathways that lead to the accumulation of sophorolipids.[112] This is usually followed by extracellular secretion of these secondary metabolites that also do not appear to be essential for cell survival.[113]

12.2.2.1 Biosynthesis and Structural Variations of SLs

Sophorolipid glycolipids comprise a long-chain hydroxy fatty acid or hydrophobic tail that is bound to a sophorose (2-*O*-β-D-glucopyranosyl-D-glucopyranose) hydrophilic head group, a disaccharide consisting of two molecules of glucose in an unusual β-1,2 linkage.[27] The SL biosynthetic pathway involves an essential step that is initiated by an NADPH-dependent cytochrome P450 monooxygenase.[27,114] Depending on the produced strain, the position of the hydroxyl groups as well as fatty acid chain length may vary; for instance, the fatty acids of *S. bombicola* sophorolipids typically have 16 or 18 carbons with hydroxylation at the terminal or subterminal carbon positions (Figure 12.2A–D).[7,27] Other species including *C. apicola* were found to produce SLs with similar conformations.[109,115] However, the SLs produced by *P. bogoriensis* contain fatty acids with a longer chain length (22 and 24 carbons) and a distinct hydroxylation near the center at the C-13 position

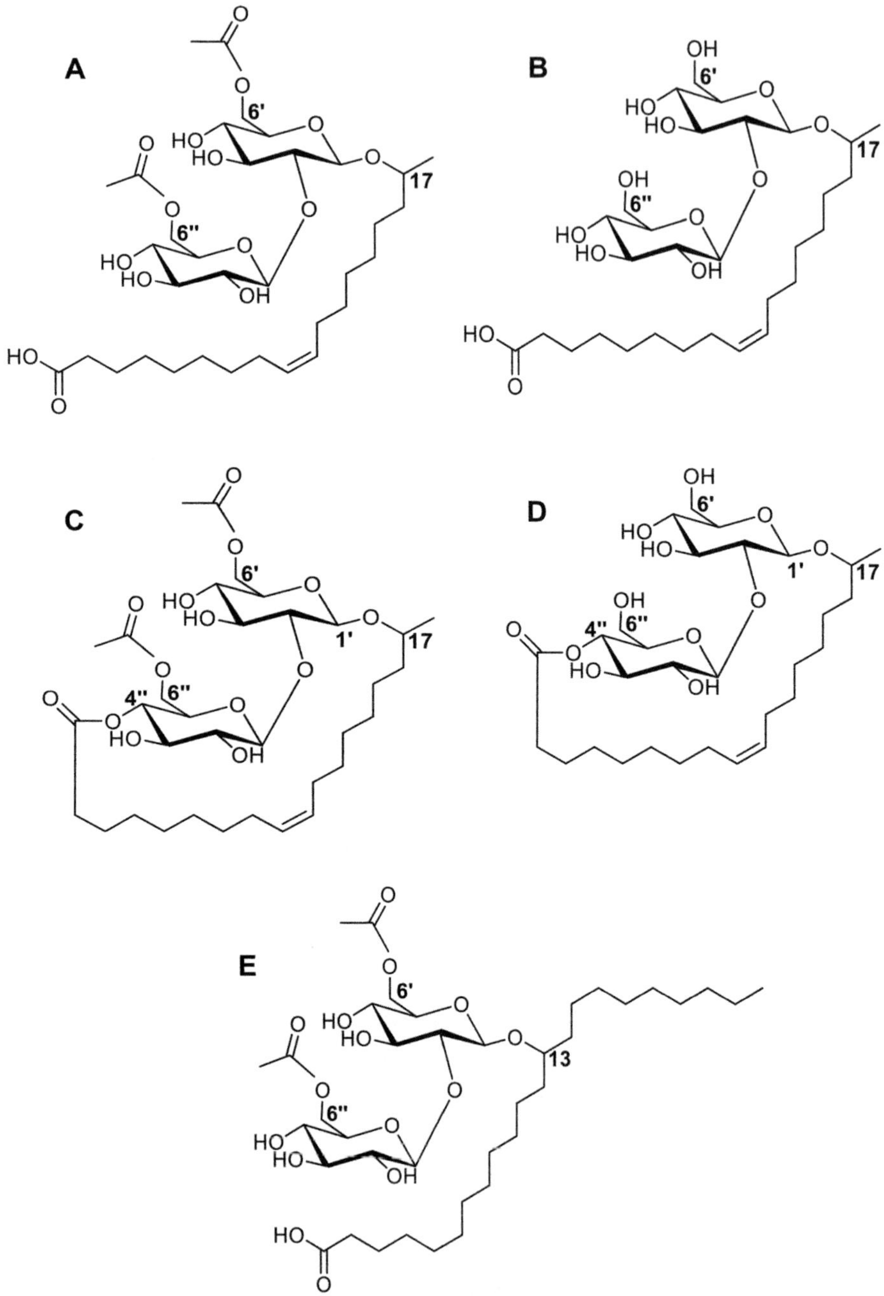

Figure 12.2 Representative chemical structures of sophorolipid biosurfactants derived from *S. bombicola* containing 17-hydroxy oleic acid in the diacetylated free acid form (A), non-acetylated free acid form (B), diacetylated lactonic form (C), and the non-acetylated lactonic form (D) and from *P. bogoriensis* containing 13-hydroxy behenic acid in the diacetylated free acid form (E).

(Figure 12.2E).[108,116,117] In the following biosynthesis steps, the hydroxyl group of the fatty acid is glycosidically linked to the first glucose residue, a reaction catalyzed by glucosyltransferase I or UgtA1 of *S. bombicola* that utilizes the uridine diphosphate-glucose or UDP-glucose as the activated donor.[118] Subsequently, glucosyltransferase II or UgtB1 catalyzes the transfer of a second glucose to create a sophorolipid molecule.[27,119,120] The hydroxyl groups at the 6 or 6′ positions of the glucose residues may be acetylated with an acetyltransferase at a single position or both positions, while deacetylated SLs were also identified.[7,117,121] In *P. bogoriensis*, acetylation is achieved by the enzyme acetyl CoA-dependent acetyltransferase.[27,122,123] Furthermore, in some yeasts SL lactones are generated from the acidic (open) forms when the carboxylic group of the fatty acid reacts with the hydroxyl group at the 4′ position on the second glucose or seldomly to the hydroxyl group at the 6 position of either glucose residue.[124] The lactonization reaction is carried out using an enzyme of the Cal-A lipase family,[125] namely the *S. bombicola* lactone esterase (SBLE) that was found capable of executing esterification reactions in aqueous media.[126,127] As a result of lactonization, hydroxyl group positions, degree of acetylation, chain length, and level of fatty acid bond saturation/unsaturation over 20 molecular forms were identified in various yeast strains with 13-[2′-*O*-β-D-glucopyranosyl-β-D-glucopyranosyloxy]-docosanoic acid 6′,6″-diacetate as the common 22 carbon SL produced by *P. bogoriensis*.[108] Among the molecules produced in this yeast, unique 24 carbon-SLs with various levels of acetylation were also identified.[117] Moreover, one of the most common SL forms produced in other yeast species is 17-L-(oxy)-octadecanoic acid 1,4″-lactone 6′,6″-diacetate, an 18-carbon-SL identified in *C. apicola* and *S. bombicola*.[27]

12.2.2.2 Genetic Regulation of SL Biosynthesis

The availability of a complete genome is essential to understand the metabolic and regulatory networks involved in sophorolipid metabolism and accumulation. Despite their commercial and economic importance, so far only a few genomes of SL-producing yeasts have been sequenced and annotated.[7] This has allowed the identification of essential genes that function in SL metabolic pathways. In the yeast *S. bombicola*, genes required for the biosynthesis and secretion of sophorolipids were found in a gene cluster organization reminiscent of bacterial operons with potentially high regulated expression. However, no genes associated with signaling or transcriptional regulations were found within the cluster.[7,27] Five identified genes were deemed required for SL biosynthesis, whereas two other genes namely the putative alcohol dehydrogenase at the 5′ end and an unknown open reading frame at the 3′ end were labeled not essential for the biosynthetic pathway or cell survival following mutational analyses.[7] Essential genes include the *CYP52M1* gene encoding for the cytochrome P450 monooxygenase, a member of the CYP52 family[128] that is capable of hydroxylating alkanes in addition to fatty acids.[129] Two *CYP52* members were identified in *C. apicola*,[130] and five different genes were identified in *S. bombicola*.[27] When the *CYP52M1* gene was inhibited in *S. bombicola*, a complete loss of the ability to synthesize SLs was observed in mutant strains,[7] thus confirming that hydroxylation is required as the first step in the biosynthetic pathway. However, a *cyp52m1* mutant was capable of restoring SL biosynthesis following chemical complementation experiments using hydroxylated fatty acids in the growth media.[7] Acetyltransferase and *UGTA1* and *UGTB1* genes, encoding for glucosyltransferase I and II respectively, were found adjacent in the gene cluster.[119–121] Another gene identified as *MDR* encodes for a multidrug resistance protein and potentially functions as a sophorolipid transporter involved in extracellular secretion.[7] Mutational analyses showed that despite a significant drop in SL secretion in *mdr* mutants, cells could still secrete the glycolipid, potentially using other transporters.[7] The *SBLE* gene encoding for lactone esterase was not identified on the same cluster with the other SL biosynthesis genes.[7] Codon-optimized *SBLE* was successfully expressed in *Pichia pastoris* as a recombinant enzyme produced for commercial applications.[111] Advances in our basic understanding of metabolic and gene regulation may provide useful hypotheses for improvements through biotechnological applications to optimize the sophorolipid biosurfactant yield and physico-chemical properties.

12.2.2.3 *Effects of Growth Substrates on the Productivity and SL Composition*

Growth substrates are key metabolic factors that can induce the accumulation of sophorolipids and affect glycolipid productivity and composition.[116] Studies have confirmed that yeast cultures can utilize or switch between different organic substrates including monosaccharide sugars and various hydrophobic compounds, such as fatty acids, alkanes, or vegetable oils.[27] However, it was found that the combination of two substrates in the growth medium, for instance a monosaccharide and a fatty acid, could improve productivity,[27] bringing new insights into commercial biosurfactant production. The yeast *C. apicola* was able to grow and increase its SL production when glucose was supplied as the sole carbon source in the growth medium.[131] Following the addition of mannose and fructose as organic substrates, *P. bogoriensis* was capable of achieving a steady growth rate and SL productivity that was comparable to glucose in the fermentation media.[117] In this experiment, the level of SL acetylation changed with growth, favoring the conversion of diacetylated SLs produced early in the culture to monoacetylated SLs following longer incubation.[132,133] Among the disaccharides tested as substrates, *S. bombicola* growth was inhibited on lactose,[27] while the yeast was able to utilize sucrose to produce sophorolipid although at lower titers compared to glucose.[134] Similarly, the SL yield was lower in *S. bombicola* when soy molasses[135] or honey[136] were supplied as growth substrates.

Several hydrophobic compounds were also assessed, and among those vegetable oils and *n*-alkanes resulted in high SL production.[27] In particular, oils that are rich in 16 or 18 carbon fatty acids (*e.g.*, olive oil, canola oil, *etc.*) had a high impact on glycolipid biosynthesis in *C. apicola* and *S. bombicola*.[27] It was found that when the level of fatty acid unsaturation was high in these oils, the resulting SLs were predominantly acidic.[27] Whereas acidic forms of SLs would be converted into lactones following longer incubation times.[112,137] Furthermore, these yeasts retain the ability to uptake and convert alkanes into hydroxy fatty acids prior to incorporation into the SL biosynthetic pathways, and the highest yields were accomplished with hexadecane and octadecane.[27] Other yeast strains including *Torulopsis gropengiesseri* could also metabolize *n*-alkanes to increase SL formation.[138] Supplementation of alkanes (*i.e.*, hexadecane, heptadecane, and octadecane) into the fermentation media also seemed to influence the level of SL molecule lactonization with the highest concentrations found as diacetylated lactones when these compounds were added.[139,140] Yeast cultures may provide a way to convert inexpensive agricultural feedstocks or other compounds into value-added sophorolipids that can be applied in nutritional, cosmetic, or medical products. However, the mechanisms underlying the uptake and utilization of several of these compounds remain poorly understood, whereas yeast species may respond differently to the substrate levels and sources.

12.2.2.4 Roles and Functional Applications of SLs

The physiological function of SLs produced by different yeast species remains mostly undetermined. However, the extracellular molecules are believed to function as energy-rich compounds that are available and accumulate when a carbon substrate is readily available.[27] When needed, these glycolipids can be assimilated and metabolized once more by the yeast.[128] Furthermore, SLs may function in defensive responses since they were found to deter the spread of other microorganisms.[141] Sophorolipid biosurfactants are also commercially produced and they have important industrial applications in several sectors.[27] They are environmentally benign due to their biodegradability and low impact on ecosystems.[27,117] So, they have been used in cleaning supplies as "green" detergents, replacing chemical surfactants.[117] Since they also showed antimicrobial activity against some bacteria and yeast species,[141] pathogenic fungi,[142] and even algae,[143] SLs have found applications in food manufacturing, processing, and storage.[144,145] In addition, they are commonly used in cosmetics and personal care products.[146] The pharmaceutical and medical sectors also benefited since SLs showed anti-inflammatory and anti-cancer potential.[115,147–151] SL lactones with acetylation on either or both glucose residues showed improved antimicrobial properties compared to the acidic forms.[152] Also, it was found that SLs with longer hydroxy fatty acids (22 or 24 carbons) showed improved surface tension lowering and emulsifying properties.[153] Thus, SLs found further applications in hydrocarbon decontamination, soil remediation, and heavy metal removal.[154–158] Finally, SLs have been a source of hydroxylated fatty acids with high industrial potential since these molecules are used to produce several types of compounds and polymers.[159,160]

12.2.3 Mannosylerythritol Lipids (MELs)

MELs are one of the most important renewable biosurfactants available from nature with multiple industrial applications due to their unique physicochemical properties.[31,161–164] Besides sophorolipids, MEL glycolipids are also abundantly synthesized by yeasts,[27,163] and similarly the accumulation of high yields is triggered by nitrogen deprivation when cultures are supplemented with glucose, glycerol, alkanes, or vegetable oils.[164] Cultures that exhibit significant accumulation usually belong to the basidiomycetous yeasts of the genera *Pseudozyma* and *Moesziomyces*,[164–168] and particularly *P. antarctica* T-34[169] and *P. parantarctica* JCM11752[170] were found to produce over 100 $g\,L^{-1}$ MEL. However, this glycolipid was originally identified and isolated from the dimorphic fungus *Ustilago maydis*, a plant pathogen that causes corn smut disease.[171,172] Various other fungi are also capable of synthesizing MELs, including *Geotrichum candidum*, *Schizonella melanogramma*, and *Kurtzmanomyces* sp.[167,173–176] Details on the biosynthetic pathways and genetic regulations are presented in the following sections.

12.2.3.1 Biosynthesis and Structure of MEL Glycolipids

MELs consist of a hydrophilic sugar-alcohol mannosylerythritol, which is acylated with hydrophobic fatty acid groups of various lengths.[177,178] These groups may be linked in two different mannosylerythritol conformations, namely the conventional 4-*O*-β-D-mannopyranosyl-(2*S*,3*R*)-erythritol, also known as the *S*-form, or the diastereomer 4-*O*-β-D-mannopyranosyl-(2*R*,3*S*)-erythritol, also known as the *R*-form.[163] MEL biosynthesis is initiated when a glycosyltransferase, identified as EMT1, catalyzes the binding of GDP-mannose and erythritol to generate mannosylerythritol.[177,179] In *U. maydis*' MELs (also known as ustilipids), the mannose residue is esterified by acyltransferase MAC1 with a short chain (C_2–C_8) fatty acid at the atom position C-2, whereas acyltransferase MAC2 esterifies the hydroxyl group at the C-3 position with a medium-chain (C_{10}–C_{18}) fatty acid.[179] Following the acylation steps, the acetyl-CoA-dependent acetyltransferase (MAT1) enzyme further acetylates the hydroxyl groups at atom position(s) C-4 and/or C-6 of the mannose residue of MELs.[179] Finally, the extracellular secretion of these glycolipids may be facilitated by the major facilitator protein MMF1.[179] The sugar-alcohol conformation, level of acetylation, and the number of acyl groups define the structural variations of mannosylerythritol lipids, characterized from various producer strains. These glycolipids are usually acylated with two fatty acids and categorized as the commonly produced diacetylated MEL-A, the less common monoacetylated MEL-B and MEL-C, with acetyl groups at the C-6 or C-4 positions of the mannose, respectively, and the nonacetylated MEL-D that may be produced *in vivo* at minor levels (Figure 12.3)[180] or synthesized by

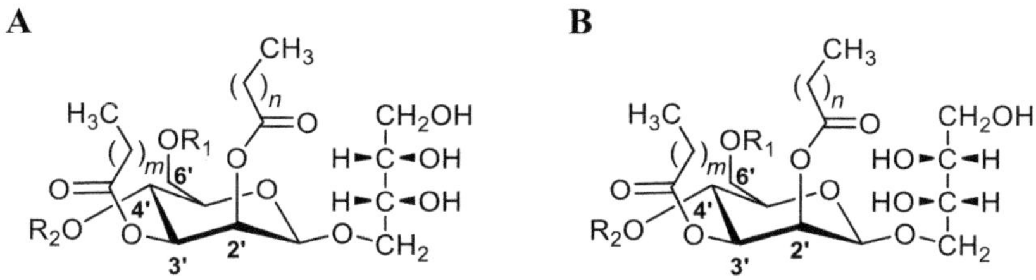

Figure 12.3 General chemical structure of *S*-form mannosylerythritol lipids (A) when R_1 and R_2 contain acetyl groups ($-COCH_3$) = MEL-A; when R_1 is acetylated and R_2 is a hydroxyl group = MEL-B; when R_1 is a hydroxyl group and R_2 is acetylated = MEL-C; and when both R_1 and R_2 are hydroxyl groups = MEL-D. In addition, structural variations can be realized through variations in fatty acid side-chain length and saturation at the C2′ and C3′ positions (typically $m = 8$–16 carbon units and $n = 2$–10 carbon units) and by the formation of the diastereometric *R*-form (B) *vs.* the more common *S*-form.

enzymatic cleavage of the acetyl group from MEL-B.[174,179,181] Like in *U. maydis*, mannosylerythritol lipids produced by most of the studied species of *Pseudozyma* (*e.g.*, *P. antarctica*, *P. hubeiensis*) and *Moesziomyces* (*e.g.*, *M. rugulosus*, *M. aphidis*, *M. parantarcticus*) have the conventional *S*-form conformation.[165,167–169,174,178] However, *P. tsukubaensis* and *P. crassa* are the only strains shown so far to produce MELs with the *R*-form conformation.[163,178,182,183] *P. tsukubaensis* produces elevated levels of mono-acetylated MEL-B (70 $g L^{-1}$) with the diastereomer 4-*O*-β-D-mannopyranosyl-(2*R*,3*S*)-erythritol as the sugar-alcohol, when supplied with olive oil in the growth media,[183] whereas *P. crassa* produces the diastereomer types of MEL-A, MEL-B, and MEL-C.[163] *P. tsukubaensis* JCM16987 could also produce MEL-B with mannosyl-L-arabitol as the sugar-alcohol moiety when supplied with L-arabitol,[184] whereas *M. parantarcticus* synthesized mannosyl-D-mannitol lipids when supplied with D-arabitol.[163] Furthermore, *P. aphidis* could produce triacylated MELs instead of the conventional diacylated lipids, with a third medium-chain fatty acid bound to the erythritol moiety, while monoacylated MELs were also identified in *Pseudozyma* sp. when solely grown on glucose,[178] thus confirming the wide structural divergence of these important glycolipid molecules.

12.2.3.2 Identification and Regulation of the MEL Gene Cluster

Since the genome of *U. maydis* has been sequenced and annotated, reverse genetic analyses coupled with enzymatic assays allowed the identification of genes involved in MEL metabolism.[179,185] Like other biosynthetic pathways of various secondary metabolites, genes encoding relevant enzymes involved in MEL biosynthesis were found to be assembled in a gene cluster composed of five different open reading frames.[179] In *U. maydis*, essential MEL genes consist of *EMT1* erythritol/mannose glycosyltransferase, *MAC1* and *MAC2* acyltransferases, *MAT1* acetyltransferase, and a putative *MMF1* gene encoding for a MEL extracellular transporter.[163,177,179,183] Other genes that encode for cytochrome P450 monooxygenases and transcription factors may be involved in MEL biosynthesis and cluster regulation, respectively.[7] Originally, microarray analyses were used to identify individual genes involved in MEL biosynthesis, following nitrogen deprivation experiments. The transcription of essential genes in the biosynthetic pathway was induced about 2.5 times upon the onset of nitrogen stress, compared to control growth conditions.[179] Additionally, mutational analyses confirmed the function of enzymes encoded by members of the gene cluster. Following *EMT1* disruption, it was found that this gene is essential for the synthesis of the mannosylerythritol moiety and thus MEL biosynthesis.[163,177,179,183] The transfer of acyl chains to mannosylerythritol is also required since inhibition of either *MAC1* or *MAC2* genes resulted in complete loss of MEL production or secretion in *U. maydis*.[179] Enzymes encoded by these genes are specific and add the acyl chain onto the mannose at the C-2 and C-3 positions, respectively.[179] Additionally, it was found that *U. maydis* can be engineered with orthologous *MAC1* and *MAC2* genes with different substrate specificities to produce MEL types with altered properties and activities.[186] *MAT1* mutation was not detrimental to MEL production, and mutant cells could produce and secrete large amounts of deacetylated MEL-D.[179] Acetyltransferase enzyme function was confirmed following *MAT1* overexpression in *Escherichia coli*.[179] Furthermore, MAT1 could also add acetyl group(s) to MELs in *in vitro* assays.[179] Finally, deletion of *MMF1* resulted in mutant cells being unable to secrete MEL extracellularly, confirming the gene function as MEL transporter in *U. maydis*.[179]

Gene clusters for MEL biosynthesis have also been determined for several *Pseudozyma* and *Moesziomyces* strains including *P. antarctica*, *P. hubeiensis*, and *M. aphidis*. MELs produced by these strains have been characterized as conventional type or *S*-form.[177,179,187–190] Using BLAST-p and sequence analyses of proteins and genes in *Ustilago maydis* and *Pseudozyma antarctica*, the gene cluster involved in the biosynthesis of the diastereomeric (*R*-form) MEL-B has been identified in the genome of *P. tsukubaensis* NBRC1940.[163] Arrangement of the genes in the cluster was found equivalent to that of *U. maydis*.[163] Moreover, essential proteins PtEMT1, PtMAC1, PtMAC2, and PtMMF1 identified in this strain were found to have over 50% homology to sequences from both *U. maydis* and *P. antarctica*.[163] *PtEMT1* is essential for the diastereomer configuration of mannosylerythritol in *P. tsukubaensis* since its introduction into the genome of mutant *P. antarctica*, unable to synthesize MELs, has resulted in the production of the *R*-form MEL-A configuration.[163] *PtMAC2* gene deletion in the strain *P. tsukubaensis* 1E5 resulted in the synthesis of monoacylated MEL with a fatty acid bound to the C-2 position of the mannose.[178] This confirmed the function of PtMAC2 as an acyltransferase capable of acylating the C-3 position of the mannose moiety.[178] Furthermore, the monoacylated MEL synthesized by the mutant strain, grown with 10% olive oil, was the diastereomer type of MEL-D with no acetyl groups.[178] As previously mentioned for *U. maydis*, disruption of the

MAC2 gene with high sequence homology to *PtMAC2* resulted in complete inhibition of MEL production, and that was in contradiction to the phenotype observed in the *P. tsukubaensis* mutant.[178,179] Moreover, when *PtMMF1* was disrupted, *P. tsukubaensis* 1E5 was able to produce and secrete monoacylated MELs.[190] Strain 1E5 was also modified by overexpression of *P. antarctica* lipases A (*PaLIPAp*) and B (*PaLIPBp*).[164] Results from these experiments showed that the biosynthesis of the diastereomer MEL-B increased about 1.6–1.9 times in recombinant *P. tsukubaensis* compared to control when grown on olive oil.[164] This class of lipases is capable of cleaving acyl groups from triglycerides when vegetable oils are supplied in the growth media, thus increasing the flux of hydrophobic substrates for MEL biosynthesis.

Genome sequence availability allowed the identification of the gene clusters for MEL biosynthetic pathways in several producer strains, as well as the development of reliable genetic engineering methods, including non-homologous recombination and gene overexpression.[178,191–194] This should improve MEL productivity and allow the synthesis of new derivatives through metabolic engineering processes.

12.2.3.3 Biological Functions and Properties of MELs

The structural variation in mannosylerythritol lipids directly affects the physico-chemical properties of these important biosurfactants, particularly the acetylation level and the sugar-alcohol conformation.[163,178] It was previously found that the hydrophobicity of the diacetylated MEL-A compounds was higher than that of monoacetylated MEL-B or MEL-C.[163] This is believed to affect the ability of MELs to solubilize organic and hydrophobic compounds. Furthermore, the diastereomer *R*-form type of MEL-B was more hydrophilic compared to the conventional *S*-form type.[195,196] The same applies for monoacylated MELs, which showed higher hydrophilicity compared to the more common diacylated forms.[178] Due to these characteristics, as well as other properties including the ability to self-assemble and function under a wide range of environmental conditions, MELs have found several applications at the industrial and commercial scales.[31,161,162] Some of these applications are in food, cleaning detergents, personal care and cosmetics, and pharmaceutical sectors.[197] For instance, the diastereomer MEL-B has been commercialized by a Japanese company (Toyobo Co., Ltd) as a skin care and damage repair compound,[198] with antioxidant and cell activation properties.[183,198,199] Like other glycolipid biosurfactants, MEL biological activities also include high antimicrobial activity against Gram-positive bacteria and anti-inflammatory and anti-cancer properties.[7,197]

12.2.4 Cellobiose Lipids (CLs)

CLs were originally described in 1950 as a class of amphipathic glycolipids produced and secreted in combination with mannosylerythritol lipids by the fungus *Ustilago maydis*.[200] Therefore, this fungus was found to simultaneously synthesize high amounts of two different classes of glycolipid biosurfactants.[177,180] CLs from *U. maydis*, designated as ustilagic acids, were produced in abundance following nitrogen deprivation, and the total glycolipid yield could reach about 23 $g\,L^{-1}$ under these conditions.[177,180] In addition to the phytopathogen *U. maydis*, CLs were identified in a few other fungi.[201,202] In particular, the biocontrol agent *Pseudozyma flocculosa* fungus was found to produce a rare CL known as flocculosin,[203] whereas CLs described from cultures of *P. fusiformata* and *P. graminicola* were similar to ustilagic acid glycolipids.[204,205] As with other biosurfactants, CLs exhibited antibiotic properties against several microorganisms with ustilagic acid showing antifungal and antibacterial activities and flocculosin acting as an antifungal compound.[7,177,203,204,206–208] The pathways and molecular and genetic regulations of CL biosynthesis are summarized below.

12.2.4.1 CL Biosynthetic Pathways

The biosynthetic pathway of cellobiose lipids was first determined for ustilagic acid in *U. maydis*.[207,208] Sequential hydroxylation at the terminal and subterminal positions of C_{16} fatty acids is carried out by cytochrome P450 monooxygenase enzymes CYP1 and CYP2, respectively, whereas the hydroxylation at the α-position is catalyzed by AHD1 α-hydroxylase.[177,207,209] The sugar moiety cellobiose (4-*O*-(β-D-glucopyranosyl)-β-D-glucopyranose) consists of two molecules of D-glucose that are linked by a β-1,4′-glycosidic bond. In ustilagic acid, cellobiose is *O*-glycosidically bound to the terminal hydroxyl group of the unusual fatty acids 15,16-dihydroxy-hexadecanoic acid or 2,15,16-trihydroxy-hexadecanoic acid (Figure 12.4),[177,179,210] a glycosylation reaction catalyzed by UGT1 glycosyltransferase.[207] The disaccharide molecule is further modified by acetylation of one glucose residue and acylation of the other glucose with a short-chain (C_6 or C_8) β-hydroxy fatty acid.[177,207,209,211] While the UAT2 enzyme was essential for the acetyl group transfer to the hydroxyl group at the C-6 position of the first glucose,[208] UHD1 enzyme catalyzes the hydroxylation of the short-chain fatty acid.[209] These structural variations, emanating from the extra hydroxylation at the α-position of the C_{16} fatty acids and the chain length of the acyl group on the sugar moiety, result in four major ustilagic acid types that are synthesized and secreted by *U. maydis*.[209] The ABC transporter ATR1, which is homologous to multidrug resistance proteins, is responsible for the extracellular transport of these molecules.[207] On the other hand, the chemical structure of flocculosin from *P. flocculosa* shows some divergence compared to ustilagic acid. Cellobiose is *O*-glycosidically bound to the terminal position of 3,15,16-trihydroxy-hexadecanoic acid following hydroxylation of 3-hydroxy-hexadecanoic acid by CYP1 and CYP2.[207,212] The fatty acid glycosylation with cellobiose is catalyzed by the FFT1 glycosyltransferase. In contrast to ustilagic acid, flocculosin is diacetylated with a second acetyl group on the disaccharide, and these reactions are catalyzed by FAT2 and FAT3 enzymes.[208]

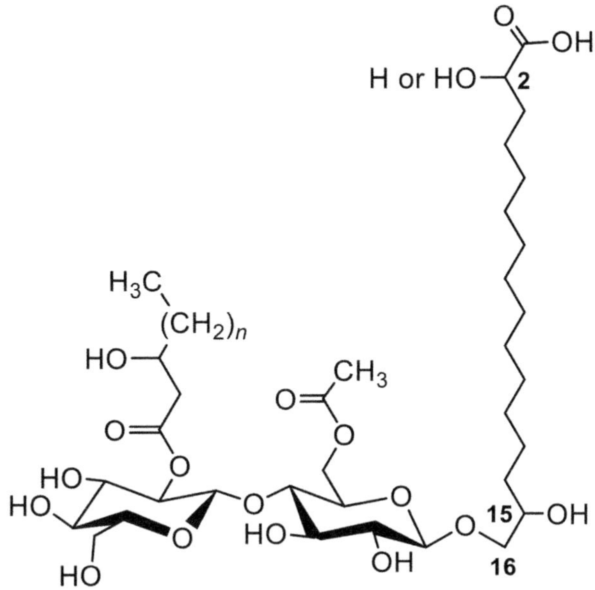

Figure 12.4 Chemical structure of ustilagic acid, a cellobiose glycolipid that is *O*-glycosidically linked to the ω-hydroxyl group of 15,16-dihydroxypalmitic acid (ustilagic acid A) or 2,15,16-trihydroxypalmitic acid (ustilagic acid B).

Furthermore, FHD1 and FAT1 catalyze the hydroxylation and transfer of β-hydroxy-octanoic acid (C_8) to the disaccharide moiety, respectively.[207–209] Although the biosynthesis of flocculosin is comparable to that of ustilagic acid and metabolic pathways are highly conserved between both fungal strains, only one type of these unique CLs was identified from *P. flocculosa* cultures.[208] Furthermore, the structural variation of the flocculosin molecule affects its functional properties and characteristics.

12.2.4.2 CL Gene Cluster Regulation

A few fungal strains of *Ustilago* and *Pseudozyma* genera were reported to be capable of producing CLs, and owing to genome sequencing and analyses they were found to contain a gene cluster with essential genes required for regulation and biosynthesis.[207–209] Comparative and mutational analyses followed by analytical determinations allowed unravelling of the function of several of these genes within each cluster.[207,209] For instance, the ustilagic acid biosynthetic cluster included 12 co-regulated open reading frames (ORFs) that covered about 45 kb near the telomeric end of chromosome 23 in *U. maydis*.[207,208] Among those identified genes, a C_2H_2 zinc finger transcription factor named *RUA1* may regulate the gene expression within the cluster,[208,213] *CYP1* and *CYP2* genes encode for the cytochrome P450 monooxygenases, *UGT1* encodes for glycosyltransferase, *UHD1* and *AHD1* encode for hydroxylases, *FAS2* encodes for fatty acid synthase, and *ATR1* encodes for a CL transporter.[207,208] Other genes in the cluster were *UAT1* involved in the short-chain fatty acid transfer to the sugar moiety, and *UAT2* required for acetylation.[208] Expression analyses confirmed that these genes were significantly upregulated under nitrogen deprivation,[177,179,207] a condition that also induced ustilagic acid production and secretion.[177,207] Gene expression was not detected under control conditions when sufficient nitrogen was supplied in the growth media.[177] Mutational analyses have confirmed that *CYP1* is required for the biosynthesis of CLs, since gene deletion resulted in complete inhibition of ustilagic acid biosynthesis in *U. maydis*.[177,207] The lack of ustilagic acid production had no effects on cell survival.[214] Moreover, the phenotype was rescued by supplementation with juniperic acid, a hydroxylated fatty acid that restored ustilagic acid biosynthesis.[207] The *CYP2* gene deletion did not affect CL biosynthesis.[177] It was also found that the *U. maydis* genome had 23 genes annotated as cytochrome P450 enzymes.[177] When the *UHD1* hydroxylase gene was disrupted, *U. maydis* was capable of producing and secreting ustilagic acid that lacked the short-chain β-hydroxy fatty acid usually bound to cellobiose.[209]

Flocculosin biosynthetic genes were identified in the sequenced genome of *Pseudozyma flocculosa*, following comparative analyses using known protein sequences encoded by *U. maydis*.[208] It was found that these genes were conserved and arranged in a large gene cluster covering about 60 kb of the *P. flocculosa* genome.[208] However, the flocculosin gene cluster contained 11 ORFs with individual candidates randomly arranged and carried a higher number of introns compared to those in *U. maydis*.[208,209] Functions of relevant genes involved in flocculosin biosynthesis were determined by genetic complementation of *U. maydis* mutants. For instance, *P. flocculosa FHD1* gene is homologous to *UHD1*. Following transformation in *U. maydis* mutant lacking UHD1 enzymatic activity, *FHD1* was capable of β-hydroxylation of the short-chain fatty acid.[209] The *P. flocculosa FAT1* gene was required for the transfer of the short-chain fatty acid to cellobiose and had a similar function to *UAT1*,[208] whereas the *FAT2* gene is homologous to *UAT2* acetyltransferase.[208] Interestingly, the flocculosin gene cluster also included another acetyltransferase *FAT3* with no homologs in *U. maydis*.[208] This gene is involved in transferring an acetyl group to the C-3 position on the second glucose molecule.[208] Conversely, *U. maydis AHD1* encoding the long-chain fatty acid α-hydroxylase was missing from the flocculosin gene cluster with no homologs found in the *P. flocculosa* genome,[208,209] which may explain the difference in the CL hydroxylation pattern between the two fungi.[209] The availability of annotated gene clusters for CL biosynthesis would allow further biotechnological applications such as trait transfer or strain improvement for optimal productivity and enhanced properties.

12.2.4.3 Biological Functions of CL Glycolipids

As with other glycolipids, several microorganisms produce and secrete CLs to potentially improve the solubility and uptake of hydrophobic substrates from their growth environment.[179] Ustilagic acids secreted by *U. maydis* were found to have several functional properties including pheromone recognition in the haploid phase prior to mating[177] and antimicrobial activities.[206] Studies have confirmed that ustilagic acid secretion inhibited the growth of the grey mold *Botrytis cinerea* when tomato leaves were co-inoculated with wild-type *U. maydis*.[207] When ustilagic acid biosynthesis was inhibited, co-culture with mutant cells could not prevent mold infection.[207] Furthermore, flocculosin secreted by *P. flocculosa* showed antifungal activity against several phytopathogens, most notably the powdery mildew fungi.[207–209,215–217] It has been suggested that CLs increased the membrane permeability resulting in ion leakage and cell death.[201] Finally, production and secretion of cellobiose lipids may help improve the survival and expansion in different environments,[209] since phylogenetically close fungi that cannot synthesize ustilagic acid (*e.g.*, *U. hordei*) were less widespread.

12.2.5 Trehalose Lipids (TLs)

Microbial TLs are glycolipid types mainly produced and characterized in bacterial genera within the phylum *Actinobacteria*.[18,38] They are amphiphilic molecules with trehalose as the hydrophilic backbone and fatty acyl chains or derivatives as the hydrophobic moiety.[18] Wide structural variations have been reported for trehalose-containing glycolipids and depending on the producing strain molecules such as trehalose mycolate, sulfolipids, trehalose lipid esters, and lipooligosaccharides, they have been elucidated.[18,218,219] For instance, in *Mycobacterium* and *Corynebacterium* trehalose is a major component of the cell envelope with trehalose mycolates possessing important immunomodulation properties.[220] Interestingly, it was found that under conditions of nutrient limitation, trehalose would also accumulate in the cells of mycobacteria.[221] In addition to their role as cell-bound structural components, these glycolipid biosurfactants can be secreted extracellularly as secondary metabolites.[219,222] TLs with various structures and chemical compositions produced and/or secreted by *Rhodococcus* sp. have been well characterized.[18,222–227] These TLs include the non-ionic trehalose mycolates as well as the anionic succinoyl TLs and trehalose tetraesters.[228,229] When grown with *n*-hexadecane, transgenic *Rhodococcus* sp. SD-74 produced high yields of TLs corresponding to about 40 $g L^{-1}$.[38] Extracellular TLs have also been reported from *Tsukamurella* cultures when supplemented with sunflower oil as a growth substrate.[219] Furthermore, *Gordonia* species could produce low molecular weight TLs when grown on different hydrocarbon and heterocyclic substrates.[219,222,230–234] TLs have been characterized in other genera including *Arthrobacter*,[235] *Brevibacterium*,[236] *Nocardia*,[237] *Nocardiopsis*,[238] and *Streptomyces*.[239] As with other glycolipids, TLs have been found to exhibit important and versatile physico-chemical properties and bioactivities, including bioremediation of oil spills[240] and antimicrobial activities.[214] Therefore, producer strains and most notably non-pathogenic bacteria (*i.e.*, *Rhodococcus*, *Tsukamurella*) have high potential for industrial and commercial applications.

The biochemical and genetic basis for TL metabolism currently available in the literature will be presented in the next sections.

12.2.5.1 Biosynthesis and Metabolic Pathways of Trehalose-based Glycolipids

In nature, trehalose disaccharide (α-D-glucopyranosyl-(1→1)-α-D-glucopyranoside) is a ubiquitous sugar with multiple physiological roles and is commonly synthesized in various microbes, plants, and invertebrates.[242–244] Being a highly stable and non-toxic molecule, it usually accumulates in response to osmotic stress, dehydration, and temperature extremes[245] and is believed to play a key role in cryoprotection, osmoregulation, and maintaining the integrity of cell membranes under stress conditions.[245] Trehalose may also function as a carbon storage and energy source, since it is hydrolyzed *in vivo* by the α,α-trehalose-glucohydrolase (trehalase) enzyme to yield two glucose molecules, which can further undergo glycolysis and ATP synthesis.[246] In various *Actinobacteria*, trehalose is required for the transport of mycolyl groups and biosynthesis of trehalose-containing glycolipids.[38,247] It may be imported from exogenous sources using specific transporters, for instance, the LpqY-SugABC transporter identified in *M. tuberculosis*, or synthesized *de novo* from glucose, fatty acids, and other substrates.[248] In these microbes (*e.g.*, *Mycobacterium* sp., *Corynebacterium* sp.), trehalose or trehalose-6-phosphate is synthesized using various routes; however, the OtsAB pathway is best studied and well characterized.[243,249] In this pathway, the enzyme trehalose-6-phosphate synthase (OtsA) catalyzes the formation of trehalose-6-phosphate using UDP-glucose and glucose-6-phosphate, whereas the enzyme trehalose-6-phosphate phosphatase (OtsB) dephosphorylates trehalose-6-phosphate to generate trehalose.[244] Furthermore, it was found that the enzyme OstA2 contributed to the increase in trehalose content in *Rhodococcus opacus* 1CP, whereas OtsA1 facilitated the increase in TL biosynthesis.[244]

Trehalose lipids characterized from different bacterial strains as cell-bound or secreted glycolipids have complex structures with multiple fatty acyl chains or their derivatives that are variable in length and level of saturation. However, our knowledge regarding the metabolic pathways and enzymes involved in the biosynthesis of various TLs remains relatively incomplete. In *M. tuberculosis*, trehalose 6,6′-dimycolate (TDM) (also known as "cord factor") (Figure 12.5) is among the early described and well-studied forms of TLs.[250,251] TDM plays a structural role as a component of the outer cell layer.[18,252,253] Biosynthesis is initiated in the cytoplasm where an esterification reaction between the hydroxyl group at the 6 position of glucose and mycolic acid forms the trehalose-6-monomycolate (TMM). Initially, the polyketide synthase Pks13 is involved in mycolic acid synthesis and maturation.[254] Subsequently, the mycolyl moiety is transferred to trehalose 6-phosphate, and the molecule is further dephosphorylated yielding TMM.[255] The precursor TMM is then transported extracellularly *via* the

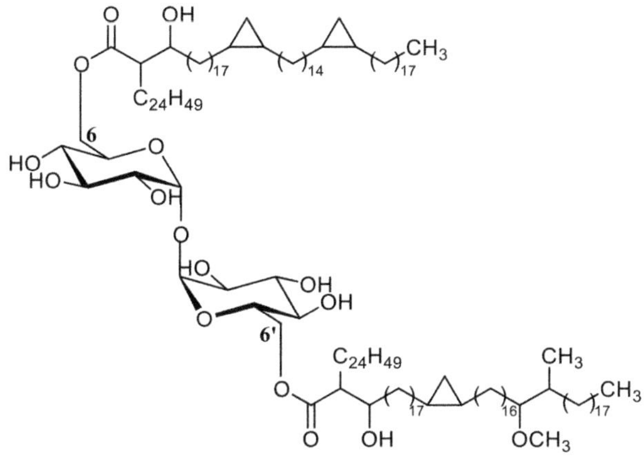

Figure 12.5 Representative chemical structure of trehalose 6,6′-dimycolate (cord factor).

MmpL3 membrane transporter[256–258] and serves as a substrate for mycolyltransferase enzymes, which transfer a mycolic acid residue either to the 6′-hydroxyl group of glucose resulting in TDM biosynthesis or to arabinogalactan resulting in the formation of the cell wall complex mycolyl-arabinogalactan-peptidoglycan.[259,260] Mycolic acids and related structures are highly hydrophobic with β-hydroxy-α-branched acyl chains that are variable in length (C_{22}–C_{92}) depending on the bacterial strains with eumycolic acid identified in *Mycobacteria*, corynemycolic acid in *Corynebacteria*, and nocardiomycolic acid in *Nocardia*.[261] Corynemycolic acids characterized by shorter and usually unsaturated chains have also been described in strains including *Arthrobacter paraffineus*, *Brevibacterium vitarumen*, *Corynebacterium* sp., and *Rhodococcus* sp.[220,262,263]

Other forms of trehalose-based glycolipids have been elucidated as antigenic glycolipids in the cell wall of virulent mycobacterial strains, including sulfolipids (Figure 12.6A) and di-, tri-, or polyacylated trehaloses, consisting of multiple (up to 5) methyl-branched fatty acyl chains that are bound to 2-, 3-, 4-, and/or 6-hydroxyl groups of glucose (Figure 12.6B).[18,264–269] In *M. tuberculosis*, the sulfolipid SL-1 (2,3,6,6′-tetraacyl-α-α′-trehalose-2′-sulfate) has been identified as the most abundant, whereas pentaacyltrehalose consisting of a saturated fatty acyl group and four mycolipenic acids was the major polyacylated trehalose. However, 30 different molecular species of polyacylated trehaloses have been identified.[268,270] SL-1 biosynthesis is initiated when the enzyme sulfotransferase (Stf0) transfers a sulfate group from 3′-phosphoadenosine-5′-phosphosulfate (PAPS) to the 2-hydroxyl group of glucose to generate trehalose-2-sulfate. An acyltransferase (PapA2) esterifies the trehalose-2-sulfate at the 2′-hydroxyl group using acyl-CoA as a donor to generate the SL659 intermediate, which undergoes further acylation by PapA1 at the 3′-hydroxyl group using a methyl-branched (hydroxy)-phthioceranoyl chain to generate the diacylated form SL1278. Polyketide

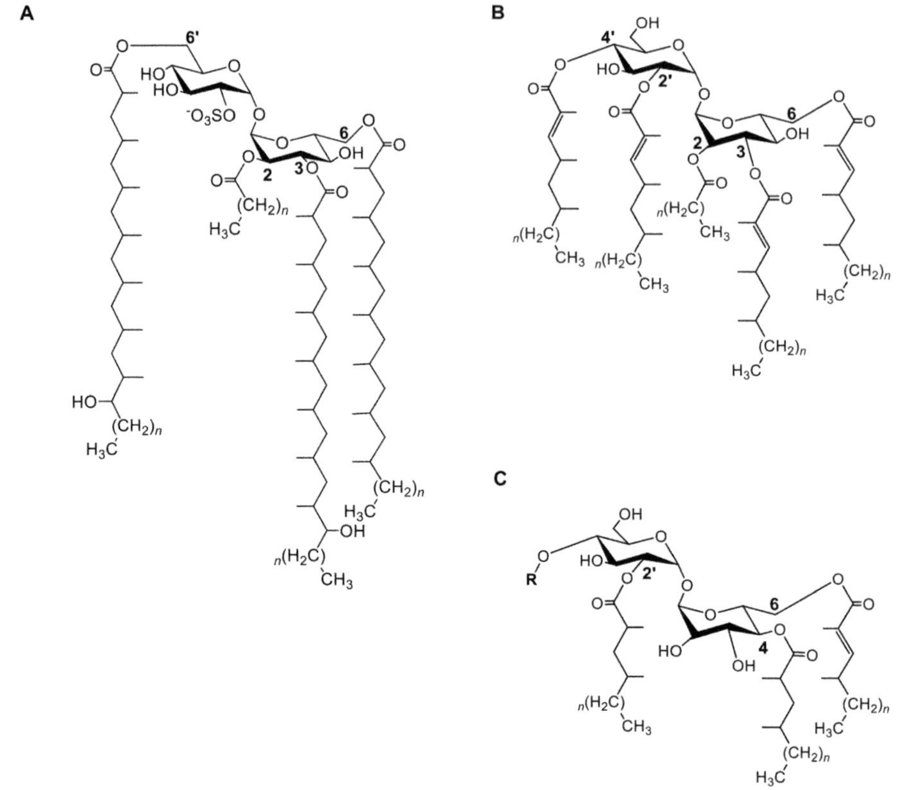

Figure 12.6 Representative chemical structures of different types of trehalolipids including sulfolipid (A), polyacyltrehalose (B), and lipooligosaccharide (C; R = complex and variable glycan moiety typically ≤6 glucose moieties) (note: types of trehalolipids and absolute chemical structures of the resulting molecules are species and process specific).

synthase (Pks2) and cutinase-like hydrolase protein (Chp1) are involved in the generation and transfer of the methyl-branched (hydroxy) phthioceranoyl chain while further acylations of 6- and 6′-hydroxyl groups generate SL-1. Furthermore, the MmpL8 transporter was found to be essential for SL-1 biosynthesis and transport across the cell membrane to its cell wall destination.[270–272] As for the biosynthesis of polyacylated trehalose, the acyltransferase PapA3 catalyzes the transfer of a saturated fatty acid (C_{10}–C_{18}) from acyl-CoA to the 2-hydroxyl group of glucose, whereas other positions on the trehalose sugar are esterified with long-chain methyl-branched fatty acids, a reaction catalyzed by PapA3 or other acyltransferases. Transfer of these molecules across the cell membrane may be facilitated by the MmpL11 transporter.[273] Mycobacterial TLs also encompass the lipooligosaccharides class identified in *M. canetti*,[274] *M. kansasii*,[275] and *M. marinum*,[276] and they not only consist of a trehalose core esterified through acylation with long-chain methyl-branched fatty acids, but also carry a polymer of (up to 6) glucose residues (Figure 12.6C).[277] Being structurally diverse and highly polar in nature, lipooligosaccharides have been linked to microbial biofilm formation but little is known about the biosynthesis of these molecules.

The class succinoyl trehalose lipids (STLs) characterized from *Rhodococcus* consists of one or two succinic acids that are bound to trehalose in addition to two fatty acids.[38] In *Rhodococcus*, TLs are major cell wall components.[244] However, some species are capable of producing high amounts of these glycolipids when grown on hydrophobic substrates.[223,244,278–282] Some of these bacteria can synthesize 2,3,4,2′-trehalose tetraesters with succinic acid and C_8 and C_{10} fatty acids when supplied with *n*-hexadecane.[244,283] Strain SD-74 was able to synthesize and secrete high amounts of two major types of STLs (1 and 2) when grown on *n*-alkanes.[38,226,229,282,284] The STL biosynthetic pathway is not fully elucidated; however, a few key enzymes were identified and found essential for production in *Rhodococcus* sp., namely an alkane monooxygenase (AlkB) that can oxidize alkane substrates as a primary step, followed by conversion to acyl-CoA or acetyl-CoA. The metabolic intermediate succinate is generated *via* the tricarboxylic acid (TCA) cycle in the mitochondria using acetyl-CoA or glucose as substrates. As a final step, TlsA (and probably other enzymes) transfers intermediates to trehalose resulting in STL formation.[38] *R. erythropolis* strains produce TLs comprising corynemycolic acids, trehalose esters, and STLs.[224,225,227,278,279,285,286] Some of those comprise β-hydroxy-α-branched acids with chain lengths between C_{20} and C_{90}.[287] When grown with long-chain *n*-alkanes, *R. opacus* 1CP produces trehalose dimycolate,[228] whereas *Rhodococcus* sp. S67 produces trehalose tetraesters with 2% *n*-hexadecane at 24 °C.[282,283] *R. fascians* strain BD8, which was isolated in the Arctic, produces TLs with one branched fatty acid with C_{16} or C_{17} chain length when *n*-hexadecane was supplied as a carbon source,[3,288,289] whereas the marine strain *Rhodococcus* sp. PML026 produces TLs with a core trehalose esterified to multiple hydroxylated and straight chain fatty acyl moieties.[290]

Compared to other actinomycetes, *Tsukamurella* spp. are capable of producing and secreting TLs with short-chain fatty acids when cultures are subject to nitrogen deprivation and when sunflower oil is supplemented as a hydrophobic substrate.[219] These glycolipids labeled TL-A and TL-B comprise a trehalose core esterified at the 2′- and 3′-hydroxyl groups of glucose with two fatty acids with short (C_4–C_6) and intermediate (C_{16}–C_{18}) chain lengths.[219,291,292] These TLs comprising shorter-chain fatty acids are less hydrophobic in comparison with other previously described classes. Furthermore, *Gordonia* sp. 1D is capable of uptake and consumption of a wide range of hydrophobic compounds, including crude oil,[222,293] while simultaneously producing and secreting trehalose-based glycolipids.[222] In the case of a heat-tolerant strain, maximum glycolipid production occurred with supplementation of *n*-hexadecane at 40 °C but it was found to tolerate temperatures as high as 50 °C.[222] Strain *G. alkanivorans* CC-JG39 is also capable of growth at high temperatures and consumption of diesel oil while simultaneously producing glycolipids.[294] Despite gaps in our knowledge of the biosynthetic pathways, several TL-producing microbes have potential for biotechnological improvements and industrial and environmental applications, most notably in the areas of biosurfactant production and oil spill cleanup.

12.2.5.2 Genetic Regulation of Trehalose-based Glycolipid Biosynthesis

Actinomycetes capable of producing glycolipids and other antimicrobial compounds have exhibited larger genomes compared to other bacteria, with relatively bigger segments encoding for enzymes involved in the biosynthesis of secondary metabolites.[219,295] As in other glycolipid-producing microbes, TL biosynthetic genes encode for fatty acid synthetic enzymes, transferases, and carbohydrate metabolism enzymes.[38] Many genes essential for key reactions remain uncharacterized. However, available knowledge on genetic regulations has offered better insights into the biosynthesis of trehalose-based glycolipids, and some of that information will be discussed next.

Several genes have been described as essential and required for trehalose disaccharide metabolism and subsequently TL biosynthesis. For instance, in *M. tuberculosis* the *Rv1235–Rv1238* genes encode for the high-affinity ABC transporter (LpqY-SugABC) that facilitates the import of exogenous trehalose,[248] whereas the *Rv2402* gene encodes for the trehalase enzyme.[246] Mutant cells unable to synthesize trehalose show slow growth and sensitivity to high temperatures when deprived of exogenous trehalose.[296] Deletion of candidate genes in the OtsAB pathway, for instance the *Rv3372* gene encoding for OtsB2, has detrimental effects on cell growth in *M. tuberculosis* and *M. bovis* despite supplementation with exogenous trehalose,[249] whereas deletion of *Rv2006* encoding for OtsB1 (homologous to trehalose-6-phosphate phosphatases) in *M. tuberculosis* has no effects on cell growth.[249] In contrast, deletion of the *Rv3490* gene encoding for OtsA has resulted in reduction in cell growth when the culture is deprived of exogenous trehalose.[249] In other bacteria, genome availability and analysis have also allowed characterization of genes required for trehalose biosynthesis.[222,244,297] For instance, *otsA*/*otsB* genes have been identified in *Gordonia* sp. 1D, as well as *treY*/*treZ* coding for enzymes that facilitate trehalose biosynthesis from maltooligosaccharides and starch.[222] Genes involved in trehalose metabolism and most notably the *ots* genes are commonly found across Rhodococci, Propionibacteria, and Streptomycetes among others.[244]

Despite current advances in *-omics* research, a great deal of information on the genetic regulation of trehalose-based glycolipid biosynthesis is available for mycobacterial strains, specifically the disease-causing agents, with near complete information on sulfolipid and polyacylated acyltrehalose biosynthetic pathways available for *M. tuberculosis*.[270] In the mycolic acid pathway, blocking the MmpL3 transporter has resulted in the accumulation of the precursor trehalose-6-monomycolate in the cells of *M. tuberculosis* and *M. smegmatis* affecting the biosynthesis of TDM.[256–258] Deletions of the *Rv0295c* gene encoding for sulfotransferase Stf0 or *pks2* encoding for the polyketide synthase enzyme inhibit the formation of essential intermediates and consequently sulfolipid SL-1 biosynthesis.[298] Inhibition of sulfolipid production in *M. tuberculosis* does not affect its pathogenicity following mice inoculation.[299,300] Furthermore, deletion of genes encoding for acyltransferases PapA1 or PapA2 has provided information on sequential acylation steps of SL-1 intermediates in *M. tuberculosis* cells.[271] *Rv3822* and *Rv3821* genes also identified on the sulfolipid biosynthetic cluster encode respectively Chp1 and the integral membrane protein Sap, which associate with MmpL8 for SL-1 transport.[272] The inhibition of the *MmpL8* gene in *M. tuberculosis* has resulted in the blocking of the SL-1 transport to the outer cell layer.[299] Similarly, genes encoding for polyacylated acyltrehalose biosynthesis are found clustered together in *M. tuberculosis*.[270] Inhibition of *Rv1182* encoding for acyltransferase PapA3 or *pks3/4* encoding for polyketide synthase results in complete loss of this type of TL.[273] Other candidates identified within the polyacylated acyltrehalose gene cluster are genes encoding for the MmpL11 transporter and FadD21 acyl-CoA synthetase/acyl-AMP ligase, which is involved in the synthesis and elongation of fatty acid precursors that are subsequently transferred to the trehalose core.[270] Furthermore, a gene cluster involved in lipooligosaccharide biosynthesis has been characterized in *M. marinum*.[301] Some of these genes have homologs in the genome of *M. tuberculosis*.[302] Mutation of *losA* gene encoding for a glycosyltransferase in *M. marinum* results in inhibition of glycosylation of the trehalose core of lipooligosaccharides.[276]

Essential genes involved in the biosynthesis of succinoyl trehalose lipids (STLs) have also been determined in *Rhodococcus* grown on *n*-alkane. Gene *alkB* is classified as an alkane monooxygenase that catalyzes the oxidation of alkane substrates to alcohol.[38] Following *alkB* inhibition, *Rhodococcus* cells are unable to synthesize STLs. However, this mutation does not affect viability, and cells have regained their ability to synthesize STLs when the alcohol hexadecanol is supplemented.[38] The *tlsA* gene encoding for acyl-CoA transferase, with sequence homology to *M. tuberculosis* acyltransferases, is required for STL biosynthesis since gene mutation results in the accumulation of intermediates affecting STL production, whereas gene overexpression leads to an increase in STL biosynthesis.[38] Furthermore, transcription of both *alkB* and *tlsA* is induced in *Rhodococcus* in the presence of *n*-hexadecane.[38] The fructose-bisphosphate aldolase encoded by the *fda* gene is also essential for glycolipid biosynthesis since this enzyme, usually involved in carbohydrate metabolism, is required for trehalose synthesis in *Rhodococcus*.[38] However, overexpression of the *fda* gene has resulted in similar STL yields when compared to the wild-type strain.[38]

Genome sequence analyses as well as mutagenesis and gene overexpression have facilitated the identification and functional characterization of multiple genes and gene clusters that are required for the biosynthesis of different types of TLs.[38,219,222,244,297] This knowledge is essential for advancing our understanding of TL biosynthesis and functions in both pathogenic and non-pathogenic bacteria, as well as for optimizing the industrial and commercial production of trehalose-based glycolipids.

12.2.5.3 Physiological Roles and Functional Applications of Trehalose Lipids

In *Actinobacteria*, trehalose-based glycolipids have been identified as major components of the cell envelope. Being highly hydrophobic molecules, they facilitate the uptake of several hydrocarbon substrates, while simultaneously inhibiting the effects of antibiotics.[18] In addition to structural roles, they have been implicated in the pathogenicity of multiple infectious agents (*e.g., M. tuberculosis* and *Corynebacterium diphtheriae*).[18,218] For instance, trehalose dimycolates affect the immunity responses in host organisms,[303] while sulfolipids identified as virulence factors in *M. tuberculosis* have also been found to contribute to pathogenicity.[304] Mycobacterial lipooligosaccharides may be implicated in biofilm formation and play a role in host cell infection.[276] Conversely, many TLs exhibit interesting properties that make the commercial production of these biosurfactants from non-pathogenic organisms highly desirable. They are ecofriendly, biodegradable, and highly stable.[305,306] TLs are naturally produced and secreted by certain bacteria when grown with specific substrates or under stress conditions (either nutrient deprivation or threat from other organisms).[18] Compared to other glycolipid biosurfactants, anionic succinoyl trehalose lipid STL-1 has shown higher surface activity and significant emulsifying potential.[229] Trehalose lipids produced by *Tsukamurella* and *Rhodococcus* have significantly reduced the surface tension immediately following culture inoculation.[3,219,282] Several studies have confirmed the ability of TL-producing bacteria to mobilize hydrophobic substrates from wastewater and oil-contaminated soils.[241,306] Oil-degrading bacteria of the genera *Rhodococcus*[282] and *Gordonia*[293] can effectively take up hydrocarbons from contaminated soils under low and high temperatures, respectively, while simultaneously synthesizing various trehalose-based glycolipid biosurfactants.[307] Furthermore, molecules produced by *R. fascians* BD8 have shown antimicrobial properties against pathogens including *Vibrio harveyi* and *Proteus vulgaris*, as well as limited growth inhibition of other bacteria and fungi.[3] Other medical properties include antiviral activity, for instance, inhibition of the influenza virus, and antitumor activity by promoting cell differentiation.[38,308,309]

12.3 Conclusions

As interest in the development of renewable glycolipids increases, the establishment of strains with significant production is becoming a critical factor for achieving commercial feasibility and sustainability. In various glycolipid producers, substrate availability seems to be a key metabolic factor affecting productivity and composition, and the carbon flux may be essential for the production of specific secondary metabolites, although the overall growth conditions may be restricted (*e.g.*, nitrogen deprivation). Indeed, excess carbon supply relative to anabolic demand appears to trigger the accumulation and secretion of glycolipid biosurfactants. In many instances, the chemical structure and content of the hydrophobic moieties are directly affected by the composition and nature of the substrates (*i.e.*, vegetable oil, *n*-alkanes) that are supplemented in the fermentation media. Productivity can be enhanced through the development of sustainable genetic strategies. Genetic information needs to be partly available to be able to design overexpression or silencing vectors that specifically induce or suppress target genes, respectively. However, transgenic improvements in yield are hampered by the lack of genome sequence availability for many producer strains, as well as the lack of understanding of the underlying signaling and molecular changes that trigger the accumulation and secretion of glycolipids. As previously mentioned, the identification of the biosynthetic gene clusters and regulatory elements in several glycolipid-producing strains is an essential step that allows the generation of novel and tailor-made glycolipid biosurfactants with improved yields.

References

1. G. S. Kiran, A. S. Ninawe, A. N. Lipton, V. Pandian and J. Selvin, *Crit. Rev. Biotechnol.*, 2016, **36**(3), 399.
2. P. Biniarz, M. Łukaszewicz and T. Janek, *Crit. Rev. Biotechnol.*, 2017, **37**, 393.
3. J. Janek, A. Krasowska, Z. Czyznikowska and M. Łukaszewicz, *Front. Microbiol.*, 2018, **9**, 2441.
4. U. A. Ochsner, A. K. Koch, A. Fiechter and J. Reiser, *J. Bacteriol.*, 1994, **176**, 2044.
5. G. S. Shreve and R. Makula, *Biomolecules*, 2019, **9**, 885.
6. P. Thakur, N. K. Saini, V. K. Thakur, V. K. Gupta, R. V. Saini and A. K. Saini, *Microb. Cell Fact.*, 2021, **20**, 1.
7. I. N. A. Van Bogaert, K. Holvoet, S. L. K. W. Roelants, B. Li, Y.-C. Lin, Y. Van de Peer and W. Soetaert, *Mol. Microbiol.*, 2013, **88**(3), 501.
8. I. M. Banat, R. S. Makkar and S. S. Cameotra, *Appl. Microbiol. Biotechnol.*, 2000, **53**, 459.
9. E. F. Santos, M. F. S. Teixeira, A. Converti, A. L. F. Porto and L. A. Sarubbo, *Biocatal. Agric. Biotechnol.*, 2019, **17**, 142.
10. S. Y. Chen, Y. H. Wei and J. S. Chang, *Appl. Microbiol. Biotechnol.*, 2007, **76**, 67.
11. I. M. Banat, *Bioresour. Technol.*, 1995, **51**, 1.
12. M. Inès and G. Dhouha, *Carbohydr. Res.*, 2015, **416**, 59.
13. G. S. Kiran, B. Sabarathnam and J. Selvin, *FEMS Immunol. Med. Microbiol.*, 2010, **59**, 432.

14. Y. Kryachko, S. Nathoo, P. Lai, J. Voordouw, E. J. Prenner and G. Vourdouw, *Inter. Biodet. Biodeg.*, 2013, **81**, 133.
15. L. A. Ławniczak, R. B. Marecik and L. Chrzanowski, *Appl. Microbiol. Biotechnol.*, 2013, **97**, 2327.
16. D. P. A. Sachdev and S. S. B. Cameotra, *Appl. Microbiol. Biotechnol.*, 2013, **97**, 1005.
17. M. M. Muller, J. H. Kugler, M. Henkel, M. Gerlitzki, B. Hormann, M. Pohnlein, C. Syldatk and R. Hausmann, *J. Biotechnol.*, 2012, **162**, 366.
18. J. H. Kügler, M. Le Roes-Hill, C. Syldatk and R. Hausmann, *Front. Microbiol.*, 2015, **6**, 212.
19. N. Christova, B. Tuleva, A. Kril, M. Georgieva, S. Konstantinov, I. Terziyski, B. Nikolova and I. Stoineva, *Appl. Biochem. Biotechnol.*, 2013, **170**, 676.
20. L. Magalhães and M. Nitschke, *Food Control*, 2013, **29**, 138.
21. T. Janek, M. Łukaszewicz and A. Krasowska, *BMC Microbiol.*, 2012, **12**, 24.
22. T. Janek, A. Krasowska, A. Radwanska and M. Łukaszewicz, *PLoS One*, 2013, **8**, e57991.
23. E. J. Gudiña, A. I. Rodrigues, E. Alves, M. R. Domingues, J. A. Teixeira and L. R. Rodrigues, *Bioresour. Technol.*, 2015, **177**, 87.
24. M. A. Díaz De Rienzo, I. M. Banat, B. Dolman, J. Winterburn and P. J. Martin, *New Biotechnol.*, 2015, **32**, 720.
25. R. M. Mann and J. R. Bidwell, *Environ. Pollut.*, 2001, **114**, 195.
26. R. M. Mann and M. R. Boddy, *Chemosphere*, 2000, **41**, 1361.
27. I. N. A. Van Bogaert, K. Saerens, C. De Muynck, D. Develter, W. Soetaert and E. J. Vandamme, *Appl. Microbiol. Biotechnol.*, 2007, **76**, 23.
28. H. Chong and Q. Li, *Microb. Cell Fact.*, 2017, **16**(1), 137.
29. G. Soberón-Chávez, and R. M. Maier, *Biosurfactants*, 2011, p. 1.
30. A. Varvaresou and K. Lakovou, *Lett. Appl. Microbiol.*, 2015, **61**, 214.
31. D. Kitamoto, T. Morita, T. Fukuoka, T. Konishi and T. Imura, *Curr. Opin. Colloid Interface Sci.*, 2009, **14**, 315.
32. D. G. De Almeida, R. C. Soares Da Silva, J. M. Luna, R. D. Rufino, V. A. Santos, I. M. Banat and L. A. Sarubbo, *Front. Microbiol.*, 2016, **7**, 1718.
33. M. Henkel, M. Geissler, F. Weggenmann and R. Hausmann, *Biotechnol. J.*, 2017, **12**, 1600561.
34. B. N. Paulino, M. G. Pessôa, M. C. Mano, G. Molina, I. A. Neri-Numa and G. M. Pastore, *Appl. Microbiol. Biotechnol.*, 2016, **100**, 10265.
35. A. Fiechter, *Trends Biotechnol.*, 1992, **10**, 208.
36. M. M. Müller and R. Hausmann, *Appl. Microbiol. Biotechnol.*, 2011, **91**, 251.
37. A. K. Mukherjee and K. Das, *Biosurfactants*, 2010, **672**, 54.
38. T. Inaba, Y. Tokumoto, Y. Miyazaki, N. Inoue, H. Maseda, T. Nakajima-Kambe, H. Uchiyama and N. Nomura, *Appl. Environ. Microbiol.*, 2013, **79**(22), 7082.
39. K. R. Meena and S. S. Kanwar, *BioMed Res. Int.*, 2015, 473050.
40. A. V. Sotirova, D. I. Spasova, D. N. Galabova, E. Karpenko and A. Shulga, *Curr. Microbiol.*, 2008, **56**, 639.

41. F. G. Jarvis and M. J. Johnson, *J. Am. Chem. Soc.*, 1949, **71**(12), 4124.
42. M. M. Müller, B. Hörmann, C. Syldatk and R. Hausmann, *Appl. Microbiol. Biotechnol.*, 2010, **87**, 167.
43. L. H. Guerra-Santos, O. Kappeli and A. Fiechter, *Appl. Environ. Microbiol.*, 1984, **48**, 301.
44. U. A. Ochsner, A. Fiechter and J. Reiser, *J. Biol. Chem.*, 1994, **269**, 19787.
45. C. Giani, D. Wullbrandt, R. Rothert and J. Meiwes, *US Pat.*, 5501966, 1996.
46. S. Haussler, M. Nimtz, T. Domke, V. Wray and I. Steinmetz, *Infect. Immun.*, 1998, **66**, 1588.
47. J. Andrä, J. Rademann, J. Howe, M. H. J. Koch, H. Heine, U. Zähringer and K. Brandenburg, *Biol. Chem.*, 2006, **387**, 301.
48. S. J. Funston, K. Tsaousi, M. Rudden, T. J. Smyth, P. S. Stevenson, R. Marchant and I. M. Banat, *Appl. Microbiol. Biotechnol.*, 2016, **100**, 7945.
49. M. M. Burger, L. Glaser and R. M. Burton, *J. Biol. Chem.*, 1963, **238**, 2595.
50. M. M. Burger, L. Glaser and R. M. Burton, *Meth. Enzymol.*, 1966, **8**, 441.
51. R. Rahim, U. A. Ochsner, M. Graninger, P. Messner, J. S. Lam and G. Soberón-Chávez, *Mol. Microbiol.*, 2001, **40**, 708.
52. A. M. Abdel-Mawgoud, F. Lepine and E. Deziel, *Appl. Microbiol. Biotechnol.*, 2010, **86**, 1323.
53. J. Toribio, A. E. Escalante and G. Soberón-Chávez, *Eur. J. Lipid Sci. Technol.*, 2010, **112**, 1082.
54. A. M. Chakrabarty, *Trends Biotechnol.*, 1985, **3**, 32.
55. B. Shao, Z. Liu, H. Zhong, G. Zeng, G. Liu, M. Yu, Y. Liu, X. Yang, Z. Li, Z. Fang, J. Zhang and C. Zhao, *Microbiol. Res.*, 2017, **200**, 33.
56. A. Wittgens, B. Santiago-Schuebel, M. Henkel, T. Tiso, L. M. Blank, R. Hausmann, D. Hofmann, S. Wilhelm, K.-E. Jaeger and F. Rosenau, *Appl. Microbiol. Biotechnol.*, 2018, **102**(3), 1229.
57. A. M. Abdel-Mawgoud, F. Lepine and E. Deziel, *Chem. Biol.*, 2014, **21**, 156.
58. C. Olvera, J. B. Goldberg, R. Sánchez and G. Soberón-Chávez, *FEMS Microbiol. Lett.*, 1999, **179**, 85.
59. R. Rahim, L. L. Burrows, M. A. Monteiro, M. B. Perry and J. S. Lam, *Microbiology*, 2000, **146**, 2803.
60. K. Zhu and C. O. Rock, *J. Bacteriol.*, 2008, **190**, 3147.
61. E. Déziel, F. Lépine, S. Milot and R. Villemur, *Microbiology*, 2003, **149**, 2005.
62. F. Lépine, E. Déziel, S. Milot and R. Villemur, *J. Mass Spectrom.*, 2002, **37**, 41.
63. L. Zhang, T. A. Veres-Schalnat, A. Somogyi, J. E. Pemberton and R. M. Maier, *Appl. Environ. Microbiol.*, 2012, **78**, 8611.
64. B. H. Rehm, N. Krüger and A. Steinbüchel, *J. Biol. Chem.*, 1998, **273**, 24044.
65. A. Wittgens, F. Kovacic, M. M. Müller, M. Gerlitzki, B. Santiago-Schübel, D. Hofmann, T. Tiso, L. M. Blank, M. Henkel, R. Hausmann, C. Syldatk, S. Wilhelm and F. Rosenau, *Appl. Microbiol. Biotechnol.*, 2017, **101**, 2865.

66. M. Hošková, R. Ježdík, O. Schreiberová, J. Chudoba, M. Šír, A. Čejková, J. Masák, V. Jirků and T. Řezanka, *J. Biotechnol.*, 2015, **193**, 45.
67. C. Syldatk, S. Lang and F. Wagner, *Z. Naturforsch.*, 1985, **40**, 51.
68. N. B. Rendell, G. W. Taylor, M. Somerville, H. Todd, R. Wilson and J. Cole, *Biochim. Biophys. Acta*, 1990, **1045**, 189.
69. E. Déziel, F. Lépine, S. Milot and R. Villemur, *Biochim. Biophys. Acta*, 2000, **1485**, 145.
70. F. Zhao, M. Yuan, L. C. L. Lei and X. Xu, *Bioresour. Technol.*, 2021, **323**, 124605.
71. M. Jadhav, S. Kalme, D. Tamboli and S. Govindwar, *J. Basic Microbiol.*, 2011, **51**, 385.
72. T. Rezanka, L. Siristova and K. Sigler, *Extremophiles*, 2011, **15**, 697.
73. D. K. Y. Solaiman, R. D. Ashby, N. W. Gunther IV and J. A. Zerkowski, *Appl. Microbiol. Biotechnol.*, 2015, **99**(10), 4333.
74. D. Dubeau, E. Déziel, D. E. Woods and F. Lépine, *BMC Microbiol.*, 2009, **9**, 263.
75. J. Y. Lim, T.-H. Lee, B. H. Nahm, Y. D. Choi, M. Kim and I. Hwang, *J. Bacteriol.*, 2009, **191**, 3758.
76. A. Latifi, M. K. Winson, M. Foglino, B. W. Bycroft, G. S. Stewart, A. Lazdunski and P. Williams, *Mol. Microbiol.*, 1995, **17**, 333.
77. A. Latifi, M. Foglino, K. Tanaka, P. Williams and A. Lazdunski, *Mol. Microbiol.*, 1996, **21**, 1137.
78. J. P. Pearson, E. C. Pesci and B. H. Iglewski, *J. Bacteriol.*, 1997, **179**, 5756.
79. A. Wittgens and F. Rosenau, *Front. Bioengin. Biotechnol.*, 2020, **8**, 594010.
80. C. E. Dulcey, Y. López de Los Santos, M. Létourneau, E. Déziel and N. Doucet, *FEBS J.*, 2019, **286**, 4036.
81. A. Germer, T. Tiso, C. Müller, B. Behrens, C. Vosse, K. Scholz, M. Froning, H. Hayen and L. M. Blank, *Appl. Environ. Microbiol.*, 2020, **86**, e002317.
82. F. Horlamus, A. Wittgens, P. Noll, J. Michler, I. Müller, F. Weggenmann, C. Oellig, F. Rosenau, M. Henkel and R. Hausmann, *GCB Bioenergy*, 2019, **11**, 260.
83. J. Beuker, A. Steier, A. Wittgens, F. Rosenau, M. Henkel and R. Hausmann, *AMB Express*, 2016, **6**, 11.
84. M. Cha, N. Lee, M. Kim, M. Kim and S. Lee, *Bioresour. Technol.*, 2008, **99**, 2192.
85. T. Tiso, A. Sabelhaus, B. Behrens, A. Wittgens, F. Rosenau, H. Hayen and L. M. Blank, *Metab. Eng. Commun.*, 2016, **3**, 234.
86. F. Zhao, R. Shi, J. Zhao, G. Li, X. Bai, S. Han and Y. Zhang, *J. Appl. Microbiol.*, 2015, **118**, 379.
87. N. Cabrera-Valladares, A. P. Richardson, C. Olvera, L. G. Treviño, E. Déziel, F. Lépine and G. Soberón-Chávez, *Appl. Microbiol. Biotechnol.*, 2006, **73**, 187.
88. F. M. Bahia, G. C. de Almeida, L. P. de Andrade, C. G. Campos, L. R. Queiroz, R. L. V. da Silva, P. V. Abdelnur, J. R. Correa, M. Bettiga and N. S. Parachin, *Sci. Rep.*, 2018, **8**, 2905.

89. D. H. Dusane, S. S. Zinjarde, V. P. Venugopalan, R. J. C. McLean, M. M. Weber and P. K. S. M. Rahman, *Biotechnol. Gen. Eng. Rev.*, 2010, **27**, 159.
90. D. H. Dusane, S. R. Damare, Y. V. Nancharaiah, N. Ramaiah, V. P. Venugopalan, A. Ravi Kumar and S. S. Zinjarde, *PLoS One*, 2013, **8**, e64501.
91. Y. Okkotsu, P. Tieku, L. F. Fitzsimmons, M. E. Churchill and M. J. Schurr, *J. Bacteriol.*, 2013, **195**, 5499.
92. Y. Zhang and R. M. Miller, *Appl. Environ. Microbiol.*, 1995, **61**, 2247.
93. R. A. Al-Tahhan, T. R. Sandrin, A. A. Bodour and R. M. Maier, *Appl. Environ. Microbiol.*, 2000, **66**, 3262.
94. W. H. Noordman and D. B. Janssen, *Appl. Environ. Microbiol.*, 2002, **68**, 4502.
95. S. Lang and D. Wullbrandt, *Appl. Microbiol. Biotechnol.*, 1999, **51**, 22.
96. M. Nitschke and G. M. Pastore, *Bioresour. Technol.*, 2006, **97**, 336.
97. I. Banat, A. Franzetti, I. Gandolfi, G. Bestetti, M. Martinotti, L. Fracchia, T. J. Smyth and R. Marchant, *Appl. Microbiol. Biotechnol.*, 2010, **87**, 427.
98. T. T. Nguyen and D. A. Sabatini, *Int. J. Mol. Sci.*, 2011, **12**, 1232.
99. K. Urum and T. Pekdemir, *Chemosphere*, 2004, **57**, 1139.
100. T. T. Nguyen, N. H. Youssef, M. J. McInerney and D. A. Sabatini, *Water Res.*, 2008, **42**, 1735.
101. G. Liu, H. Zhong, X. Yang, Y. Liu, B. Shao and Z. Liu, *Biotechnol. Bioeng.*, 2018, **115**, 796.
102. Q. Wang, X. Fang, B. Bai, X. Liang, P. J. Shuler, W. A. I. Goddard and Y. Tang, *Biotechnol. Bioeng.*, 2007, **98**, 842.
103. R. Sharma, J. Singh and N. Verma, *3 Biotech*, 2018, **8**, 20.
104. F. Zhao, H. Jiang, H. Sun, C. Liu, S. Han and Y. Zhang, *RSC Adv.*, 2019, **9**(6), 2885.
105. V. A. L. Rocha, L. V. A. Castilho, R. P. V. de Castro, D. B. Teixeira, A. V. Magalhaes, J. G. C. Gomez and D. M. G. Freire, *Biotechnol. Prog.*, 2020, **36**(4), e2981.
106. M. Robineau, S. L. Guenic, L. Sanchez, L. Chaveriat and E. A. Barka, *Molecules*, 2020, **25**, 3108.
107. P. A. J. Gorin, J. F. T. Spencer and A. P. Tulloch, *Can. J. Chem.*, 1961, **39**, 846.
108. A. P. Tulloch and J. F. T. Spencer, *Can. J. Chem.*, 1968, **46**(3), 345.
109. J. F. T. Spencer, P. A. J. Gorin and A. P. Tulloch, *Antonie van Leeuwenhoek*, 1970, **36**, 129.
110. H.-J. Daniel, M. Reuss and C. Syldatk, *Biotechnol. Lett.*, 1998, **20**, 1153.
111. S. De Waele, I. Vandenberghe, B. Laukens, S. Planckaert, S. Verweire, I. N. A. Van Bogaert, W. Soetaert, B. Devreese and K. Ciesielska, *Protein Expression Purif.*, 2018, **143**, 62.
112. A. M. Davila, R. Marchal and J. P. Vandecasteele, *Appl. Microbiol. Biotechnol.*, 1992, **38**, 6.
113. A. Albrecht, U. Rau and F. Wagner, *Appl. Microbiol. Biotechnol.*, 1996, **46**, 67.
114. D. F. Jones, *J. Chem. Soc. C*, 1968, 2827.

115. J. Chen, X. Song, H. Zhang, Y. B. Qu and J. Y. Miao, *Appl. Microbiol. Biotechnol.*, 2006, **72**, 52.
116. A. Nuñez, R. Ashby, T. A. Foglia and D. K. Y. Solaiman, *Biotechnol. Lett.*, 2004, **26**, 1087.
117. I. A. Ribeiro, M. R. Bronze, M. F. Castro and M. H. L. Ribeiro, *J. Mol. Recognit.*, 2012, **25**, 630.
118. T. B. Breithaupt and R. J. Light, *J. Biol. Chem.*, 1982, **257**, 9622.
119. K. M. J. Saerens, S. L. K. W. Roelants, I. N. A. Van Bogaert and W. Soetaert, *FEMS Yeast Res.*, 2011, **11**, 123.
120. K. M. J. Saerens, J. Zhang, S. Saey, I. N. A. Van Bogaert and W. Soetaert, *Yeast*, 2011, **28**, 279.
121. K. M. J. Saerens, L. Saey and W. Soetaert, *Biotechnol. Bioeng.*, 2011, **108**, 2923.
122. T. W. Esders and R. J. Light, *J. Biol. Chem.*, 1972, **247**, 1375.
123. M. L. Bucholtz and R. J. Light, *J. Biol. Chem.*, 1976, **251**, 424.
124. H. J. Asmer, S. Lang, F. Wagner and V. Wray, *J. Am. Oil Chem. Soc.*, 1988, **65**, 1460.
125. M. Widmann, P. B. Juhl and J. Pleiss, *BMC Genomics*, 2010, **11**, 123.
126. K. Ciesielska, I. N. A. Van Bogaert, S. Chevineau, B. Li, S. Groeneboer, W. Soetaert, Y. Van de Peer and B. Devreese, *J. Proteomics*, 2014, **98**, 159.
127. K. Ciesielska, S. L. Roelants, I. N. A. Van Bogaert, S. De Waele, I. Vandenberghe, S. Groeneboer, W. Soetaert and B. Devreese, *Appl. Microbiol. Biotechnol.*, 2016, **100**, 9529.
128. I. N. A. Van Bogaert, M. De Mey, D. Develter, W. Soetaert and E. J. Vandamme, *FEMS Yeast Res.*, 2009, **9**, 87.
129. D. R. Nelson, *Methods Mol. Biol.*, 1998, **107**, 15.
130. K. Lottermoser, W. H. Schunck and O. Asperger, *Yeast*, 1996, **12**, 565.
131. R. K. Hommel, L. Weber, A. Weiss, U. Himmelreich, O. Rilke and H. P. Kleber, *J. Biotechnol.*, 1994, **33**, 147.
132. T. W. Esders and R. J. Light, *J. Lipid Res.*, 1972, **13**, 663.
133. A. J. Cutler and R. J. Light, *J. Biol. Chem.*, 1979, **254**(5), 1946.
134. V. Klekner, N. Kosaric and Q. H. Zhou, *Biotechnol. Lett.*, 1991, **13**, 345.
135. D. K. Y. Solaiman, R. D. Ashby, A. Nunez and T. A. Foglia, *Biotechnol. Lett.*, 2004, **26**, 1241.
136. G. Pekin, F. Vardar-Sukan and N. Kosaric, *Eng. Life Sci.*, 2005, **5**, 357.
137. A. M. Davila, R. Marchal and J. P. Vandecasteele, *Appl. Microbiol. Biotechnol.*, 1997, **47**, 496.
138. D. F. Jones and R. Howe, *J. Chem. Soc.*, 1968, 2801.
139. A. M. Davila, R. Marchal and J. P. Vandecasteele, *J. Ind. Microbiol.*, 1994, **13**, 249.
140. D. A. Cavalero and D. G. Cooper, *J. Biotechnol.*, 2003, **103**, 31.
141. S. Ito, M. Kinta and S. Inoue, *Agric. Biol. Chem.*, 1980, **44**, 2221.
142. D. S. Yoo, B. S. Lee and E. K. Kim, *J. Microbiol. Biotechnol.*, 2005, **15**, 1164.
143. K. E. Gi, *Korean Pat.*, 0083614, 2004.
144. S. Akari and Y. Akari, *Jp. Pat.*, 61205449, 1987.

145. K. Masaru, N. Takashi, A. Yoji, N. Kazuo, N. Tatsu, T. Sumiko and N. Jotaro, *Jp. Pat.*, 2001131538, 2001.
146. H. Mager, R. Röthlisberger and F. Wagner, *EP Pat.*, 0209783, 1987.
147. V. Shah, G. F. Doncel, T. Seyoum, K. M. Eaton, I. Zalenskaya, R. Hagver, A. Azim and R. Gross, *Antimicrob. Agents Chemother.*, 2005, **49**(10), 4093.
148. V. Shah, M. Jurjevic and D. Badia, *Biotechnol. Prog.*, 2007, **23**, 512.
149. J. Chen, X. Song, H. Zhang, Y. B. Qu and J. Y. Miao, *Enzyme Microb. Technol.*, 2006, **39**, 501.
150. M. H. Bluth, E. Kandil, C. M. Mueller, V. Shah, Y. Y. Lin, H. Zhang, L. Dresner, L. Lempert, M. Nowakowski, R. Gross, R. Schulze and M. E. Zenilman, *Crit. Care Med.*, 2006, **34**, 188.
151. L. M. Napolitano, *Crit. Care Med.*, 2006, **34**, 258.
152. S. Lang, E. Katsiwela and F. Wagner, *Fat Sci. Technol.*, 1989, **91**, 363.
153. J. D. Shin, J. Lee, Y. B. Kim, I. Han and E.-K. Kim, *Bioresour. Technol.*, 2010, **101**, 3170.
154. M. Baviere, D. Degouy and J. Lecourtier, *US Pat.*, 5326407, 1994.
155. R. Marchal, J. Lemal, C. Sulzer and A. M. Davila, *US Pat.*, 5900366, 1999.
156. L. Pesce, *World Pat.*, 02/062495, 2002.
157. J. Ducreux, D. Ballerini, M. Baviere, C. Bocard and N. Monin, *US Pat.*, 5654192, 1997.
158. C. N. Mulligan, R. N. Yong and B. F. Gibbs, *J. Hazard. Mater.*, 2001, **85**, 111.
159. S. Inoue and N. Miyamoto, *US Pat.*, 4201844, 1980.
160. J. A. Zerkowski and D. K. Y. Solaiman, *J. Am. Oil Chem. Soc.*, 2006, **83**, 621.
161. D. Kitamoto, H. Isoda and T. Nakahara, *J. Biosci. Bioeng.*, 2002, **94**, 187.
162. T. Imura, H. Yanagishita and D. Kitamoto, *J. Am. Chem. Soc.*, 2004, **126**, 10804.
163. A. Saika, H. Koike, T. Fukuoka, S. Yamamoto, T. Kishimoto and T. Morita, *PLoS One*, 2016, **11**(6), e0157858.
164. A. Saika, H. Koike, S. Yamamoto, T. Kishimoto and T. Morita, *Appl. Microbiol. Biotechnol.*, 2017, **101**, 8345.
165. T. Morita, M. Konishi, T. Fukuoka, T. Imura, H. K. Kitamoto and D. Kitamoto, *FEMS Yeast Res.*, 2007, **7**, 286.
166. T. Morita, T. Fukuoka, T. Imura and D. Kitamoto, *J. Oleo Sci.*, 2015, **64**, 133.
167. U. Rau, L. A. Nguyen, S. Schulz, V. Wray, M. Nimtz, H. Roeper, H. Koch and S. Lang, *Appl. Microbiol. Biotechnol.*, 2005, **66**, 551.
168. M. Konishi, T. Morita, T. Fukuoka, T. Imura, K. Kakugawa and D. Kitamoto, *Appl. Microbiol. Biotechnol.*, 2007, **75**, 521.
169. D. Kitamoto, T. Ikegami, G. T. Suzuki, A. Sasaki, Y. Takeyama, Y. Idemoto, N. Koura and H. Yanagishita, *Biotechnol. Lett.*, 2001, **23**, 1709.
170. T. Morita, M. Konishi, T. Fukuoka, T. Imura, H. Sakai and D. Kitamoto, *J. Oleo Sci.*, 2008, **57**, 557.
171. R. H. Haskins, J. A. Thorn and B. Boothroyd, *Can. J. Microbial.*, 1955, **1**, 749.

172. R. Kahmann, C. Basse and M. Feldbrügge, *Curr. Opin. Microbiol.*, 1999, **2**, 647.
173. G. Deml, T. Anke, F. Oberwinkler, B. M. Giannetti and W. Steglich, *Phytochemistry*, 1980, **19**, 83.
174. D. Kitamoto, S. Akiba, C. Hioki and T. Tabuchi, *Agric. Biol. Chem.*, 1990, **54**, 31.
175. M. Kurz, C. Eder, D. Isert, Z. Li, E. F. Paulus, M. Schiell, L. Toti, L. Vertesy, J. Wink and G. Seibert, *J. Antibiot.*, 2003, **56**, 91.
176. K. Kakugawa, M. Tamai, K. Imamura, K. Miyamoto, S. Miyoshi, Y. Morinaga, O. Suzuki and T. Miyakawa, *Biosci., Biotechnol., Biochem.*, 2002, **66**, 188.
177. S. Hewald, K. Josephs and M. Bolker, *Appl. Environ. Microbiol.*, 2005, **71**(6), 3033.
178. A. Saika, Y. Utashima, H. Koike, S. Yamamoto, T. Kishimoto, T. Fukuoka and T. Morita, *Appl. Microbiol. Biotechnol.*, 2018, **102**, 1759.
179. S. Hewald, U. Linne, M. Scherer, M. A. Marahiel, J. Kamper and M. Bolker, *Appl. Environ. Microbiol.*, 2006, **72**(8), 5469.
180. S. Spoeckner, V. Wray, M. Nimtz and S. Lang, *Appl. Microbiol. Biotechnol.*, 1999, **51**, 33.
181. T. Fukuoka, Y. Yanagishita, T. Imura, T. Morita, H. Sakai, M. Abe and D. Kitamoto, *Carbohydr. Res.*, 2011, **346**(2), 266.
182. T. Fukuoka, T. Morita, M. Konishi, T. Imura and D. Kitamoto, *Carbohydr. Res.*, 2008, **343**(3), 555.
183. T. Morita, M. Takashima, T. Fukuoka, M. Konishi, T. Imura and D. Kitamoto, *Appl. Microbiol. Biotechnol.*, 2010, **88**(3), 679.
184. T. Morita, T. Fukuoka, A. Kosaka, T. Imura, H. Sakai, M. Abe and D. Kitamoto, *Appl. Microbiol. Biotechnol.*, 2015, **99**, 5833.
185. J. Kämper, R. Kahmann, M. Bölker, L. J. Ma, T. Brefort and B. J. Saville, *et al.*, *Nature*, 2006, **444**, 97.
186. F. Becker, T. Stehlik, U. Linne, M. Bolker, J. Freitag and B. Sandrock, *Metab. Eng. Commun.*, 2021, **12**, e00165.
187. T. Morita, H. Koike, Y. Koyama, H. Hagiwara, E. Ito, T. Fukuoka, T. Imura, M. Machida and D. Kitamoto, *Gen. Announ.*, 2013, **1**, e00064.
188. A. Saika, H. Koike, T. Hori, T. Fukuoka, S. Sato, H. Habe, D. Kitamoto and T. Morita, *Gen. Announ.*, 2014, **2**, e00878.
189. S. Lorenz, M. Guenther, C. Grumaz, S. Rupp, S. Zibek and K. Sohn, *Genome Announc.*, 2014, **2**, e00053.
190. A. Saika, T. Fukuoka, H. Koike, S. Yamamoto, T. Sugahara, A. Sogabe, D. Kitamoto and T. Morita, *Appl. Microbiol. Biotechnol.*, 2020, **104**, 10105.
191. M. Konishi, Y. Yoshida, M. Ikarashi and J. Horiuchi, *Biotechnol. Lett.*, 2015, **37**(8), 1679.
192. G. Marchand, E. Fortier, B. Neveu, S. Bolduc, F. Belzile and R. R. Bélanger, *J. Microbiol. Methods*, 2007, **70**(3), 519.
193. T. Morita, H. Habe, T. Fukuoka, T. Imura and D. Kitamoto, *J. Biosci. Bioeng.*, 2007, **104**(6), 517.

194. T. Watanabe, T. Morita, H. Koike, T. Yarimizu, Y. Shinozaki, Y. Sameshima-Yamashita, S. Yoshida, M. Koitabashi and H. Kitamoto, *Appl. Microbiol. Biotechnol.*, 2016, **100**(7), 3207.
195. J. Yoshikawa, T. Morita, T. Fukuoka, M. Konishi, T. Imura, K. Kakugawa and D. Kitamoto, *Appl. Microbiol. Biotechnol.*, 2014, **98**, 823.
196. T. Fukuoka, T. Yanagihara, T. Imura, T. Morita, H. Sakai, M. Abe and D. Kitamoto, *Carbohydr. Res.*, 2012, **351**, 81.
197. N. A. Inge, V. Bogaert, J. Zhang and W. Soetaert, *Proc. Biochem.*, 2011, **46**, 821.
198. H. Yamamoto, T. Morita, T. Fukuoka, T. Imura, S. Yanagidani, A. Sogabe, D. Kitamoto and M. Kitagawa, *J. Oleo Sci.*, 2012, **64**, 407.
199. M. Takahashi, T. Morita, T. Fukuoka, T. Imura and D. Kitamoto, *J. Oleo Sci.*, 2012, **61**, 457.
200. R. H. Haskins, *Can. J. Res.*, 1950, **28**, 213.
201. T. V. Kulakovskaya, E. V. Kulakovskaya and W. I. Golubev, *FEMS Yeast Res.*, 2003, **3**, 401.
202. E. O. Puchkov, U. Zahringer, B. Lindner, T. V. Kulakovskaya, U. Seydel and A. Wiese, *Biochim. Biophys. Acta*, 2002, **1558**, 161.
203. Y. Cheng, D. J. McNally, C. Labbe, N. Voyer, F. Belzile and R. R. Belanger, *Appl. Environ. Microbiol.*, 2003, **69**, 2595.
204. T. V. Kulakovskaya, A. S. Shashkov, E. V. Kulakovskaya and W. I. Golubev, *FEMS Yeast Res.*, 2005, **5**, 919.
205. V. I. Golubev, T. V. Kulakovskaya, A. S. Shashkov, E. V. Kulakovskaya and N. V. Golubev, *Mikrobiologiia*, 2008, **77**, 201.
206. R. H. Haskins and J. A. Thorn, *Can. J. Bot.*, 1951, **29**, 585.
207. B. Teichmann, U. Linne, S. Hewald, M. A. Marahiel and M. Bölker, *Mol. Microbiol.*, 2007, **66**(2), 525.
208. B. Teichmann, C. Labbé, F. Lefebvre, M. Bölker, U. Linne and R. R. Bélanger, *Mol. Microbiol.*, 2011, **79**(6), 1483.
209. B. Teichmann, F. Lefebvre, C. Labbe, M. Bolker, U. Linne and R. R. Belanger, *Appl. Environ. Microbiol.*, 2011, **77**(21), 7823.
210. R. U. Lemieux, *Can. J. Chem.*, 1953, **31**, 396.
211. R. U. Lemieux and R. Charanduk, *Can. J. Chem.*, 1951, **29**, 759.
212. B. Mimee, C. Labbe, R. Pelletier and R. R. Belanger, *Antimicrob. Agents Chemother.*, 2005, **49**, 1597.
213. B. Teichmann, L. Liu, K. O. Schink and M. Bölker, *Appl. Environ. Microbiol.*, 2010, **76**, 2633.
214. L. C. Vining, *Annu. Rev. Microbiol.*, 1990, **44**, 395.
215. T. C. Paulitz and R. R. Belanger, *Annu. Rev. Phytopathol.*, 2001, **39**, 103.
216. T. J. Avis and R. R. Belanger, *FEMS Yeast Res.*, 2002, **2**, 5.
217. R. R. Belanger, A. J. Dik and J. G. Menzies, *Plant–Microbe Interactions and Biological Control*, Marcel Dekker, New York, 1998, pp. 89–109.
218. M. S. Kuyukina and I. B. Ivshina, *Biology of Rhodococcus*, Springer, Berlin, 2010, pp. 231–262.

219. J. H. Kügler, C. Muhle-Goll, B. Kühl, A. Kraft, R. Heinzler, F. Kirschhöfer, M. Henkel, V. Wray, B. Luy, G. Brenner-Weiss, S. Lang, C. Syldatk and R. Hausmann, *Appl. Microbiol. Biotechnol.*, 2014, **98**, 8905.
220. P. J. Brennan, D. P. Lehane and D. W. Thomas, *Eur. J. Biochem.*, 1970, **13**, 117.
221. A. D. Elbein and M. Mitchell, *J. Bacteriol.*, 1973, **113**, 863.
222. Y. Delegan, A. Sargsyan, N. Hovhannisyan, B. Babayan, K. Petrikov and M. Vainstein, *J. Basic Microbiol.*, 2019, **60**, 14.
223. P. Rapp, H. Bock, V. Wray and F. Wagner, *J. Gen. Microbiol.*, 1979, **115**, 491.
224. M. E. V. Singer and W. R. Finnerty, *Can. J. Microbiol.*, 1990, **36**, 741.
225. M. J. Espuny, S. Egido, M. E. Mercadé and A. Manresa, *Toxicol. Environ. Chem.*, 1995, **48**, 83.
226. S. Lang and J. C. Philp, *Antonie van Leeuwenhoek*, 1998, **74**, 59.
227. F. Peng, Z. Liu, L. Wang and Z. Shao, *J. Appl. Microbiol.*, 2007, **102**, 1603.
228. S. Niescher, V. Wray, S. Lang, S. R. Kaschabek and M. Schlömann, *Appl. Microbiol. Biotechnol.*, 2006, **70**(5), 605.
229. Y. Tokumoto, N. Nomura, H. Uchiyama, T. Imura, T. Morita, T. Fukuoka and D. Kitamoto, *J. Oleo Sci.*, 2009, **58**(2), 97.
230. K. Iwahori, T. Tokutomi, N. Miyata and M. Fujita, *J. Biosci. Bioeng.*, 2001, **92**(1), 77.
231. F. T. Shen, L. S. Young, M. F. Hsieh, S. Y. Lin and C. C. Young, *Syst. Appl. Microbiol.*, 2010, **33**, 53.
232. L. Lo Piccolo, C. De Pasquale, R. Fodale, A. M. Puglia and P. Quatrini, *Appl. Environ. Microbiol.*, 2011, **77**, 1204.
233. H. F. Santos, F. L. Carmo, J. E. Paes, A. S. Rosado and R. S. Peixoto, *Water, Air, Soil Pollut.*, 2011, **216**, 329.
234. A. B. Jackisch-Matsuura, L. S. Santos, M. N. Eberlin, A. Fonseca de Faria, T. Matsuura, M. J. Grossman and L. R. Durrant, *Braz. Arch. Biol. Technol.*, 2014, **57**(1), 138.
235. M. Morikawa, H. Daido, T. Takao, S. Murata, Y. Shimonishi and T. Imanaka, *J. Bacteriol.*, 1993, **175**(20), 6459.
236. H. Okazaki, H. Sugino, T. Kanzaki, H. Fukuda and K. Isobe, *Agric. Biol. Chem.*, 1969, **33**, 764.
237. M. Powalla, S. Lang and V. Wray, *Appl. Microbiol. Biotechnol.*, 1989, **31**(5–6), 473.
238. G. S. Kiran, T. A. Thomas and J. Selvin, *Colloids Surf., B*, 2010, **78**(1), 8.
239. M. Richter, J. M. Willey, R. Süßmuth, G. Jung and H. P. Fiedler, *FEMS Microbiol. Lett.*, 1998, **163**(2), 165.
240. M. Arenskötter, D. Bröker and A. Steinbuchel, *Appl. Environ. Microbiol.*, 2004, **70**, 3195.
241. A. Franzetti, I. Gandolfi, G. Bestetti, T. J. P. Smyth and I. M. Banat, *Eur. J. Lipid Sci. Technol.*, 2010, **112**, 617.
242. J. C. Arguelles, *Arch. Microbiol.*, 2000, **174**, 217.
243. A. D. Elbein, Y. T. Pan, I. Pastuszak and D. Carroll, *Glycobiology*, 2003, **13**, 17R.

244. D. Tischler, S. Niescher, S. R. Kaschabek and M. Schlomann, *FEMS Microbiol. Lett.*, 2013, **342**, 113.
245. S. Vanin, L. Bubacco and M. Beltramini, *Cryo. Lett.*, 2008, **29**, 485.
246. J. D. Carroll, I. Pastuszak, V. K. Edavana, Y. T. Pan and A. D. Elbein, *FEBS J.*, 2007, **274**, 1701.
247. K. A. L. De Smet, A. Weston, I. N. Brown, D. B. Young and B. D. Robertson, *Microbiology*, 2000, **146**, 199.
248. R. Kalscheuer, B. Weinrick, V. Veeraraghavan, G. S. Besra and W. R. J. Jacobs, *Proc. Natl. Acad. Sci. U. S. A.*, 2010, **107**, 21761.
249. H. N. Murphy, G. R. Stewart, V. V. Mischenko, A. S. Apt, R. Harris, M. S. B. McAlister, P. C. Driscoll, D. B. Young and B. D. Robertson, *J. Biol. Chem.*, 2005, **280**, 14524.
250. G. Middlebrook, R. J. Dubos and C. Pierce, *J. Exp. Med.*, 1947, **86**, 175.
251. H. Noll, H. Bloch, J. Asselineau and E. Lederer, *Biochim. Biophys. Acta*, 1956, **20**, 299.
252. C. Hoffmann, A. Leis, M. Niederweis, J. M. Plitzko and H. Engelhardt, *Proc. Natl. Acad. Sci. U. S. A.*, 2008, **105**, 3963.
253. B. Zuber, M. Chami, C. Houssin, J. Dubochet, G. Griffiths and M. Daffe, *J. Bacteriol.*, 2008, **190**, 5672.
254. G. S. Besra, T. Sievert, R. Lee, R. A. Slayden, P. J. Brennan and K. Takayama, *Proc. Natl. Acad. Sci. U. S. A.*, 1994, **91**, 12735.
255. K. Takayama, C. Wang and G. S. Besra, *Clin. Microbiol. Rev.*, 2005, **18**, 81.
256. A. E. Grzegorzewicz, H. Pham, V. A. Gundi, M. S. Scherman, E. J. North, T. Hess, V. Jones, V. Gruppo, S. E. Born, J. Kordulakova, S. S. Chavadi, C. Morisseau, A. J. Lenaerts, R. E. Lee, M. R. McNeil and M. Jackson, *Nat. Chem. Biol.*, 2012, **8**, 334.
257. K. Tahlan, R. Wilson, D. B. Kastrinsky, K. Arora, V. Nair, E. Fischer, S. W. Barnes, J. R. Walker, D. Alland, C. E. Barry and H. I. Boshoff, *Antimicrob. Agents Chemother.*, 2012, **56**, 1797.
258. C. Varela, D. Rittmann, A. Singh, K. Krumbach, K. Bhatt, L. Eggeling, G. S. Besra and A. Bhatt, *Chem. Biol.*, 2012, **19**, 498.
259. J. T. Belisle, V. D. Vissa, T. Sievert, K. Takayama, P. J. Brennan and G. S. Besra, *Science*, 1997, **276**, 1420.
260. V. Puech, C. Guilhot, E. Perez, M. Tropis, L. Y. Armitige, B. Gicquel and M. Daffe, *Mol. Microbiol.*, 2002, **44**, 1109.
261. J. Asselineau and G. Lanéelle, *Front. Biosci.*, 1998, **3**, e164.
262. T. Suzuki, K. Tanaka, I. Matsubara and S. Kinoshita, *Agric. Biol. Chem.*, 1969, **33**, 1619.
263. M.-A. Lanéelle and J. Asselineau, *Biochim. Biophys. Acta*, 1977, **486**, 205.
264. D. E. Minnikin, G. Dobson, D. Sesardic and M. Ridell, *J. Gen. Microbiol.*, 1985, **131**, 1369.
265. M. Daffe, C. Lacave, M. A. Laneelle, M. Gillois and G. Laneelle, *Eur. J. Biochem.*, 1988, **172**, 579.
266. A. Lemassu, M. S. Laneelle and M. Daffe, *FEMS Microbiol. Lett.*, 1991, **62**, 171.

267. G. S. Besra, R. C. Bolton, M. R. McNeil, M. Ridell, K. E. Simpson, J. Glushka, H. van Halbeek, P. J. Brennan and D. E. Minnikin, *Biochemistry*, 1992, **31**, 9832.
268. M. Munoz, M. A. Laneelle, M. Luquin, J. Torrelles, E. Julian, V. Ausina and M. Daffe, *FEMS Microbiol. Lett.*, 1997, **157**, 251.
269. A. A. Khan, B. L. Stocker and M. S. Timmer, *Carbohyd. Res.*, 2012, **356**, 25.
270. R. Kalscheuer and K. Koliwer-Brandl, *Molecular Genetics of Mycobacteria*, American Society for Microbiology, Washington, DC, 2nd edn, 2014.
271. P. Kumar, M. W. Schelle, M. Jain, F. L. Lin, C. J. Petzold, M. D. Leavell, J. A. Leary, J. S. Cox and C. R. Bertozzi, *Proc. Natl. Acad. Sci. U. S. A.*, 2007, **104**, 11221.
272. J. C. Seeliger, C. M. Holsclaw, M. W. Schelle, Z. Botyanszki, S. A. Gilmore, S. E. Tully, M. Niederweis, B. F. Cravatt, J. A. Leary and C. R. Bertozzi, *J. Biol. Chem.*, 2012, **287**, 7990.
273. S. K. Hatzios, M. W. Schelle, C. M. Holsclaw, C. R. Behrens, Z. Botyanszki, F. L. Lin, B. L. Carlson, P. Kumar, J. A. Leary and C. R. Bertozzi, *J. Biol. Chem.*, 2009, **284**, 12745.
274. M. Daffe, M. McNeil and P. J. Brennan, *Biochemistry*, 1991, **30**, 378.
275. S. W. Hunter, R. C. Murphy, K. Clay, M. B. Goren and P. J. Brennan, *J. Biol. Chem.*, 1983, **258**, 10481.
276. A. Burguiere, P. G. Hitchen, L. G. Dover, L. Kremer, M. Ridell, D. C. Alexander, J. Liu, H. R. Morris, D. E. Minnikin, A. Dell and G. S. Besra, *J. Biol. Chem.*, 2005, **280**, 42124.
277. M. Gilleron and G. Puzo, *Glycoconjugate J.*, 1995, **12**, 298.
278. J. S. Kim, M. Powalla, S. Lang, F. Wagner, H. Lunsdorf and V. Wray, *J. Biotechnol.*, 1990, **13**, 257.
279. M. J. Espuny, S. Egido, I. Rodon, A. Manresa and M. E. Mercade, *Biotechnol. Lett.*, 1996, **18**, 521.
280. T. P. Pirog, T. A. Shevchuk, I. N. Voloshina and E. V. Karpenko, *Appl. Biochem. Mirobiol.*, 2004, **40**, 470.
281. C. Zheng, L. Shuguang, Y. Li, H. Lixin and W. Qinghong, *Ann. Microbiol.*, 2009, **59**, 771.
282. T. M. Luong, O. N. Ponamoreva, I. A. Nechaeva, K. V. Petrikov, Y. A. Delegan, A. K. Surin, D. Linklater and A. E. Filonov, *World J. Microbiol. Biotechnol.*, 2018, **34**, 20.
283. P. Rapp and L. H. E. Gabriel-Jurgens, *Microbiology*, 2003, **149**, 2879.
284. Y. Uchida, R. Tsuchiya, M. Chino, J. Hirano and T. Tabuchi, *Agric. Biol. Chem.*, 1989, **53**, 757.
285. A. Kretschmer and F. Wagner, *Biochim. Biophys. Acta*, 1983, **753**, 306.
286. Y. Ishigami, S. Suzuki, T. Funada, M. Chino, Y. Uchida and T. Tabuchi, *J. Jpn. Oil Chem. Soc.*, 1987, **36**, 847.
287. R. Kurane, K. Hatamochi, T. Kakuno, M. Kiyohara, T. Tajima, M. Hirano and Y. Taniguchi, *Biosci., Biotechnol., Biochem.*, 1995, **59**, 1652.
288. S. Ueda, N. Fujiwara, T. Naka, I. Sakaguchi, Y. Ozeki, I. Yano, T. Kasama and K. Koboyashi, *Microb. Pathog.*, 2001, **30**, 91.

289. A. M. Marqués, A. Pinazo, M. Farfan, F. J. Aranda, J. A. Teruel, A. Ortiz, A. Manresa and M. J. Espuny, *Chem. Phys. Lipids*, 2009, **158**, 110.
290. D. White, L. Hird and S. Ali, *J. Appl. Microbiol.*, 2013, **115**, 744.
291. C. Asselineau and J. Asselineau, *Prog. Chem. Fats Other Lipids*, 1978, **16**, 59.
292. E. Vollbrecht, R. Heckmann, V. Wray, M. Nimtz and S. Lang, *Appl. Microbiol. Biotechnol.*, 1998, **50**(5), 530.
293. Y. A. Delegan, A. A. Vetrova, V. N. Akimov, M. A. Titok, A. E. Filonov and A. M. Boronin, *Appl. Biochem. Microbiol.*, 2016, **52**, 389.
294. C. C. Young, T. C. Lin, M. S. Yeh, F. T. Shen and J. S. Chang, *World J. Microbiol. Biotechnol.*, 2005, **21**, 1409.
295. R. H. Baltz, *Curr. Opin. Pharmacol.*, 2008, **8**(5), 557.
296. P. J. Woodruff, B. L. Carlson, B. Siridechadilok, M. R. Pratt, R. H. Senaratne, J. D. Mougous, L. W. Riley, S. J. Williams and C. R. Bertozzi, *J. Biol. Chem.*, 2004, **279**, 28835.
297. Y. A. Delegan, L. N. Valentovich, S. M. Shafieva, K. G. Ganbarov, A. E. Filonov and M. B. Vainstein, *Folia Microbiol.*, 2019, **64**, 41.
298. J. D. Mougous, C. J. Petzold, R. H. Senaratne, D. H. Lee, D. L. Akey, F. L. Lin, S. E. Munchel, M. R. Pratt, L. W. Riley, J. A. Leary, J. M. Berger and C. R. Bertozzi, *Nat. Struct. Mol. Biol.*, 2004, **11**, 721.
299. S. E. Converse, J. D. Mougous, M. D. Leavell, J. A. Leary, C. R. Bertozzi and J. S. Cox, *Proc. Natl. Acad. Sci. U. S. A.*, 2003, **100**, 6121.
300. C. Rousseau, O. C. Turner, E. Rush, Y. Bordat, T. D. Sirakova, P. E. Kolattukudy, S. Ritter, I. M. Orme, B. Gicquel and M. Jackson, *Infect. Immun.*, 2003, **71**, 4684.
301. H. Ren, L. G. Dover, S. T. Islam, D. C. Alexander, J. M. Chen, G. S. Besra and J. Liu, *Mol. Microbiol.*, 2007, **63**, 1345.
302. P. Brodin, Y. Poquet, F. Levillain, I. Peguillet, G. Larrouy-Maumus, M. Gilleron, F. Ewann, T. Christophe, D. Fenistein, J. Jang, M. S. Jang, S. J. Park, J. Rauzier, J. P. Carralot, R. Shrimpton, A. Genovesio, J. A. Gonzalo-Asensio, G. Puzo, C. Martin, R. Brosch, G. R. Stewart, B. Gicquel and O. Neyrolles, *PLoS Pathog.*, 2010, **6**, e1001100.
303. Z. Shao, *Biosurfactants-from Genes to Applications*, Springer, Berlin, 2011.
304. M. Gilleron, S. Stenger, Z. Mazorra, F. Wittke, S. Mariotti, G. Böhmer, J. Prandi, L. Mori, G. Puzo and G. De Libero, *J. Exp. Med.*, 2004, **199**, 649.
305. R. K. Hommel, *Biodegradation*, 1990, **1**, 107.
306. N. Christofi and I. B. Ivshina, *J. Appl. Microbiol.*, 2002, **93**, 915.
307. J. Philp, M. Kuyukina, I. Ivshina, S. Dunbar, N. Christofi, S. Lang and V. Wray, *Appl. Microbiol. Biotechnol.*, 2002, **59**, 318.
308. H. Isoda, D. Kitamoto, H. Shinmoto, M. Matsumura and T. Nakahara, *Biosci., Biotechnol., Biochem.*, 1997, **61**, 609.
309. T. Sudo, X. X. Zhao, Y. Wakamatsu, M. Shibahara, N. Nomura, T. Nakahara, A. Suzuki, Y. Kobayashi, C. Y. Jin, T. Murata and K. K. Yokoyama, *Cytotechnology*, 2000, **33**, 259.

Subject Index

β-hydroxyl esters, 50

acidic media, mixed natural acids and esters in, 135
 computational evaluation, 139–141
 methyl coumalate from dimethyl malate, 142
 synthesis of **4**, procedure for, 141
 synthesis of **6**, procedure for, 141
acrylated epoxidized soybean oil (AESO), 24, 46–48, 51–53
adhesives, 85
AESO. *See* acrylated epoxidized soybean oil (AESO)
algae, 12
aliphatic diisocyanates, 18
alkenes, 24
amidation, 65–66
amidation/amination, 179
aminolysis kinetics study, 95–96
animal fats, 2
anionic polyurethane dispersions, 77–78
aviation fuel, 151–153
azomethine-functionalized chitosan, 185

BHT. *See* butylated hydroxytoluene (BHT)
bio-based isocyanate, 73
bio-content in polymeric materials, 14
biological oils, 2–7
blown oils, 8
bodied oil, 8
Breeze software, 29
11-bromoundecanoic acid, 7
Burkholderia pseudomallei, 217
Burkholderia sp., 215
Burkholderia thailandensis, 217
butylated hydroxytoluene (BHT), 186

calendula, 4
Calophyllum inophyllum oil, 177
canola oil, 2
carbonated plant oils, 79
carbonation, 95, 100–101
carbon-containing fatty acid derivative, epoxy resins from, 118–119
carbon dioxide, EuO natural epoxy ring opening polymerization in, 30–31
carbon neutrality, 75
carboxylated nitrile rubber (XNBR), 199
cardanol, 48, 122
castor bean lipase (CBL), 178
castor oil, 61, 70
catalytic partial hydrogenation, 178
cationic polyurethane dispersions, 75–77
CBL. *See* castor bean lipase (CBL)
CCs. *See* cyclic carbonates (CCs)
cellobiose lipids (CLs), 227
 biological functions of CL glycolipids, 230–231
 biosynthetic pathways, 228–229
 gene cluster regulation, 229–230

Cellvibrio japonicus, 218
chain extenders from plant oils, 71–73
citric acid, 135
CLs. *See* cellobiose lipids (CLs)
copolymerization of functionalized oils, 18
crystallization, 12
cyclic carbonates (CCs), 92
cycloaliphatic diisocyanates, 18
CYP52M1 gene, 222

DAC. *See* dialdehyde cellulose (DAC)
decarbonylation, 147–148
decarboxylation
 iridium-based, 151–153
 ruthenium-based, 149–151
designer oils by genetic modification, 12–13
DGEBA. *See* diglycidyl ether of bisphenol A (DGEBA)
DHA. *See* docosahexaenoic acid (DHA)
diacid, 137
dialdehyde cellulose (DAC), 51
Dickeya dadantii, 218
Diels–Alder reaction, 160
differential scanning calorimetry (DSC), 27, 30
diglycidyl ether of bisphenol A (DGEBA), 115, 122
9,10-dihydro-9-oxa-10-phosphaphenanthrene-10-oxide (DOPO) ester, 118, 122
dimethylacetamide (DMA), 137
dimethyl malate, methyl coumalate from, 142
2,2-dimethylolbutanoic acid, 68
2,2-dimethylolbutyric acid, 77–78
2,2-dimethylolpropionic acid, 68, 70, 77–78
dioctyl phthalate (DOP), 196–197
Dixoniella grisea, 185
DMA. *See* dimethylacetamide (DMA)
docosahexaenoic acid (DHA), 93
DOP. *See* dioctyl phthalate (DOP)
DOPO ester. *See* 9,10-dihydro-9-oxa-10-phosphaphenanthrene-10-oxide (DOPO) ester
double bond effect on polymer properties, 9, 10
drug delivery, 86
drying oils, 8, 160
DSC. *See* differential scanning calorimetry (DSC)

EAME. *See* eleostearic acid methyl ester (EAME)
ECH. *See* epichlorohydrin (ECH)
EcoLabel, 171
eicosapentaenoic acid (EPA), 93
Elektrionization™, 181–182
eleostearic acid, 160
eleostearic acid methyl ester (EAME), 159
 characterization, 161
  FT-IR system, 161
  NMR, 161
 methods
  lubricity determination, 162
  preparation of EAME/MA and esterification with methanol and butanol, 162
  tung oil, preparation of EAME from, 161–162
 results and discussion
  characterization of EAMA/MA/ME and EAMA/MA/BU, 163–165
  lubricity, 165–167
EMI. *See* 2-ethyl-4-methylimidazole (EMI)
ENR. *See* epoxidized natural rubber (ENR)
Environment Protection Agency (EPA), 171
EOC. *See* epoxy oxygen content (EOC)

EPA. *See* eicosapentaenoic acid (EPA)
EPA. *See* Environment Protection Agency (EPA)
epichlorohydrin (ECH), 114
EPO. *See* epoxidized palm oil (EPO)
epoxidation
and carbonation, 179–181
of *Nannochloropsis* oil by *m*-CPBA, 99–100
of soybean, linseed and *Schizochytrium* triglyceride oils by H_2O_2, 97–99
epoxidation/oxirane ring opening, 61–62
epoxidation/ring-opening method, 66–69
epoxide-ring opening, 43
epoxide rings, natural plant oils with, 24–27
epoxidized algae-DHA triglyceride oil, methylation of, 100
epoxidized linseed, methylation of, 95
epoxidized methyl oleate glyceride, 180
epoxidized natural rubber (ENR), 199
epoxidized oils, 17
epoxidized palm oil (EPO), 115, 196
epoxidized plasticizers, 195
plant oil-based, 195–197
epoxidized soybean oil (ESO), 24–26, 49–50, 51, 53
epoxidized VOs, 48–49
epoxy oxygen content (EOC), 95
epoxy resins, 20, 48–51
from carbon-containing fatty acid derivative, 118–119
from halogen-containing fatty acid derivative, 115–118
from intumescent flame retardants (IFRs) and nanoscale flame retardants, 130–131
from nitrogen-containing fatty acid derivative, 123–127
from phosphorus-containing fatty acid derivative, 119–123
preparation of epoxy resins from plant oils, 112–115
from silicon-containing fatty acid derivative, 127–130
epoxy soybean oil (ESO), 195–196, 197
erucic acid, 2
Escherichia coli, 81
ESO. *See* epoxidized sunflower oil (ESO); epoxy soybean oil (ESO)
ester bonds, 19–20
estolide formation/oligomerization, 182–184
2-ethyl-4-methylimidazole (EMI), 50
ethyl butyrate, 139
EuO
analysis and composition of, 31
extraction, 28
EuO natural epoxy ring opening polymerization in CO_2, 30–31
Euphorbia lagascae, 25
Euphorbia lathyris oil, 179
Euphorbia resinifera seeds, 25
euphorbia seeds, 27

FAAE. *See* fatty acid alkyl esters (FAAE)
FAME. *See* fatty acid methyl ester (FAME)
FAs. *See* fatty acids (FAs)
FAS II complex. *See* fatty acid synthase II (FAS II) complex
fatty acid alkyl esters (FAAE), 176
fatty acid composition determination by gas chromatography (GC), 28
fatty acid methyl ester (FAME), 177
fatty acid profile, 96–97
fatty acids (FAs), 4, 5, 173

fatty acid synthase II (FAS II) complex, 216
FFAs. *See* free fatty acids (FFAs)
fish fats, 2
flame-retardant epoxy resins from fatty acid derivatives
  carbon-containing fatty acid derivative, epoxy resins from, 118–119
  halogen-containing fatty acid derivative, epoxy resins from, 115–118
  intumescent flame retardants (IFRs), epoxy resins from, 130–131
  nanoscale flame retardants, epoxy resins from, 130–131
  nitrogen-containing fatty acid derivative, epoxy resins from, 123–127
  phosphorus-containing fatty acid derivative, epoxy resins from, 119–123
  silicon-containing fatty acid derivative, epoxy resins from, 127–130
flame retardant plasticizers, plant oil-based, 201–206
foams, 85–86
Fourier-transform infrared spectroscopy (FTIR), 28, 96
free fatty acids (FFAs), 94
FTIR. *See* Fourier-transform infrared spectroscopy (FTIR)
fully bio-based polyurethanes, 75
fumaric acid, 50
functional groups, position of, 11–12
functionalized oils, copolymerization of, 18

gas chromatography (GC), fatty acid composition determination by, 28
GC. *See* gas chromatography (GC)
gelation theory, 15
gel permeation chromatography (GPC), 27, 30
genetic modification, designer oils by, 12–13
Gibbs free energy change, 32
glycerol, 72
glyceryl ester plasticizer, 206
glycidol, 199
GPC. *See* gel permeation chromatography (GPC)
growth substrates, 222–223

H_2O_2
  epoxidation by, 94
  epoxidation of soybean, linseed and *Schizochytrium* triglyceride oils by, 97–99
halogenated flame retardants (HFRs), 115
halogen-containing fatty acid derivative, epoxy resins from, 115–118
HFRR method. *See* high-frequency reciprocating rig (HFRR) method
HFRs. *See* halogenated flame retardants (HFRs)
high-frequency reciprocating rig (HFRR) method, 162
high-performance liquid chromatography (HPLC), 29
HPLC. *See* high-performance liquid chromatography (HPLC)
hydroformylation/reduction method, 63–64
hydrogenated oil, 178
hydrolysis, 177–178
hydrolytic stability of oil products, 19–20
4-hydroxy-6-methylpyrone, 139
2-(2-hydroxyethylamino)ethanol, 71
hydroxy fatty acid, 8

IFRs. *See* intumescent flame retardants (IFRs)
IM. *See* isosorbide dimethacrylate (IM)

industrial chemicals *via* decarboxylation of natural carboxylic acids, 144
  iridium-based decarboxylation and aromatization reactions, 151–153
  itaconic acid, methacrylic acid from selective decarboxylation of, 153–155
  natural carboxylic acids, chemicals from decarboxylation of, 149
  oxygen removal from natural carboxylic acids, 147–148
  plant feedstocks for industrial use, 144–146
  ruthenium-based decarboxylation, 149–151
initiator concentration, effect of, 31–34
internal emulsifiers from plant oils, 66
intumescent flame retardants (IFRs), epoxy resins from, 130–131
iridium-based decarboxylation and aromatization reactions for aviation fuel, 151–153
isocyanates from plant oils, 73–74
isosorbide, 48
isosorbide dimethacrylate (IM), 48
isostearic acids, 178
itaconic acid, methacrylic acid from selective decarboxylation of, 153–155

karanja oil (KO), 180
KO. *See* karanja oil (KO)

Lesquerella oil, 4
light scattering (LS), 7
lignin, 48
lignocellulose-derived furan compounds, 186
Lindlar catalyst, 178
linoleic acid, 11
linseed oil, 26
liquid carbon dioxide–green solvents, natural epoxy oil polymerization in, 23
  chemicals, 28
  differential scanning calorimetry (DSC), 30
  EuO, analysis and composition of, 31
  EuO extraction, 28
  EuO natural epoxy ring opening polymerization in CO_2, 30–31
  euphorbia seeds, 27
  fatty acid composition determination by gas chromatography (GC), 28
  Fourier-transform infrared spectroscopy (FTIR), 28
  high-performance liquid chromatography (HPLC), 29
  initiator concentration, effect of, 31–34
  monomer concentration, effect of, 34
  MW determination of polymers, 30
  natural plant oils with epoxide rings, 24–27
  nuclear magnetic resonance (NMR) spectroscopy, 29
  plant/natural oil-based polymers, 24
  polymers, thermal properties of, 29
  RPEuO, thermal stability of, 38–39
  Soxhlet extraction and characterization of the extracted RPEuO-soluble substance, 34–38
  spectroscopic characterization, 28
  thermogravimetric analysis (TGA), 29
LS. *See* light scattering (LS)

lubricants (base oil/additives) from other natural resources, 185–186
lubricity, 165–167
  determination, 162

MA. *See* maleic anhydride (MA)
macrodiols, 8
macromolecular plasticizers, plant oil-based, 197–200
major facilitator superfamily (MFS), 217
maleic anhydride (MA), 18, 50, 160
malic acid, 135
mannosylerythritol lipids (MELs), 224
  biological functions and properties of, 227
  biosynthesis and structure of MEL glycolipids, 224–225
  identification and regulation of MEL gene cluster, 225–227
marigold seed oil, 4
mass spectrometry techniques, 7
MCC. *See* microcrystalline cellulose (MCC)
m-CPBA, epoxidation by, 95
MELs. *See* mannosylerythritol lipids (MELs)
3-mercaptopropionic acid, 70
methacrylated fatty acid esters (MFA), 47–48
methacrylic acid from selective decarboxylation of itaconic acid, 153–155
methanol and butanol, EAME/MA esterification with, 162
methyl coumalate from dimethyl malate, 142
methyl oleate (MO), 177, 180
methyl palmitoleate diol, 72
MFA. *See* methacrylated fatty acid esters (MFA)
MFS. *See* major facilitator superfamily (MFS)
MFSP. *See* minimum fuel selling price (MFSP)
microbially-produced citric acid, 135
microcrystalline cellulose (MCC), 51
minimum fuel selling price (MFSP), 93
MO. *See* methyl oleate (MO)
Mo-DTC. *See* molybdenum dithiocarbamate (Mo-DTC)
molecular weight (MW) determination of polymers, 30
molybdenum dithiocarbamate (Mo-DTC), 173
monoglycerides, 72
monomer-11-ω-amino undecanoic acid, 7
monomer concentration, effect of, 34
monomers, 48
monomers prepared from plant oils, 60
  chain extenders from plant oils, 71–73
  epoxidation/ring-opening method, 66–69
  internal emulsifiers from plant oils, 66
  isocyanates from plant oils, 73–74
  polyols from plant oils, 61
    epoxidation/oxirane ring opening, 61–62
    hydroformylation/reduction method, 63–64
    ozonolysis/reduction method, 65
    thiol–ene coupling reaction, 64
    transesterification/amidation, 65–66
  thiol–ene photo-click method, 70–71

Nannochloropsis crude lipid, 94
Nannochloropsis oil
  methylation of, 94
  short path distillation of, 94

nanoscale flame retardants, epoxy resins from, 130–131
National Fire Protection Association (NFPA), 171
natural carboxylic acids
  chemicals from decarboxylation of, 149
  oxygen removal from, 147–148
natural epoxy oil polymerization in liquid carbon dioxide–green solvents, 23
  chemicals, 28
  differential scanning calorimetry (DSC), 30
  EuO, analysis and composition of, 31
  EuO extraction, 28
  EuO natural epoxy ring opening polymerization in CO_2, 30–31
  euphorbia seeds, 27
  fatty acid composition determination by gas chromatography (GC), 28
  Fourier-transform infrared spectroscopy (FTIR), 28
  high-performance liquid chromatography (HPLC), 29
  initiator concentration, effect of, 31–34
  monomer concentration, effect of, 34
  MW determination of polymers, 30
  natural plant oils with epoxide rings, 24–27
  nuclear magnetic resonance (NMR) spectroscopy, 29
  plant/natural oil-based polymers, 24
  polymers, thermal properties of, 29
  RPEuO, thermal stability of, 38–39
  Soxhlet extraction and characterization of the extracted RPEuO-soluble substance, 34–38
  spectroscopic characterization, 28
  thermogravimetric analysis (TGA), 29
natural oil-based polymers, electrical properties of, 20
natural oils, 1
  bio-content in polymeric materials, 14
  biological oils, 2–7
  designer oils by genetic modification, 12–13
  functional groups, position of, 11–12
  hydrolytic stability of oil products, 19–20
  natural oil-based polymers, electrical properties of, 20
  oil heterogeneity on the properties of polyurethanes, 10–11
  planning polymeric materials from, 7–8
  polymer properties, effect of double bonds on, 9, 10
  polymers, technologies for converting oils to, 16–19
  saturated fatty acids, use of oils and fats with a high content of, 14–15
  thermosets, effect of cross-linking density on the properties of, 15–16
  vegetable oils, functionalization of, 8–9
natural oils, sustainable lubricant formulations from, 170
  lubricants (base oil/additives) from other natural resources, 185–186
  vegetable oils, characteristics of, 173–174
  vegetable oils, chemical modification of, 174

amidation/amination, 179
catalytic partial hydrogenation, 178
Elektrionization™, 181–182
epoxidation and carbonation, 179–181
estolide formation/oligomerization, 182–184
hydrolysis, 177–178
polymerization, 184–185
transesterification, 176–177
natural plant oils with epoxide rings, 24–27
natural raw materials, 4
NBR. *See* nitrile rubber (NBR)
neopentylglycol (NPG), 177
NFPA. *See* National Fire Protection Association (NFPA)
NIPUs. *See* non-isocyanate polyurethanes (NIPUs)
nitrile rubber (NBR), 199
nitrogen-containing fatty acid derivative, epoxy resins from, 123–127
NMR spectroscopy. *See* nuclear magnetic resonance (NMR) spectroscopy
non-ionic polyurethane dispersions, 78
non-isocyanate polyurethane from vegetable and microalgal oils, 92
aminolysis kinetics study, 95–96
analytical methods, 96
carbonation, 95, 100–101
epoxidation of *Nannochloropsis* oil by *m*-CPBA, 99–100
epoxidation of soybean, linseed and *Schizochytrium* triglyceride oils by H_2O_2, 97–99
epoxidized algae-DHA triglyceride oil, methylation of, 100
epoxidized linseed and *Schizochytrium* oils, methylation of, 95
fatty acid profile, 96–97
H_2O_2, epoxidation by, 94
m-CPBA, epoxidation by, 95
Nannochloropsis oil
methylation of, 94
short path distillation of, 94
plant algal oil feedstocks, characterization of NIPU polymers from, 106–108
polymer precursor fractionation, 95, 102
reaction kinetics study, 102–108
resin preparation, 96
short path distillation, 97
non-isocyanate polyurethanes (NIPUs), 78, 79, 81, 92
carbonated plant oils, 79
NIPU foams prepared from plant oils, 81–84
solvent-based NIPUs prepared from plant oils, 79–81
waterborne NIPUs prepared from plant oils, 81
Novozyme 435, 177
NPG. *See* neopentylglycol (NPG)
nuclear magnetic resonance (NMR) spectroscopy, 29
N-vinyl pyrrolidone (NVP), 52
NVP. *See* *N*-vinyl pyrrolidone (NVP)

OAP. *See* oleic acid polyesters (OAP)
oil-based polyurethanes, 20
oil heterogeneity on the properties of polyurethanes, 10–11
oil products, hydrolytic stability of, 19–20
olefinic terminal double bonds, 18
oleic acid, 74, 152
oleic acid polyesters (OAP), 198
ortho-phosphoric acid, 68
oxygen removal from natural carboxylic acids, 147–148

ozonolysis, 9
ozonolysis/reduction method, 65

palmitoleic acid methyl ester, 72
palm oil, 71, 72
palm oil monoglyceride-based polyester plasticizer (POMP), 200
palm stearin methoxy ester ethoxylate, 180
palm stearin oil methyl ester (PSME), 179
partially bio-based polyurethanes, 74–75
PBA. *See* poly(butylene adipate)s (PBA)
PBS. *See* polybutylene succinate (PBS)
PE. *See* pentaerythritol (PE); polyethylene (PE)
pentaerythritol (PE), 177
petrochemical products, 1
phosphaphenanthrene group-containing soybean-oil-based plasticizer (PSOPE), 205
phosphate esters, 201
phosphorus-based flame retardants, 120
phosphorus-containing fatty acid derivative, epoxy resins from, 119–123
Pichia pastoris, 222
plant algal oil feedstocks, characterization of NIPU polymers from, 106–108
plant feedstocks for industrial use, 144
  plant oil production and availability, 145–146
plant/natural oil-based polymers, 24
plant oil-based plasticizers, 194, 207–209
  glyceryl ester plasticizer, 206
  plant oil-based epoxidized plasticizer, 195–197
  plant oil-based flame retardant plasticizers, 201–206
  plant oil-based polyester plasticizers and macromolecular plasticizers, 197–200
  plant oil-based reactive plasticizers and internally plasticized PVC, 206–207
plant oil-based polyurethanes, 59
  application of, 84
    adhesives, 85
    drug delivery, 86
    foams, 85–86
    protective coatings, 84–85
    slow-release fertilizers, 86–87
  monomers prepared from plant oils, 60
    chain extenders from plant oils, 71–73
    epoxidation/ring-opening method, 66–69
    internal emulsifiers from plant oils, 66
    isocyanates from plant oils, 73–74
    polyols from plant oils, 61–66
    thiol–ene photo-click method, 70–71
  non-isocyanate polyurethanes, 78
    carbonated plant oils, 79
    NIPU foams prepared from plant oils, 81–84
    solvent-based NIPUs prepared from plant oils, 79–81
    waterborne NIPUs prepared from plant oils, 81
  plant oil-based waterborne polyurethane dispersions, 75
    anionic polyurethane dispersions, 77–78
    cationic polyurethane dispersions, 75–77
    non-ionic polyurethane dispersions, 78

solvent-based plant oil-based polyurethanes, 74
fully bio-based polyurethanes, 75
partially bio-based polyurethanes, 74–75
plant oil-based waterborne polyurethane dispersions, 75
anionic polyurethane dispersions, 77–78
cationic polyurethane dispersions, 75–77
non-ionic polyurethane dispersions, 78
plant oils
carbonated, 79
chain extenders from, 71–73
internal emulsifiers from, 66
isocyanates from, 73–74
NIPU foams prepared from, 81–84
polyols from, 61
epoxidation/oxirane ring opening, 61–62
hydroformylation/reduction method, 63–64
ozonolysis/reduction method, 65
thiol–ene coupling reaction, 64
transesterification/amidation, 65–66
preparation of epoxy resins from, 112–115
solvent-based NIPUs prepared from, 79–81
waterborne NIPUs prepared from, 81
polyamide 11, 7
poly(butylene adipate)s (PBA), 198
polybutylene succinate (PBS), 195
polyester plasticizers, plant oil-based, 197–200
polyesters, 7
polyethylene (PE), 198
poly(hydroxycarbamate) foams, 83
polyhydroxy compounds, 199
polyitaconic acid, 12
polymeric materials
bio-content in, 14
polymeric materials from natural oils, 7–8
polymerization, 184–185
polymerized polyol, 8
polymer precursor fractionation, 95, 102
polymer properties, effect of double bonds on, 9, 10
polymers
molecular weight (MW) determination of, 30
soybean oil-based
epoxy resins, 48–51
polyurethane, 43–45
vinyl resins, 45–48
technologies for converting oils to, 16–19
thermal properties of, 29
polyols from plant oils, 61
epoxidation/oxirane ring opening, 61–62
hydroformylation/reduction method, 63–64
ozonolysis/reduction method, 65
thiol–ene coupling reaction, 64
transesterification/amidation, 65–66
polypropylene (PP), 198
polyunsaturated fatty acids (PUFA), 93
polyurethane (PU), 9, 43–45, 81, 85, 92
oil heterogeneity on the properties of, 10–11
plant oil-based. *See* plant oil-based polyurethanes
polyvinyl chloride (PVC), 194
POMP. *See* palm oil monoglyceride-based polyester plasticizer (POMP)
PP. *See* polypropylene (PP)

pressure-sensitive adhesives (PSAs), 120
Prileschajew reaction, 24
protective coatings, 84–85
PSAs. *See* pressure-sensitive adhesives (PSAs)
Pseudomonas aeruginosa, 215
Pseudomonas desmolyticum, 217
Pseudomonas fluorescens, 218
Pseudomonas putida, 218
Pseudomonas sp., 215
Pseudomonas stutzeri, 218
Pseudozyma flocculosa, 230
PSME. *See* palm stearin oil methyl ester (PSME)
PSOPE. *See* phosphaphenanthrene group-containing soybean-oil-based plasticizer (PSOPE)
PU. *See* polyurethane (PU)
PUFA. *See* polyunsaturated fatty acids (PUFA)
PVC. *See* polyvinyl chloride (PVC)
pyrones, 138

Qualitas algal oil, 96

radical polymerization of oils, 17
rapeseed oil, 2
RBO. *See* rice bran oil (RBO)
regional plant oil availability, 146
resin preparation, 96
rhamnolipids (RLs), 215
    functional properties and characteristics of RL biosurfactants, 218–219
    genetic regulation and heterologous expression of RL genes, 217–218
    rhamnolipid biosynthetic pathway, 216–217
Rhodococcus, 237
rice bran oil (RBO), 180
ricinoleic acid, 8
ring-opening polymerization of epoxidized soybean oil (RPESO), 25
RLs. *See* rhamnolipids (RLs)
RPESO. *See* ring-opening polymerization of epoxidized soybean oil (RPESO)
RPEuO
    -soluble substance, 34–38
    thermal stability of, 38–39
ruthenium, 152
ruthenium-based decarboxylation, 149–151

Saccharomyces cerevisiae, 218
saturated branched chain fatty acids (sbc-FA), 178
saturated fatty acids, use of oils and fats with a high content of, 14–15
sbc-FA. *See* saturated branched chain fatty acids (sbc-FA)
SBO. *See* soybean oil (SBO)
$scCO_2$. *See* supercritical carbon dioxide ($scCO_2$)
Schizochytrium oils, methylation of, 95
scuffing load ball-on-cylinder lubricity evaluator (SL-BOCLE), 162
SEC. *See* size exclusion chromatography (SEC)
select glycolipid biosurfactants, synthesis and regulation of, 215
    cellobiose lipids (CLs), 227–231
    mannosylerythritol lipids (MELs), 224–227
    rhamnolipids (RLs), 215–219
    sophorolipids (SLs), 219–224
    trehalose lipids (TLs), 231–238
short path distillation (SPD), 94, 97
silicon-containing fatty acid derivative, epoxy resins from, 127–130
size exclusion chromatography (SEC), 7
SL-BOCLE. *See* scuffing load ball-on-cylinder lubricity evaluator (SL-BOCLE)
slow-release fertilizers, 86–87

SLs. *See* sophorolipids (SLs)
solvent-based NIPUs prepared from plant oils, 79–81
solvent-based plant oil-based polyurethanes, 74
  fully bio-based polyurethanes, 75
  partially bio-based polyurethanes, 74–75
sophorolipids (SLs), 219
  biosynthesis and structural variations of, 219–221
  effects of growth substrates on productivity and SL composition, 222–223
  genetic regulation of SL biosynthesis, 221–222
  roles and functional applications of, 223–224
Soxhlet extraction and characterization of extracted RPEuO-soluble substance, 34–38
soybean oil (SBO), 4, 24, 42
  characteristics of, 42–43
  chemical structure of, 43
soybean oil-based matrix, composites from, 51–52
soybean oil-based polymers
  epoxy resins, 48–51
  polyurethane, 43–45
  vinyl resins, 45–48
soybean oil derivatives, composites compatibilized by, 52–53
SPD. *See* short path distillation (SPD)
spectroscopic characterization, 28
Staphylococcus aureus, 81
STLs. *See* succinoyl trehalose lipids (STLs)
styrene, 47
succinoyl trehalose lipids (STLs), 235
sulfonate-type hydrophilic chain extender, 78
supercritical carbon dioxide ($scCO_2$), 25, 26
sustainable lubricant formulations from natural oils, 170
  lubricants (base oil/additives) from other natural resources, 185–186
  vegetable oils, characteristics of, 173–174
  vegetable oils, chemical modification of, 174
    amidation/amination, 179
    catalytic partial hydrogenation, 178
    Elektrionization™, 181–182
    epoxidation and carbonation, 179–181
    estolide formation/oligomerization, 182–184
    hydrolysis, 177–178
    polymerization, 184–185
    transesterification, 176–177

TAG. *See* triacylglycerols (TAG)
tall oil, 2
TBAB. *See* tetrabutylammonium bromide (TBAB)
TCA cycle. *See* tricarboxylic acid (TCA) cycle
TDM. *See* trehalose 6,6'-dimycolate (TDM)
terminal hydroxy groups, 12
tetrabutylammonium bromide (TBAB), 95
TGA. *See* thermogravimetric analysis (TGA)
thermogravimetric analysis (TGA), 27, 29
thermosets, 7
  effect of crosslinking density on the properties of, 15–16
thermosetting polymers, 15
thiol–ene coupling reaction, 64
thiol–ene photo-click method, 70–71
TLs. *See* trehalose lipids (TLs)
TME. *See* trimethylolethane (TME)
TMH. *See* trimethylolhexane (TMH)

TMM. *See* trehalose-6-monomycolate (TMM)
TMP. *See* trimethylolpropane (TMP)
TMTE. *See* tung-maleic triglycidyl ester (TMTE)
Torulopsis gropengiesseri, 223
transesterification, 176–177
transesterification/amidation, 65–66
trehalose 6,6′-dimycolate (TDM), 232–233
trehalose-6-monomycolate (TMM), 232
trehalose-based glycolipids
  biosynthesis and metabolic pathways of, 232–236
  genetic regulation of biosynthesis of, 236–238
trehalose lipids (TLs), 231
  physiological roles and functional applications of, 238
triacylglycerols (TAG), 4
tricarboxylic acid (TCA) cycle, 235
triglycerides, 100
trimethylolethane (TME), 177
trimethylolhexane (TMH), 177
trimethylolpropane (TMP), 177
Tsukamurella spp., 235
tung-maleic triglycidyl ester (TMTE), 197
tung oil, preparation of EAME from, 161–162
tung oil fatty acid esters, 161

vapor pressure osmometry (VPO), 6–7
vegetable and microalgal oils, non-isocyanate polyurethane from, 92
vegetable oil (VO), 23, 43, 170–171, 174
  -based plasticizer, 195
  characteristics of, 173–174
  chemical modification of, 174
    amidation/amination, 179
    catalytic partial hydrogenation, 178
    Elektrionization^{TM}, 181–182
    epoxidation and carbonation, 179–181
    estolide formation/oligomerization, 182–184
    hydrolysis, 177–178
    polymerization, 184–185
    transesterification, 176–177
  epoxidation of, 49
  functionalization of, 8–9
Vernicia fordii, 160
vernolic acid, 25
vernonia seed oil, 4
Vessel Incidental Discharge Act (VIDA), 171
VIDA. *See* Vessel Incidental Discharge Act (VIDA)
vinyl resins, 45–48
vitrimer, 50
VO. *See* vegetable oil (VO)
VOCs. *See* volatile organic compounds (VOCs)
volatile organic compounds (VOCs), 25
VPO. *See* vapor pressure osmometry (VPO)

waste cooking oils (WCO), 171
waterborne NIPUs prepared from plant oils, 81
waterborne polyurethane, 75
waterborne polyurethane dispersion systems, 86
water-soluble citric acid, 50
WCO. *See* waste cooking oils (WCO)

XNBR. *See* carboxylated nitrile rubber (XNBR)

ZDDP. *See* zinc dialkyldithiophosphate (ZDDP)
zinc dialkyldithiophosphate (ZDDP), 173